ECOLOGY
OF THE
ENGLISH CHALK

ECOLOGY
OF THE
ENGLISH CHALK

C. J. Smith

Merrist Wood Agricultural College,
Worplesdon, Guildford, Surrey, England

1980

ACADEMIC PRESS

A Subsidiary of Harcourt Brace Jovanovich, Publishers

London • New York • Toronto • Sydney • San Francisco

ACADEMIC PRESS INC. (LONDON) LTD
24–28 Oval Road,
London NW1

U.S. Edition published by
ACADEMIC PRESS INC.
111 Fifth Avenue
New York, New York 10003

British Library Cataloguing in Publication Data

Smith, C. J.
  Ecology of the English Chalk.
  1. Ecology — England 2. Chalk — England
  I. Title
  574.5′264   QH138.A1   79-41168

  ISBN 0-12-651850-5

Photoset in Times by Rugcliff Ltd., Cuckfield, Sussex
Printed by Whitstable Litho Ltd., Whitstable, Kent

Foreword

by

PROFESSOR A. H. BUNTING

. . . the man who should know the true history of the bit of chalk which every carpenter carries about in his breeches pocket, though ignorant of all other history, is likely, if he will think his knowledge out to its ultimate results, to have a truer, and therefore a better, conception of this wonderful universe, and of man's relation to it, than the most learned student who is deep-read in the records of humanity and ignorant of those of nature.

T. H. Huxley "On a Piece of Chalk", Norwich, 1868

The open and undulating landscapes, the sensuous curving horizons, the grasslands and woods, the commanding escarpments, the bold cliffs and headlands confronting the challenging ocean, the tumuli and rings, the dolmens, circles and standing stones, the wind rippling the barley, the song of larks in the luminous air—all these, and many more, evoke deep-felt emotions in all who know and love the English Chalk, feelings of magical enchantment, of antiquity and permanence amid ceaseless change "without haste, but without rest", yet of an inner freedom and peace. It is small wonder that the chalk country has commanded the devotion, down the centuries, of so large a company of countrymen, naturalists and scholars, whose labours and writings have added so much to our science, to our agriculture and to our sense of the unity of nature, gods and man.

Dr Smith is the latest of this distinguished line, and he is far from the least. His book is a substantial work of scholarship in which he ranges, with breadth and with judgement, over the many fields of learning which contribute to the synoptic science of ecology, and to its application to agriculture, environmental management, conservation and other important and urgent problems of practice. He has also mastered the frequently voluminous productions of large numbers of earlier writers, in several countries, on the biology, ecology and agriculture of the Chalk. More, he has brought the ample harvest of his endeavour into the barn, where he has set it out in a logical and intellectually satisfying pattern, to synthesise understanding in many areas and to point out lacunae in others. As a result, "Ecology of the

v

English Chalk" is at once a significant contribution to hard ecology and an instructive guide to further enquiry. Though its style is relaxed, even informal, this is an important learned work for which research workers and students of all ages will be grateful.

Dr Smith began his work on the ecology of the Chalk in this University, which, surrounded as it is by chalk country, had been to the fore in chalkland studies for many years. He now offers a generous reward to all who encouraged his effort, many of whom he has evidently outstripped.

It is at once an honour to introduce, and a pleasure to commend this book.

University of Reading, January 1980

Preface

The Chalk formation of southern and eastern England forms a very distinctive block of country which has sustained and inspired its inhabitants for over 5000 years. Indeed, the importance of this long association with man in the ecology of the chalklands can hardly be overstated. On the other hand, it is only within the last 200 years that others—archaeologists, naturalists, geologists and, still more recently, ecologists and conservationists (not to mention the increasing number of people bent simply on spending their leisure time there)—have come to the Downs and Wolds to stake their own particular claims for study or relaxation. This book is an attempt to collate and review, for all those interested or involved in any of the many facets of chalkland ecology in its broadest sense, the present state of knowledge of the geological and archaeological history, climate, soils, flora, fauna and husbandry of the English chalk country, and finally to consider perhaps the most challenging and potentially controversial aspect of all—its conservation. All these subjects are covered in such a way as to provide an informative foundation for the undergraduate and interested layman, but the main objective has been a detailed, critical handbook, well laced with references, for the more specialised reader. In such a broad field, moreover, it ought to help geologists, botanists, entomologists, agricultural advisors and other specialists to know a little more about each other's disciplines, for it is desirable that these people get together more than they have done in the past. Finally, I hope that the book will stimulate continued constructive interest in the chalklands, for we are still astonishingly short even of the most fundamental descriptive data, and numerous tantalising problems and paradoxes of applied ecology await solution.

Of course, the subject matter encompassed is far too great for any normal mortal to accommodate single-handed, and I have many friends, colleagues and correspondents to thank for so generously giving me the benefit of their specialist knowledge, without which this book could never have been written. For the hours of work involved in reading my early drafts, I am especially indebted to B. W. Avery, Dr T. Batey, H. C. Bowen, Mrs Ursula Bowen, Dr J. G. Evans, D. F. Fourt, Dr P. J. Grubb, Dr J. F. Hope-Simpson, Professor Eric Kerridge, Dr M. G. Morris, Dr G. F. Peterken,

Professor C. D. Pigott, Dr Francis Rose, Professor E. W. Russell, J. H. P. Sankey, Dr R. J. Small, L. P. Smith, D. T. Streeter, Dr Lena K. Ward, T. C. E. Wells, Dr A. S. Watt and C. J. Wood. My former chief, Professor Hugh Bunting, who first encouraged and consolidated my interest in chalkland ecology, kindly read the whole draft, and agreed to write the foreword. Many others, too numerous alas to mention by name, guided me on points of detail and permitted me to reproduce tables, diagrams and other data from their published works. I only hope that I have done my distinguished mentors justice.

I must, too, extend my warmest appreciation to my family, for whom "The Book" has become something of an institution during its seven years' gestation. My wife undertook the bulk of the typing, copied most of the diagrams, and drew the original illustrations of plants and fossils. Mrs Kate Ibberson and my daughter Sandie helped with the diagrams and with the daunting task of indexing and checking nomenclature. Denys Ovenden was responsible for the drawings of invertebrates, birds and mammals. Vincent Morris worked wonders with my attempts at photography. Christina Ashman, Ann Crossley, Mrs D. Murdoch and Mrs M. Murray helped type the final draft, and Isobel Elsey the index. And the whole venture might never have evolved had Mrs I. M. Howden not placed at my disposal her charming cottage at Lyme Regis where, during some of the most intensive periods of creative study I have ever forced myself to endure, "Ecology of the English Chalk" took shape. To all of these, my grateful thanks.

C. J. Smith   *January, 1980*

Acknowledgements and Nomenclature

Acknowledgements

The following publishers and organisations kindly permitted the reproduction of copyright material: George Allen and Unwin Ltd (Fig. 9.4); Applied Science Publishers Ltd (Tables 6.3, 10.1, 10.10); Bell and Hyman Ltd (Fig. 1.2); Berks., Bucks. and Oxon. Naturalists' Trust Ltd (Fig. 10.4b); Blackwell Scientific Publications Ltd (Figs 6.2, 6.11, 6.13, 6.20); British Ecological Society (Figs 2.16c,d, 2.18, 3.6–3.9, 3.11–3.13, 5.8, 5.10, 5.14, 5.15, 5.19–5.26, 6.3, 6.4, 6.18, 6.19, 7.2, 7.8a, 8.10, 8.12; Tables 3.3, 4.1, 5.6, 5.9–5.12, 6.5–6.9, 6.11–6.13, 7.3, 7.4, 8.1, 8.2); British Grassland Society (Figs 9.9, 9.10; Tables 6.1, 9.6–9.8); British Trust for Ornithology (Figs 7.12, 7.13, 10.5; Tables 7.6, 7.7); Cambridge University Press (Figs 1.16, 1.23, 1.28, 6.1; Tables 9.2–9.4); Cement and Concrete Association (Fig. 9.2); Chapman and Hall Ltd (Figs 7.8b, 10.4a; Tables 5.5, 7.5); Wm Collins Sons and Co. Ltd (Figs 1.12, 1.13, 1.18, 9.6; Table 5.2); Dorset County Council (Tables 10.2, 10.3); Faber and Faber Ltd (Table 1.1); VEB Gustav Fischer Verlag (Fig. 2.4); Forestry Commission (Figs 9.18–9.20); Geologists' Association (Table 1.2); The Hamlyn Group (Fig. 9.7); Harvard University Press (Figs 2.16a,b); Imperial Chemical Industries Ltd (part of Fig. 4.7); Institute of Geological Sciences (Figs 1.10a, 1.12, 1.13; Table 1.1); Kent Trust for Nature Conservation Ltd (Fig. 1.10b; Table 5.1); Linnean Society of London (Fig. 8.6); Longman Group Ltd (Figs 1.8, 1.9, 1.14, 1.15, 2.11, 2.14, 2.15; Tables 1.4, 2.1, 3.7); Meteorological Office (Figs 2.1–2.3); Methuen and Co. Ltd (Fig. 6.10); Ministry of Agriculture, Fisheries and Food (Tables 9.1, 9.10); *Nature*/Macmillan Journals Ltd (Figs 4.1, 4.2); Nature Conservancy Council (Table 10.9); Newnes–Butterworth (Fig. 2.6); Martinus Nijhoff BV (Fig. 3.10); Oliver and Boyd Ltd (Fig. 2.16e); Oxford University Press (Fig. 3.5); Pergamon Press Ltd (Fig. 1.17); G. Philip and Son Ltd (Fig. 1.11); Royal Meteorological Society (Figs 2.5, 2.7, 2.13, 2.19); Seminar Press Ltd (Figs 1.29, 6.8); Soil Survey of England and Wales (Figs 3.3, 3.4; Tables 3.1, 3.5, 3.6, 3.8, 9.11); Unesco (Fig. 2.17); J. Wiley and Sons Ltd (Table 10.4).

Regarding verbatim quotations, the lines by Thomas Hardy (Section 1.3.2.4) are from "The Revisitation": see "The Complete Poems of Thomas Hardy" (New Wessex Edition), published by Macmillan, London and Basingstoke. The apt description of the habits of landslips (Section 1.2.3) is quoted by permission of Serendip Fine Books, Lyme Regis. The sources of other quotations are indicated in the text.

Nomenclature

A major problem in a work of this scope is the standardisation of scientific names. Here, nomenclature of the grasses follows Hubbard (1968), other flowering plants and ferns Clapham *et al.* (1962), and mosses Watson (1968). The names used for other lower plants are the ones adopted by authorities cited in the text. Among the animals, mammals are named according to Corbet and Southern (1977), birds Sharrock (1976), insects Kloet and Hincks (1964–1972), spiders Locket *et al.* (1951–1974), woodlice Sutton (1972) and molluscs Waldén (1976). In a few cases, especially among the invertebrates, alternative names are used, but these, together with other synonyms (as well as common names), can all be cross-checked in the index.

The cinnabar moth is referred to in the index and on pp. 151, 153, 161 and 293 as *Callimorpha dominula*. This should read *Callimorpha* (or *Tyria*) *jacobaeae*.

Contents

Chapter 1. The Making of the Chalk Landscape

Chapter 2. The Climate of the English Chalklands

Chapter 3. Soils of the Chalk

Chapter 4. **The Colonisation of Exposed Chalk**

Chapter 5. **Chalk Grassland: I. Climatic and Edaphic Influences**

Chapter 6. **Chalk Grassland: II. Biotic Influences**

Chapter 7. **Chalk Scrub**

Chapter 8. **Woodlands on the Chalk**

Chapter 9. **Economic Aspects of the Chalklands**

Chapter 10. **Conservation**

To Sandie and Jenny

1

The Making of the
Chalk Landscape

1.1 The Geology of Chalk

1.1.1 The origin of chalk

The story of chalk begins no less than one hundred million years ago. One hundred million years! A million centuries. There is no way of saying it that makes this vast expanse of time in any way conceivable to the human mind. Yet, this is but a fraction of the probable age of the Earth, which is currently put at 4650 million years. And in a mere 100 million years, the spiralling galaxy to which our solar system belongs has made just half a revolution.

At about this time the Earth was mid-way through the Cretaceous Period in its geological history (Fig. 1.1). In fact it was from the latin word for chalk (*creta*) that the name for this period was derived, although in the thirty-five million years that had so far elapsed (the Lower Cretaceous), Wealden Beds, Greensand and Gault Clay rather than chalk were characteristic deposits. But a vast change was imminent. We know, for example, from recent developments in tectonics, that the huge single super-continent of Pangaea had separated into its northern (Laurasia) and southern (Gondwanaland) components, separated by the Tethys Sea, and that both continental masses had been drifting steadily northwards from a position at one time almost entirely in the southern hemisphere (Fig. 1.2). The ancient beginnings of what is now Britain lay somewhere near the centre of Laurasia, lapped along its eastern flank by a long, narrow strip of sea extending from the far north, known as the Boreal Ocean, and ancestor of the modern North Sea.

Now there was a marked subsidence to the east of Britain, perhaps associated with the imminent break-up of Laurasia, which allowed inundation by the sea on a vast scale. In the region which now underlies north-western Europe, the sea deepened and spread westwards and northwards (Fig. 1.3) in what came to be called the great Cenomanian

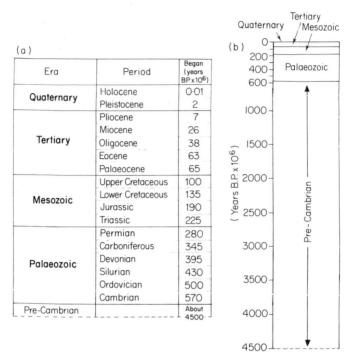

Era	Period	Began (years B.P x 10⁶)
Quaternary	Holocene	0·01
	Pleistocene	2
Tertiary	Pliocene	7
	Miocene	26
	Oligocene	38
	Eocene	63
	Palaeocene	65
Mesozoic	Upper Cretaceous	100
	Lower Cretaceous	135
	Jurassic	190
	Triassic	225
Palaeozoic	Permian	280
	Carboniferous	345
	Devonian	395
	Silurian	430
	Ordovician	500
	Cambrian	570
Pre-Cambrian		About 4500

Fig. 1.1. The geological column, displayed (a) conventionally and (b) to scale. The Quaternary and Tertiary are alternatively regarded as periods within the Cenozoic Era, with Palaeocene, Eocene, etc. given epoch status.

Transgression, although this is now known to have occurred in several distinct episodes. Nevertheless, the following thirty-five million years (the Upper Cretaceous) witnessed the deposition of the familiar white chalk, found now not only in England, Belgium and northern France where it is most extensive and well known, but also in Denmark, East and West Germany, Poland, Russia, parts of the Middle East, the Gulf Coast of the U.S.A. (the Alabama, Austin and Selma Chalks), and western Australia (the Gingin Chalk). The Chalk accumulated at an incredibly slow rate—on average about 2–4mm per century—but the sea floor continued to subside so that an enormous depth, more than 1 km in places, eventually resulted. The Chalk of the Upper Cretaceous Period is virtually unique in the whole of geological time. Apart from the Albian red chalks (see below) and Tertiary chalks such as the Danian of Denmark and some Middle East deposits, nothing quite like it had ever been deposited before, and none has since. What was it about the Chalk Sea, as it came to be called by geologists, that was so special?

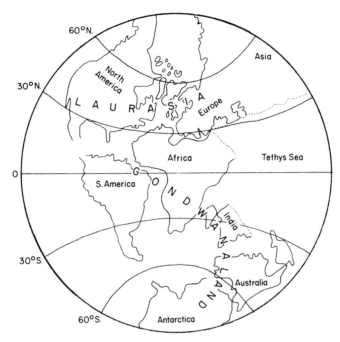

Fig. 1.2. The approximate positions of the continents in Upper Cretaceous times. Britain (which is shown as it appears now, of course) was still fully 10° further south than at the present time. After Tarling and Tarling (1971).

The essential feature of the Chalk Sea was that its deposits were derived almost entirely from the marine organisms which inhabited it. The proportion of sediments originating from land (the terrigenous materials such as sand, silt and clay) was very small indeed and, because the shape of such particles as are present indicates that they were carried there by wind, rather than washed there by water (Double, 1927), the Chalk Sea has been pictured as rather like an ancient Aegean, clear, warm and tranquil, but bordered by a desolate low-lying, arid landscape (Bailey, 1924). An alternative suggestion is that tropical rainforest (which could account no less plausibly for the absence of erosion) clothed the islands and larger landmasses in and around the Chalk Sea (Montford, 1970). At the present time, the arid-climate theory tips the balance, but only just (Hancock, 1975).

1.1.2 The composition of chalk

Opening his book on chalkland ecology, Sankey (1966) quotes an apposite reflection by G.M. Davies, that "chalk is the best-known rock in England,

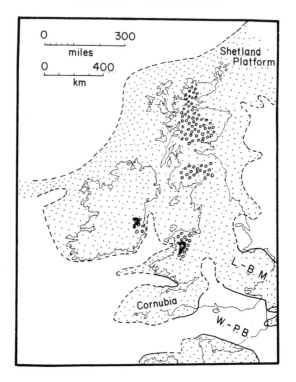

☷ Massifs and structural highs.

☷ Probable land at highest sea-level during Upper Campanian.

Fig. 1.3. The geological setting of the Upper Cretaceous in Britain. There is still some uncertainty over how much land was left unsubmerged during late Campanian times: the areas shown here are the probable maxima. L.–B. M., London–Brabant Massif; W.–P. B., Wessex–Paris Basin. After Hancock (1975), simplified.

and inability to distinguish it is proverbially a sign of mental deficiency", though in fact the familiar "chalk from cheese" maxim relates to the contrast between chalk and cheese *country* (see Chapter 9). Apart from its colour, the softness of chalk is perhaps its most familiar characteristic, at least to southerners. In soft chalk, the component particles have never become hardened or cemented together (indurated), hence the various adjectival connotations such as earthy, dull, pulverulent, weakly cohesive and even meagre (Fig. 1.4). Nevertheless not all chalk is soft. As we shall see below, harder bands occur throughout the chalk deposits of southern England,

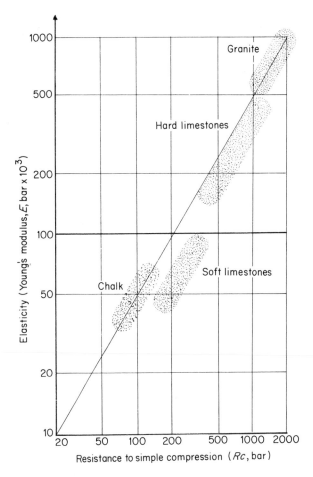

Fig. 1.4. Elasticity and resistance to compression of chalk and other rocks. From Pedro (1972).

while the chalk of the Wolds (and of Ulster) is altogether harder. Exactly what it is that accounts for these differences in chalk lithology has kept geologists debating hotly for many decades since Sorby and Jukes-Browne in England, and Cayeux in France, first began to look closely at it towards the end of the nineteenth century.

Chalk is a particularly pure form of limestone, so although it is conventional to say "chalk and limestone", technically this is not quite correct: "chalk and other limestones" would be more proper, if a shade pedantic. Limestones are sedimentary rocks, formed by the deposition in water of solid calcium carbonate, in some cases by inorganic chemical precipitation (see Tarr, 1925), but mainly as organic (bioclastic) remains. The forms in which bioclastic calcium carbonate normally occur are as low- or high-Mg calcite or aragonite. The calcium carbonate of chalk consists almost entirely of low-Mg calcite (Hancock, 1975).

This calcite derives largely from marine algae of the class Haptophyceae which abounded in the plankton of the Chalk Sea. The cell walls of these algae are reinforced by calcite tablets (laths) arranged in rings (coccoliths), sometimes with a stalk (rhabdoliths). Complete and partially dismembered coccoliths and rhabdoliths, as well as individual laths (Fig. 1.5a), account for

Fig. 1.5a. Scanning electron micrograph of Upper Campanian chalk (*Belemnitella langei* zone) from Norfolk, showing a complete coccolith and detached laths (×12 000). (J. M. Hancock)

75–90% of the bulk of most white chalk, endowing it with a large excess of particles in the size-range 0·5–4 μm, with important consequences for its water relations (Sections 1.2.2.3 and 3.6.2). A second peak in the particle size graph, in the range 4–10 μm, is accounted for mainly by the shells (tests) of foraminifera (Fig. 1.5b), and by fragments of bivalves (particularly *Inoceramus*), echinoderm (sea-urchin) plates and bryozoans.

The proportion of impurities in chalk—the insoluble residue left behind

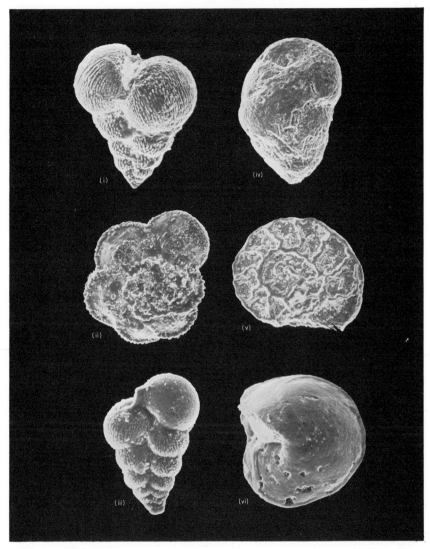

Fig. 1.5b. Scanning electron micrographs of foraminifera from chalk. (i)–(iii) Planktonic species from a borehole in Suffolk; (iv)–(vi) benthonic species from a chalk-pit near Arundel. Approximate magnifications: (i) × 160; (ii) × 110; (iii) × 200; (iv) and (v) × 115; (vi) × 135. (B. M. Funnell)

when the chalk is dissolved away with either hydrochloric or acetic acid—varies considerably. The white chalks are often very pure indeed. Thirteen samples of white chalk analysed by Dr N. Walsh from a wide geographical range of sites in England contained between 96·77% and 99·09% CaCO₃, mean 98·17% (see Hancock, 1975, Table 1 for full details). An example from Newhaven cited by Shepherd (1972) is shown in Table 1.1. On the other hand, up to 50% of the bulk of the more clayey (argillaceous) chalks may be made up of montmorillonitic, micaceous and other clays. Chalks

Table 1.1

Composition of white chalk from near Newhaven [a]

	Component	Concentration (%)
Matrix	Calcium carbonate	97·89
	Magnesium carbonate	0·75
	Silica	0·65
	Calcium phosphate	0·22
	Iron and aluminium oxides	0·14
	Water (combined)	0·35
Flints	Silica	98·00
	Water (combined)	2·00

[a] Geological Survey quoted by Shepherd (1972).

which contain as much clay as this are known as marls, and although they occur in seams throughout the Chalk, they are especially typical of the Lower (Cenomanian) Chalk (Avery, 1964; Jeans, 1968). The clay content of chalk will clearly modify its properties of water retention and cation exchange, as well as its resistance to erosion. These aspects are returned to later in this chapter and in Chapter 3. Along the old Chalk Sea shorelines there is often an increase in the proportion of sand, as in the gritty Cenomanian "chalks" of Devon (now regarded as limestones), which in places are essentially calcareous sands (Edmonds *et al.*, 1969).

The flints so characteristic of the higher horizons of the Chalk are usually found in seams as tabular sheets following bedding planes (Fig. 1.6), but less frequently along vertical joints and fissures. They can persist, of course, long after the chalk beds of which they were a part have been removed by erosion. Occasionally gigantic flints (paramoudras) up to several metres across are found. Flints are essentially lumps of dark-coloured, microscopically crystalline quartz which develop a white rind on exposure due to loss of water. Most geologists agree that they are biogenic, and formed by the redistribution of colloidal silica long after the chalk itself was deposited, but there are arguments and counter-arguments about the complex mechanisms involved in flint formation (see Shepherd, 1972; Hancock, 1975). Where flints are absent, as in the Lower Chalk and lower part of the

Middle Chalk, the concentration of silica in the chalk itself is correspondingly greater. To put it another way, in the flinty chalks there has been an almost perfect separation of the matrix into pure silica and pure calcium carbonate (Table 1.1). Perrin (1965) has demonstrated the abundance of soluble silica in chalk ground water.

Some layers of the Chalk contain impregnations, coatings, nodules and

Fig. 1.6. Alum Bay and the Needles, at the western extremity of the Isle of Wight. The steep Alpine folding of the Upper Chalk is strikingly shown by the bands of flints. (Cambridge University Collection, copyright reserved)

pellets of calcium phosphates, sometimes loosely referred to as collophane, but consisting mainly of carbonate–apatite. As a rule phosphatic chalks are discontinuous in their distribution and are mainly confined to hardgrounds (see below), but they occur in northern France and are particularly extensive in Belgium where they have long been quarried for fertiliser. In England they have been found at widely separated localities in Berkshire: Taplow in the east, and Kintbury in the west, for example. The greenish mineral glauconite, a silicate of iron and potassium, occurs generally throughout the Chalk, but with the exception of certain strata such as the Glauconitic Marl of the Lower Chalk (Section 1.1.3.2), is found only in minute quantities. Pyrite, a form of ferrous sulphide, is commonly found in the Lower Chalk, either dispersed through the matrix where its oxidation to ferric hydroxide stains the chalk yellow, or as rusty nodules ("thunderbolts") containing an attractive radiating crystal structure. It is inappropriate here to delve further into chalk mineralogy. Further information can be found in Weir and Catt (1965), Jeans (1968) and Hancock (1975).

Harder horizons in the Chalk can often be explained by shallower phases during which subsidence and chalk deposition were interrupted. The watery ooze which would otherwise have formed soft chalk was subjected to diagenesis, and the sea floor was cemented, normally to a depth of up to $0\cdot5$ m, and in extreme cases to as much as 2 m. In places, a succession of these so-called *hardgrounds* makes up a single hard stratum, as in the Chalk Rock of the Chilterns. Hardgrounds frequently contain concentrations of phosphate and glauconite. An additional point of interest is that these horizons may contain extensive fossilised burrows of contemporary crustacean inhabitants (Thalassinidea) of the Chalk Sea floor (Bromley, 1967). Rarely, the burrows are found hollow, but more usually they contain unconsolidated fill in which very delicate fossils are sometimes preserved. Alternatively, fully silicified twig-like or reticulate flint casts of the burrows may be found; these were at one time given the name *Spongia paradoxica* because of their superficial resemblance to certain sponges. Simultaneous cementation and burrowing (bioturbation) may have led to the formation of the nodular chalk so frequently encountered.

Elsewhere diagenesis took place much later. Thus some local hardening of chalk (as in parts of Dorset) may be a consequence of tectonic stress, while the much more general hardness of the northern chalks, notably in Yorkshire and Ulster, has resulted largely from infilling and crushing of the original porous matrix by overburden weight and associated phenomena (see Hancock and Kennedy, 1967; Hancock, 1975). It might be noted here that *tufa* has quite a different origin, and is formed by the redeposition of calcium carbonate by the evaporation of saturated drainage water, as in springs.

1.1.3 Chalk stratigraphy

1.1.3.1 Fossil indicators

It is well known that the decline of the great reptiles and the almost explosive evolution of the flowering plants took place during the Upper Cretaceous, but remains of terrestrial organisms are few and far between in the Chalk, amounting to the occasional bone or tooth, rarely even the whole skeleton of an errant flying reptile. Stones of alien composition (erratics) occasionally found buried in chalk may have been carried on rafts of seaweeds or trees blown far out from the shore, or they could even be stomach stones (gastroliths) of marine reptiles (Gallois, 1965). On the whole it is evident that evolution proceeded at a far slower rate in the Chalk Sea than on the land. Yet, thirty-five million years is a long time. Changes did take place and contemporary marine organisms evolved in their own relatively modest way to provide a valuable key to the different chalk strata. The use of foraminifera and coccoliths in stratigraphy has already been noted, while among the macrofauna certain echinoderms (sea-urchins), for example, progressively adapted themselves to burrow deeper into the seabed (Black, 1970).

The abundance of echinoderms in the Upper Cretaceous is well known, but there were still some ammonites to be found in the Chalk Sea, as well as brachiopods, bivalves, gastropods and belemnites. Remains of free-swimming stalkless crinoids (sea lilies), sponges, corals, worm-tubes and the teeth of sharks and rays may also be found. Figure 1.7 illustrates a selection of fossils either common in, or characteristic of the chalk strata. For our purposes, it is adequate to be acquainted with only the main zones of the Chalk, and these, together with their fossil indicators, are summarised in Table 1.2.

1.1.3.2 The main Chalk strata

First a word on terminology. It is conventional to use the proper noun for the main geological formation of the Chalk as a whole, and for recognised subdivisions like the Cenomanian Chalk, the Chalk Marl, and so on, and to regard these as being made up of chalk (or chalks) of varying lithology. Hence the sometimes baffling interchange of capital and lower-case c's.

It is worth brief mention that the earliest chalk to be laid down in Britain was red, not white, and of Lower Cretaceous (specifically Albian) origin, not Upper: its fossils match those of the Gault Clay and Upper Greensand. This was the Red Chalk, or Hunstanton Red Rock, described by Wilson (1948) as a ferruginous, hard, nodular chalky limestone, ranging in colour from pink to brick red, and containing large, rounded grains of quartz. It is confined mainly to Norfolk, and to Lincolnshire and Yorkshire where it

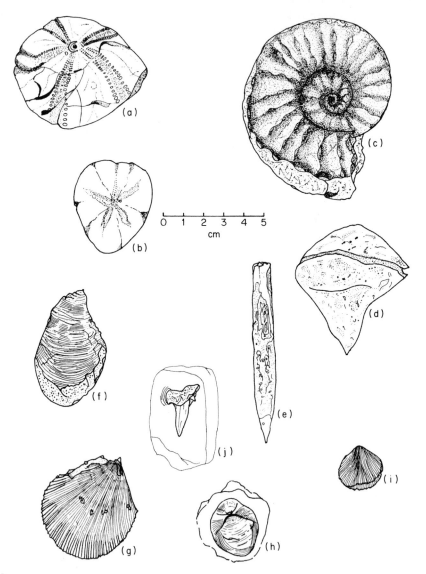

Fig. 1.7. Some characteristic fossils of the Chalk, drawn from specimens in the County Museum, Aylesbury: (a) *Echinocorys* sp. and (b) *Micraster cor-anguinum* (echinoids); (c) *Schloenbachia ?subtuberculata* (an ammonite); (d) *Ventriculites* sp. (a sponge); (e) *Actinocamax plenus* (a belemnite); (f) *Inoceramus mytiloides* and (g) *Spondylus spinosus* (bivalves); (h) *Terebratula carnea* and (i) *Rhynchonella pucatilis* (brachypods); (j) shark's tooth.

underlies the Wolds. It is well seen, as its alternative name implies, underlying white chalk at Hunstanton, but it reaches its greatest thickness at Speeton, North Yorkshire.

Most ecologists familiar with the English Chalk are aware of its three main subdivisions: the Lower, Middle and Upper Chalk. Some will know that these divisions correspond approximately (though not exactly) with the Cenomanian, Turonian and Senonian stages named by continental stratigraphers. Among geologists, however, the English names are now regarded as obsolete and even the continental scheme has recently been much modified, with the Senonian stage replaced by Coniacian, Santonian and Campanian, and the adoption of alternative zones and indicator fossils. The uppermost layers of the Chalk exposed in Britain are now recognised as Lower Maastrichtian. Table 1.2, taken from Hancock (1975), is a helpful guide to the revised stratigraphy of the Chalk, but as the author himself states, this should not be regarded as final. A more detailed discussion will be found in Rawson *et al.* (1978). For all this, however, English names are retained here, not with the deliberate intention of perpetuating obsolescence, but simply because no work on soils or vegetation with a geological emphasis has yet made use of the new scheme and, as can be seen from Table 1.2, it is not possible simply to translate old to new.

Two main subdivisions of the Lower (Cenomanian) Chalk are recognised. The lowermost is the Chalk Marl, including at its base the Glauconitic Marl (originally misnamed the Chloritic Marl), containing pelletal glauconite and phosphatised pebbles. Above this is the massive Grey Chalk, mainly grey, but white and even pink in parts of Lincolnshire and Yorkshire. Although in England the Lower Chalk becomes more sandy towards the west and less clayey (argillaceous) north of Norfolk, its predominant feature in the south is its marliness: it is the "white clay" of the English agricultural historians and, usually forming a landscape of subdued relief, is exploited widely for arable agriculture. It makes, too, an ideal mixture of chalk and clay for use in the manufacture of cement (see Chapter 9).

Nevertheless, there are harder seams (the rags), often visible as breaks of slope, within the marly chalks (Fig. 1.8). For example, a notable feature of the Lower Chalk from Buckinghamshire and Bedfordshire northwards is a band of hard, grey, sandy chalk, the Totternhoe Stone, named after the village of that name in Bedfordshire, and which has been widely used for building. This is known as the Burwell Rock in Cambridgeshire. There is locally a thin layer of glauconitised chalk pebbles at its base containing phosphatic nodules. The Marl Rock is another hard bed locally developed in the Lower Chalk which often betrays its presence as a spring line. The possible significance of these local variations in hardness and marliness for landscape evolution, as well as for soil formation, is obvious. Indeed,

Table 1.2

Table 1.2
Divisions of the Upper Cretaceous in south-east England [a,b]

BIOSTRATIGRAPHICAL UNITS			LITHOSTRATIGRAPHICAL UNITS	
STAGES	ZONES	Subzones or other zones		
MAASTRICHTIAN — Upper	*Belemnella kazimiroviensis*		White-chalk; absent on land in British Isles, but present in North Sea	
	Belemnitella junior			
MAASTRICHTIAN — Lower	*Belemnella occidentalis*			
	Belemnella lanceolata			
CAMPANIAN — Upper	*Belemnitella langei*			UPPER CHALK—largely white-chalk with flints
	Belemnitella minor			
	Belemnitella mucronata			
CAMPANIAN — Middle	*Gonioteuthis quadrata*			
CAMPANIAN — Lower	*Offaster pilula*	*Echinocorys cincta*	Planoconvexa Bed (1m white-chalk enclosed in two beds of marl)	
		Echinocorys depressula		
SANTONIAN	*Marsupites testudinarius*			
	Uintacrinus socialis			
	Micraster coranguinum			
CONIACIAN	*Micraster cortestudinarium*	*Micraster decipiens*		
		Micraster normanniae		
TURONIAN — Upper	*Holaster planus*		Top Rock	
			Chalk Rock	
TURONIAN — Middle	*Terebratulina lata*		Spurious Chalk Rock	MIDDLE CHALK
TURONIAN — Lower	*Inoceramus labiatus*		Melbourn Rock	
CENOMANIAN — Upper	*Sciponoceras gracile*		Plenus Marls	LOWER CHALK—partly developed in chalk – marl facies
	Calycoceras naviculare			
CENOMANIAN — Middle	*Acanthoceras rhotomagense*		Totternhoe Stone	
CENOMANIAN — Lower	*Mantelliceras mantelli*		Glauconitic Marl	

[a] From Hancock (1975).

[b] The zonal succession is complete, but there are many other possible zones and subzones which can be, and have been, used, particularly based on species of *Echinocorys* and *Inoceramus*, quite apart from microfossil zones. The lithological units listed are not all present in any one district. This table is not drawn to scale; for thickness of stages see Table III of Hancock's paper. For a more detailed treatment of Chalk stratigraphy, see Rawson *et al.* (1978).

distinct topographical features can sometimes result from very small litho-logical differences (see below). Likewise, local concentrations of phosphate and glauconite might be expected to show up in soils developed on these beds, although their very hardness may lessen their agronomic and ecological significance. It is impractical to deal here with every local geological variation of this kind, and the reader should consult the appropriate Memoirs and Handbooks of the Geological Survey. Useful summaries of information on the Chalk can be found in Wilson (1948), Chatwin (1960, 1961), Sherlock *et al.* (1960), Gallois (1965) and Edmonds *et al.* (1969).

The Middle Chalk is much purer, and so whiter, than the Lower Chalk. It tends, too, to be both harder and more permeable, so resisting erosion more effectively and often forming important topographical features where the angle of dip allows (Fig. 1.14). This property is enhanced by the occurrence at the base of the Middle Chalk of a hard, nodular bed, the Melbourn Rock (named after the village of Melbourn in Cambridgeshire), rendered parti-cularly hard and gritty towards the top by an abundance of shell fragments of bivalves of the genus *Inoceramus*. In Devon, where the Melbourn Rock is not developed, a similar hard, gritty layer forms the well-known Beer Stone, long quarried and used extensively for building. The general lack of flints in the Middle Chalk has given it the name of the "white chalk without flints" in southern England, but it does contain flints in its uppermost layers, and is altogether more flinty further north.

The Upper Chalk is, as we have seen, white and very pure, but with abundant flints. It is particularly strongly fissured. A hard, creamy coloured band, the Chalk Rock, actually a complex of hardgrounds, is found at the bottom of the Upper Chalk. Here is an example where it is important to be careful with the capital letters, for many authors use this name without capitals—chalk rock—to describe any chalk parent material, or C horizon, in soil profile analysis, regardless of whether or not the true Chalk Rock is involved. The Top Rock is another named hard band in the Upper Chalk. Locally, as we noted previously, chalk rich in phosphate occurs, as at Taplow and Kintbury. North and south of the London Basin a greater thickness of the Upper Chalk has been preserved, particularly in Norfolk, Purbeck and the Isle of Wight, where the fossil record shows beds higher in the sequence than any others in England.

The lower layers of the Upper Chalk, sometimes called the Echinoid Chalk after their indicator fossils, tend to be more resistant to erosion than the uppermost strata, the Belemnite Chalk (Fig. 1.9), but the reasons for this are not always obvious. In some places a slight increase in marl content seems to be the cause (Sparks, 1960); elsewhere local hard bands such as the Whitway Rock of north Hampshire, and others unnamed, may be re-sponsible (Hawkins, 1931). In Northern Ireland the Upper Chalk (the

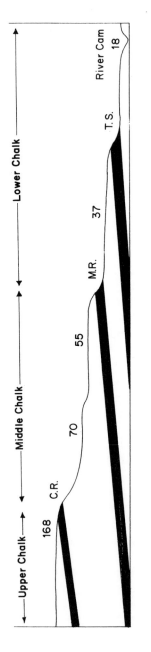

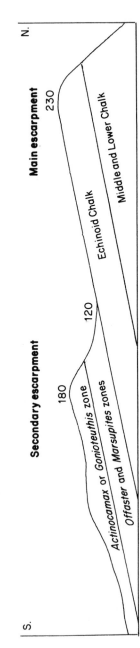

Fig. 1.8 (top). The influence of hard strata on the topography of the Chalk in south-west Cambridgeshire. C.R. = Chalk Rock; M.R. = Melbourn Rock; T.S. = Totternhoe Stone. Altitude is in metres O.D. and the vertical scale is exaggerated. From Sparks (1960).

Fig. 1.9 (bottom). The South Downs escarpments in relation to the divisions of the Chalk. Altitude is in metres O.D. and the vertical scale is exaggerated. From Sparks (1960).

White Limestone), which rests directly on Greensand, has little ecological significance, for, as we shall see shortly, it has been covered by basalt. Indeed, where these basalts extruded, the surface of the chalk was baked (metamorphosed) to marble, although this was not the main cause of the hardness of the chalk there. A recent discovery of Senonian Chalk in south-west Ireland (Walsh, 1966) is of great interest, but, like the Albian and Ulster chalks, is of limited ecological significance because so little is exposed at the surface.

The fourth and most recent series of European chalk deposits is the Danian, so called from its extensive occurrence in Denmark, though, as was noted previously, this is of Tertiary origin. In any case, Danian Chalk is absent from Britain, other than as a submarine outcrop on the bed of the Western Approaches, and this is hardly relevant to the terrestrial ecologist. The relationship of the Chalk outcrops of mainland Europe to their distribution in Britain is shown in Fig. 1.10.

1.2 The Evolution of the Chalk Landscape

1.2.1 The Tertiary Era*

1.2.1.1 The Palaeocene, Eocene and Oligocene Periods

Towards the close of the Upper Cretaceous Period, earth movements were renewed, and about sixty-five million years ago the chalk seabed began to rise above water level, heralding the opening of the Tertiary Era (Fig. 1.1). Uplift began in the north and west sooner than in the south and east, and it is possible that marine erosion started to plane off the surface of the Chalk before it actually emerged. In any event, the sea did finally recede, exposing the chalk landmass for the first time to the forces of subaerial erosion. As much as 170 m of chalk was removed in the vicinity of what is now the London Basin to form the so-called Sub-Eocene surface, subsequently preserved by renewed sedimentation (Fig. 1.16). Further north, the Chalk was to be eroded away entirely, except in the far north-west of Britain. There, during early Tertiary times (possibly even before the end of the Upper Cretaceous), intense volcanic activity developed, pouring out lava over the Chalk of Northern Ireland and north-west Scotland, and protecting some of it from erosion there to the present day.

Alternating marine and estuarine conditions returned to south-east England in due course, depositing the sands, clays and marls which are now a feature of the London and Hampshire Basins. Palaeogeographic maps of

---

* Alternatively referred to as the Tertiary Period within the Cenozoic Era, with its subdivisions given Epoch status (e.g. Eocene Epoch).

the extent of these successive sedimentations have featured in several classic textbooks on stratigraphy, but more recent work has indicated the need for revision of these ideas (Owen, 1976). Here, suffice it to say that Palaeocene and Eocene deposits overlie the Chalk in both the London and Hampshire Basins, and Oligocene deposits in the latter only. The nature of these deposits is summarised in Table 1.3.

Much work has been done to reconstruct the contemporary terrestrial flora and fauna (see, for example, Pennington, 1974), but, tempting as it is to speculate on the ecology of the tropical Tertiary chalklands, this is another highly controversial field, and is not pursued here.

1.2.1.2 The Alpine Orogeny

Meanwhile, the continents were drifting on. Africa had been steadily catching up with Europe and as the ocean between them closed, they finally met in the protracted tectonic collision which formed the Alps some time early in the Miocene Period. Southern England took a considerable buckling, and a marked series of east–west folds was formed (Fig. 1.11), well seen in the Isles of Wight and Purbeck (Figs 1.6 and 1.12), and of enormous significance to subsequent landscape development. Anticlines and synclines formed ridges and valleys, usually but not necessarily respectively, for where anticlinal warping weakened a chalk crest to expose a stratum less resistant to weathering, a valley would be formed. This is known as inverted relief, classically seen in the Weald of Kent and Sussex, though visible on a smaller scale in the Greensand Vale of the Isle of Wight (Fig. 1.12). The subject is discussed in detail by Small (1970) who has made a close study of this aspect of chalk landscape evolution.

The influence of the tilt of a stratum on escarpment formation and recession has already been mentioned. Clearly, where local folding reverses

Fig. 1.10. Outcrops of Chalk in (a) Britain and (b) north-west Europe. Drift not shown. (a) Adapted by permission of the Institute of Geological Sciences from the Geological Survey's 1:625 000 map (1957 Edition, Sheet 2), and including approximate positions of main topographic features and key localities mentioned in the text. BA = Barton Hills, BD = Berkshire Downs (now officially in Oxfordshire), BE = Beer Head, BH = Beachy Head, BL = Breckland, CC = Cranborne Chase, CH = Chiltern Hills, DO = Dorset Downs, DU = Dunstable Downs, DV = Devonshire outliers, EA = East Anglian Heights, FH = Flamborough Head, GM = Gog-Magog Hills, HB = Hampshire Basin, HC = Hampshire Chalk, HG = Hog's Back, HU = Hunstanton, IP = Isle of Purbeck, IT = Isle of Thanet, IW = Isle of Wight, LB = London Basin, LW = Lincolnshire Wolds (now partly in South Humberside), MD = Marlborough Downs (sometimes incorporated with the Berkshire Downs as the North Wessex Downs), ND = North Downs, PD = Portsdown Hill, SD = South Downs, SP = Salisbury Plain, VA = Vale of Aylesbury, VO = Vale of Oxford, VP = Vale of Pewsey, WD = Weald of Kent and Sussex, YW = Yorkshire Wolds (now mainly within North Humberside). (b) From Shimwell (1973).

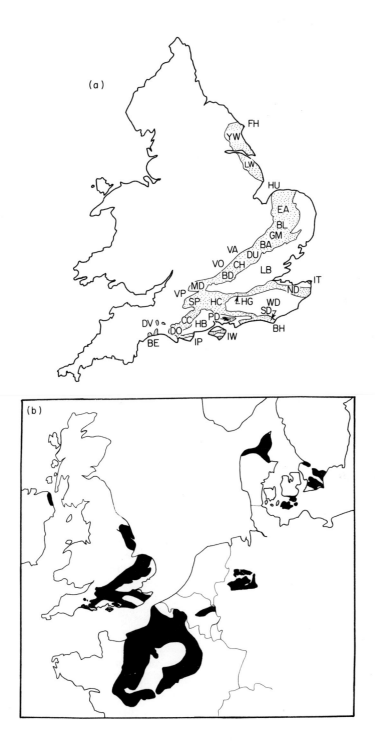

(a)

FH
YW
LW
HU
EA
BL
GM
BA
VA
DU
VO
CH
BD
LB
MD
VP
SP
HC
HG
WD
ND
IT
SD
DV
CC
PD
BH
DO
HB
BE
IP
IW

(b)

Table 1.3

Tertiary deposits overlying the Chalk (more recent strata towards the top)

PLIOCENE	Lenham Beds	Sands; North and South Downs, especially towards the east[a]
OLIGOCENE	Hamstead Beds	Clays, loams, sands and shales
	Bembridge Marl	Marls
	Bembridge Limestone	Limestone
	Osborne Beds	Clays and marls
	Headon Beds	Marls and clays with bands of limestone and sand
EOCENE	Barton Beds	Clays
	Bracklesham Beds	Clays and sands
	Bagshot Beds	Sands
	Claygate Beds	Sandy clay, with sandy intercalations
	London Clay	Stiff clay, sandy at top
	Oldhaven and Blackheath Beds	Sand and shingle; overlie Chalk locally in North Downs
	Reading Beds	Clays and sands with pebble beds;[b] main source of sarsens and puddingstones
	Woolwich Beds	Clays, loams, sands and pebbles, locally a source of conglomerate; overlie Chalk in North Downs
PALAEOCENE	Thanet Beds	Sands and silts;[b] form a narrow outcrop along the margin of the North Downs Chalk

[a] Also certain high-level gravels, e.g. in the Chilterns.
[b] Often fill solution pipes and hollows in Chalk.

the angle of dip, as in the London Basin syncline (Fig. 1.13), two mirror-image scarps will be formed. Thus while the Chiltern escarpment continues to recede south-eastwards, the North Downs escarpment is actually re-treating north-westwards to meet it: a fact of no small significance to future inhabitants of the London commuter belt! Another point not always appreciated is that the steepest angle of tilt does not lead to the boldest topographical features: where the strata are presented edge-on to the forces of subaerial erosion, the outcrop is more readily worn away. This can be seen by comparing the altitude of the steeply inclined Hog's Back of Surrey and central ridge of the Isle of Wight with adjacent shallower-angled outcrops of the same formations (Fig. 1.14). Of course, these are all features

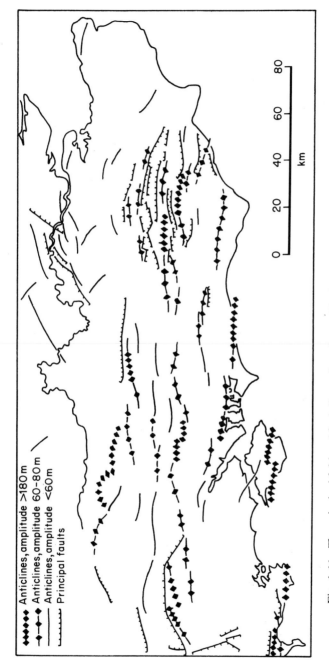

Anticlines, amplitude >180 m
Anticlines, amplitude 60–80 m
Anticlines, amplitude <60 m
Principal faults

0   20   40   60   80
km

Fig. 1.11. The principal fold and fault lines of south-east England. From Wooldridge and Linton (1955).

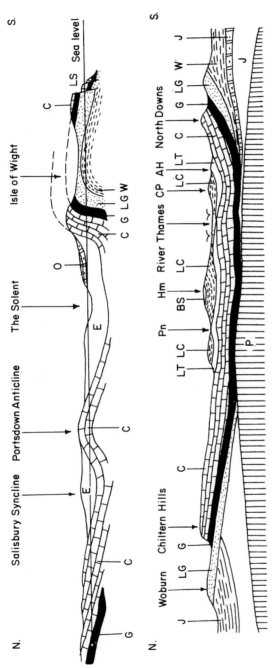

Fig. 1.12 (top). Geological cross-section of the Hampshire Basin and Isle of Wight. Vertical scale exaggerated. Strata shown are (oldest first): Wealden (W), Lower Greensand (LG), Gault and Upper Greensand (G), Chalk (C), Eocene (E), Oligocene (O) and the landslip (LS). From Stamp (1967).

Fig. 1.13 (bottom). Geological cross-section of the London Basin. Strata shown are the Palaeozoic floor (P), various Jurassic rocks (J), Wealden Beds (W), Lower Greensand (LG), Gault and Upper Greensand (G), Lower London Tertiaries (LT), London Clay (LC) and Bagshot Sand (BS). Localities indicated by letters are Crystal Palace (CP), Hampstead (Hm), Pinner (Pn) and the Addington Hills (AH). From Stamp (1967).

of a relatively youthful landscape. The longer subaerial erosion can continue uninterrupted by tectonic upheavals or changes in sea-level, the greater is the likelihood that the ultimate stage in landscape evolution, the peneplain, will be reached. Thus in parts of Salisbury Plain such as the Great Ridge, the trained eye can pick out remnants of the old chalk peneplain of Miocene-Pliocene times (Wooldridge and Linton, 1955).

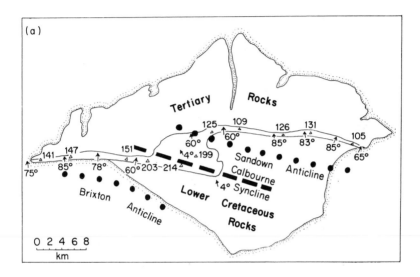

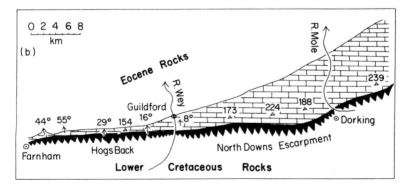

Fig. 1.14. The relationship between angle of dip and elevation of the Chalk outcrops of (a) the central ridge of the Isle of Wight and (b) the North Downs and Hog's Back, Surrey. Arrows indicate angle of dip (degrees from horizontal) and triangles O.D. in metres. From Sparks (1960).

1.2.1.3 The Pliocene Sea

It seems generally less widely appreciated that much of the southern chalklands were subject to another marine incursion by that fickle ancestor of the modern North Sea early in the Pliocene Period, about 12 million years ago. This, the Pliocene or Calabrian Sea, left the shelly Crag deposits of East Anglia, and possibly the sandy Lenham Beds of the eastern South and North Downs, as well as some of the higher-level gravels to be found scattered about the Downs and Chilterns further west (Fig. 1.15), though this is another area of controversy.

Most of the Pliocene deposits have been worn away, and the bench of the sea shore, or surface of the seabed, are all that remain (Fig. 1.16), but these elusive imprints may have local ecological significance through their influence on soils and vegetation through microtopography and micro-climate (Chapters 2 and 3), just like the prehistoric earthworks to be described in due course.

1.2.2 The Pleistocene glaciations

1.2.2.1 The advance of the ice

Because of changes in the Earth's climate as a result of oscillations in the intensity of the sun's radiation which have yet to be fully explained (see Zeuner, 1959; West, 1968; Crowe, 1971), temperatures now plunged, seas froze, glaciers formed and the great northern ice sheets of the Pleistocene Period spread southwards to influence, directly or indirectly, all the chalklands of Britain for over a million years. Directly, ice from indigenous glaciers in highland Britain and from the Scandinavian ice sheet which crossed the North Sea, repeatedly ground and scraped its way across the Wolds of Yorkshire and Lincolnshire and the East Anglian Heights, perhaps even carving out the missing stretch of escarpment east of the Hitchin Gap (Linton, 1963). Each time, it left behind its great burden of rubble, clay and other solids (till) as it melted and retreated (Fig. 1.17). Indirectly, the land to the south of the ice cap was held in the grip of *periglacial* conditions, which froze the ground to a great depth. Only the surface layer thawed in the brief summer, and in hilly areas slope wastage occurred on a large scale. There was probably no period throughout geological history which had a more profound effect over such a short time upon the topography of the chalk landscape than this million years of the Pleistocene.

1.2.2.2 The direct effects of glaciation

It is in East Anglia that the most extensive deposits of glacial till are found, formerly known as boulder clay, but in fact consisting of every combination

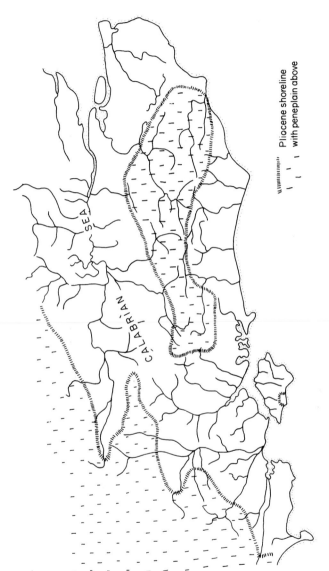

Fig. 1.15. The Pliocene (Calabrian) Sea in south-east England. From Sparks (1960).

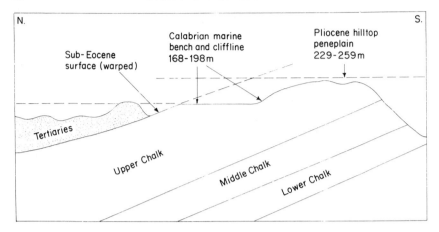

Fig. 1.16. The relationship between the Sub-Eocene surface, the Pliocene peneplain and the Calabrian bench in the central North Downs. From Small (1970).

of chalk, sands and clays, rocks, pebbles and gravel, including erratics and "far-travelled stones", all testifying to the repeated advance and retreat of glacial ice of subtly differing provenance and properties. The geology of Pleistocene East Anglia is extraordinarily difficult to unravel, and largely outside the terms of reference of this book, but it is worth noting that the last (Devensian or Hunstanton) glaciation, which ended about 11 500 years ago, was not the most extensive. This distinction is held by the second main surge, the Anglian or Great Eastern, which reached its peak about one million years ago. Between these two was the Wolstonian or Little Eastern episode, and of course long spells of much warmer conditions (interstadials and interglacials) interspersed the glaciations. The possibility that we may now be experiencing the beginnings of a return to a further glacial period has been well aired in recent years.

Mention ought to be made in passing of the Chalky Boulder Clay of Norfolk, Suffolk, Hertfordshire, Cambridgeshire and Lincolnshire, for this "composite till of tough and unstratified stony clay with many fragments of chalk and flint" (Small, 1970) demonstrates well both erosive and dumping powers of ice with specific reference to the Chalk. The properties of some of the facies of the Chalky Boulder Clay have been reviewed by Baden-Powell (1948). Curious contortions in the Chalk and other strata in East Anglia have been ascribed to glacial disturbance (Chatwin, 1961; see also Section 9.4.2.5).

Reference to Fig. 1.17 will show that the uplands of Lincolnshire and Yorkshire had their share of glaciation, but in fact the Wolds failed to accumulate the depth of till to be found in East Anglia. Partly, this was due

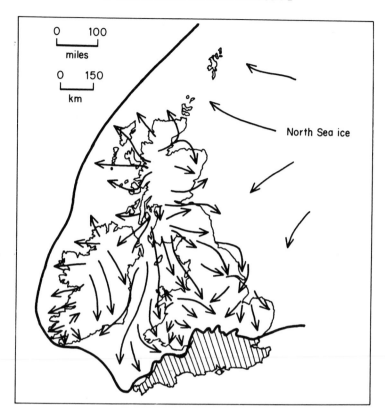

Fig. 1.17. The long-established picture of the probable maximum extent of the Pleistocene ice cap over Britain. From Owen (1976). Evidence for more widespread glaciation in southern England continues to accumulate: for example, ice from the west may have reached Salisbury Plain and even the Marlborough Downs during Anglian times (Kellaway, 1971).

to altitude which was insufficient to generate an ice cap but great enough to ensure a less prolonged receipt of glacial deposits than the lower areas of East Anglia to the south. But in addition to this, the Wolds were apparently free from ice for much of the latter half of the Pleistocene Period (Fig. 1.18). They received none of the tills collectively termed the Newer Drift, and there was a very long period over which the cover of Older Drift of the earlier glaciations could be steadily removed again (Wilson, 1948). The result is that much of the Wolds, particularly of Yorkshire, are today remarkably free from glacial drift.

1.2.2.3 The effects of periglacial conditions on exposed chalk
During the Pleistocene glaciations, periglacial conditions prevailed

throughout the southern chalk country, and alternated further north with glacial conditions as the ice cap oscillated and finally retreated altogether. Many areas, especially in the south and in the Yorkshire Wolds, were deeply dissected during this period, and retain to the present day what in effect is a fossilised late glacial landform arrested in its development in a relatively youthful state. The mainspring of periglacial erosion was a subsoil frozen to a considerable depth (permafrost), and a surface layer (active layer) which, though frozen and snow-covered in winter, thawed rapidly in the spring for as long as the temperature remained above freezing point. It seems appropriate here to digress briefly in order to consider the properties of chalk which determined its response to these conditions.

Chalk has an enormous capacity to absorb water because it is highly

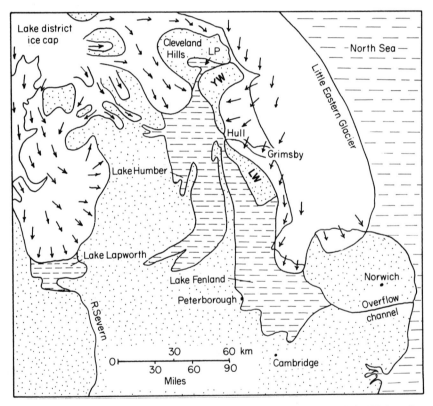

Fig. 1.18. The Lincolnshire and Yorkshire Wolds (LW and YW) in Wolstonian times, strikingly isolated by ice to the east, and the ponded-up meltwaters of Lakes Pickering (LP), Humber and Fenland to the north, west and south. After Stamp (1967).

porous. White chalk typically has a pore space of between 35% and 47% of its total volume (Hancock, 1975), so that a block of chalk 1 m³ in volume may hold nearly 0·5 m³ (500 l) of water. However, the voids of the chalk are very small (Fig. 1.19), and transmissivity of water in the matrix is extremely low. On the other hand, numerous joints and fissures endow white chalk with a high permeability, so that rain water not drawn into the matrix drains rapidly down to the water table. These properties have important implications for plant–water relations which we consider in later chapters, but the point here is that, physically, water has very little direct effect upon white chalk, which thus tends to resist erosion and to be left upstanding, as we have already seen. Even quite soft chalk may be left essentially intact as long as it is pure. The more marly bands are both less porous and less permeable, and so more prone to erosion. These differences are scrutinised more closely in due course.

Chemically, the direct effect of water is likewise small, for calcium carbonate is only slightly soluble in water. Were the chalk to be acted upon by moderately warm distilled water, then it might be expected to show essentially no signs of erosion at all. But percolating water typically contains carbon dioxide in solution as carbonic acid, picked up as it fell as rain and

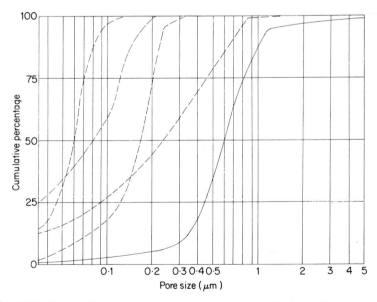

Fig. 1.19. Cumulative pore-size curves for white chalk (solid line) and other lime-stones (broken lines) from Normandy and the Charentes, France. Data from the *Centre de Géomorphologie de Caen*. After Pedro (1972).

particularly by trickling through foliage, roots and humus, and this greatly enhances its properties as a solvent, for calcium bicarbonate is formed, which is far more soluble in water than calcium carbonate:

$$CO_2 + H_2O \rightleftharpoons H_2CO_3$$

$$CaCO_3 + H_2CO_3 \rightleftharpoons Ca(HCO_3)_2.$$

This process, *carbonation*, is the prime agent in chemical weathering of chalk, but a fact often overlooked is that as the temperature falls the solubility of carbon dioxide, and so calcium carbonate, in water increases (Table 1.4). Moreover, as is well known, water expands as it cools and freezes. Only one freeze–thaw cycle is needed to turn a block of saturated chalk, particularly of the soft kind, into a heap of wet powder, and it is plain to see that even without invoking any other processes, a change to cold conditions makes chalk very vulnerable indeed to erosion.

Table 1.4

The effect of carbon dioxide (as carbonic acid) on the solubility of calcium carbonate at different temperatures [a]

Carbon dioxide content of water	Temperature (°C)	Solubility of $CaCO_3$ in water (g l^{-1})
Water and adjacent air free from CO_2	25 50	0·014 0·015
Water exposed to normal atmospheric concentration of CO_2 (300 vpm)	0 10 20	0·081 0·070 0·065
Water saturated with CO_2 at 1 atm pressure	0 15	1·500 1·175

[a] From Parkes (1956).

In that periglacial climate there were processes aplenty to ensure the constant removal of the products of frost shattering and low-temperature carbonation, which were to mould the characteristic (though not unique) convexo-concave slopes and valleys (coombes) of the chalk country which we see today. It is not difficult to imagine the effects of a sudden spring thaw on a long winter's accumulation of snow and ice, with or without accompanying rainfall, in undulating terrain still saturated and frozen solid less than half a metre below the surface. Following in the wake of this annual spring scouring of the valleys, other phenomena, less dramatic but no less important, played their part, particularly nivation, spring-sapping and solifluxion.

Nivation refers to the curious ability of snow patches, typically those persisting in summer on shaded, north-facing slopes, to carve out shallow, corrie-like hollows into the hillside in a way still not fully understood. Bull (1940) invoked nivation-hollowing to explain certain features of the north-facing escarpment of the South Downs near Eastbourne, Sussex. *Spring-sapping* occurred where percolating water encountered the freeze-line, or an impervious marly layer, in the chalk. The bed of chalk surrounding the spring, sometimes initially turned into tufa, would ultimately collapse to form a small cave, and the process could go on repeating itself.

In *solifluxion*, melting ice within the interstices of the surface waste expanded and lubricated the whole saturated mass of chalk rubble, clay, gravel and flints, which could then flow downhill to accumulate and con-solidate on the valley sides and in the bottoms as Head, or Coombe Rock (Fig. 1.20). According to Small (1970), solifluxion debris might typically

Fig. 1.20. Solifluxion debris (Coombe Rock or Head) exposed in the cliffs at Fresh-water Bay, Isle of Wight. (C.J.S.)

flow at a rate of 2–5 cm/year down a 10–15° slope. This is far slower than seems to be generally appreciated, though still of great significance over a period of many thousands of years. During times of less severe climate, particularly when soil could form and vegetation establish, movement would have been much less rapid and limited in the main to soil creep, though after a cold winter, spring meltwater could still be a potent agent of erosion. Lord Kennet's (1940) letter to *Nature* about the stream which flowed clean through his house describes a classic example of this process on a small scale in modern times, though altogether more dramatic were the Hawkley land-slip of March 1774 (Section 1.2.3), and the great Till flood of January 1841. Here, a rapid thaw and heavy rain after prolonged frost and snow brought floodwater from a still-frozen Salisbury Plain sweeping down the Till and Wylye valleys into the Avon. Three people were drowned and 72 cottages (built of cob—Section 9.2) destroyed (Cross, 1967).

A curious feature of chalk country, long recognised by European geomor-phologists (e.g. Pinchemel, 1954), is the asymmetry of the valleys (Figs 1.21 and 9.8). In the Chilterns, for example, slopes facing south or west are typically steeper than those facing north or east (Ollier and Thomasson, 1957), although elsewhere this pattern can be reversed. It can be logically argued that the weathering processes outlined above have been more active on the steep slopes, exposed as they are to maximum insolation and the greatest extremes of temperature and humidity, than on the shallower slopes. The trouble is, valley conformation can be equally plausibly explained by the very opposite argument: that excessive weathering results in a long, shallow valley side rather than a shorter, steeper one. So the true mode of origin of the asymmetrical valleys has to remain an open question.

Locally, combinations of the more intensive forms of erosion joined forces to carve out gigantic gashes in the chalk scarps, sometimes with a curious right-angled turn, and seemingly out of all proportion to the gentle landscape surrounding them. A good example is Incombe Hole, near Iving-hoe, shown in Fig. 1.22. Rake Bottom near Butser Hill, Hampshire (Small, 1958), and the valleys in the Hertfordshire chalk scarp near Pegsdon (Sparks and Lewis, 1957) are examples which have been documented. The impressive Happy Valley at Great Kimble in the Buckinghamshire Chilterns is so steep that a true turf is unable to form on the loose scree (Fig. 8.14).

1.2.3 Modern exposures of raw chalk

The very nature of chalk and the way it erodes makes natural steep cliffs rare features of chalk landscapes, but these can be formed where the chalk is undercut by a river, or by the sea. Such features are, of course, in no way confined to periglacial climates, and provide modern examples of natural

Fig. 1.21. An asymmetrical dry valley in the Chilterns near West Wycombe, Bucking-hamshire. The shallower east-facing slopes and valley floor are under arable cultivation. The steep west-facing bank in the foreground has been derelict for over 40 years and provides classic examples of the ungrazed chalk grassland and scrub communities described in Chapters 6 and 7. Of great botanical and entomological interest, this site, known locally as Buttlers Hangings, is now leased from the West Wycombe Estate as a Nature Reserve by the Berks., Bucks. and Oxon. Naturalists' Trust (see Chapter 10). (C.J.S.)

exposure of the raw chalk. River cliffs are rare and modest in extent in the English Chalk (the Mole cliff below Box Hill, Surrey, is a well-known exception), though they occur widely in northern France: the Seine Valley is particularly spectacular. Sea cliffs of chalk form some of our most impressive coastal scenery. The cliffs of Dover, the Seven Sisters of the Sussex coast and the Needles of the Isle of Wight (Fig. 1.6) are among the more famous examples. The white chalks form the boldest cliffs, especially if they contain flints, and the cliff face may be quite vertical where jointing favours undercutting. Cliffs of the more marly chalks typically slope more gently. Liger (1956) reports that slumping is more frequent on the Cenomanian Chalk of the Normandy coast than where Turonian or Senonian deposits form the cliffs.

Landslips occur on an altogether larger scale where an underlying

Fig. 1.22. Incombe Hole, Ivinghoe, in the Buckinghamshire Chilterns near the Hertfordshire border. The floor of the valley exhibits the curious right-angled bends characteristic of these features. Some slumping can be seen on the opposite bank. (C.J.S.)

impermeable stratum, such as Gault Clay, dips seaward to form a lubricated platform (see Figs 1.12 and 1.23), so that the ground "fidgets to be in the sea", forming broken terrain of great biological interest (see Chapter 4). Landslips involving chalk occur along the southern coast of the Isle of Wight, at Folkestone Warren, Kent (Hutchinson, 1969), and between Lyme Regis and Axmouth (Arber, 1940, 1973; Fig. 1.24). It was here, at Dowlands Chasm, that the great slip of 1839 wrenched away a huge chunk of farmland, leaving the season's winter wheat crop perched on lofty pinnacles, from which precarious stations it was harvested, amid great celebrations, the following summer (Fig. 1.25).

Such phenomena are not always confined to the coast. Gilbert White (1788) describes a slip at Hawkley Hanger, a precipitous stretch of the western North Downs escarpment in Hampshire which so impressed Cobbett nearly half a century later. White's observations, recorded in his forty-fifth letter to Thomas Pennant, vividly describe the effects of an excessively wet spring:

> The months of January and February, in the year 1774, were remarkable for great melting snows and vast gluts of rain, so that by the end of the latter month the land-springs or lavants, began to prevail, and to be near as high as in the

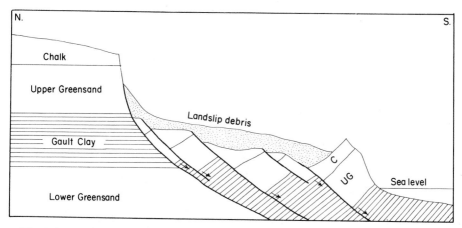

Fig. 1.23. Rotational slipping where Chalk (C) and Upper Greensand (UG) overlie impermeable Gault Clay. From Small (1970).

memorable winter of 1764. The beginning of March also went on in the same tenor; when, in the night between the 8th and 9th of that month, a considerable part of the great woody hanger at Hawkley was torn from its place, and fell down, leaving a high freestone cliff naked and bare, and resembling the steep side of a chalk-pit. It appears that this huge fragment, being perhaps sapped and undermined by waters, foundered, and was engulfed, going down in a perpendicular direction; for a gate which stood in the field, on the top of the hill, after sinking with its posts for thirty or forty feet, remained in so true and upright a positon as to open and shut with great exactness, just as in its first situation. Several oaks also are still standing, and in a state of vegetation, after taking the same desperate leap.

1.2.4 The superficial deposits

Having digressed to present-day geomorphological processes, we return here to our more strictly sequential account of the evolution of the chalk landscape. All through the millions of years that the chalk was exposed to subaerial erosion, a residue of its less-weatherable constituents, mainly clay and flints, had steadily accumulated on the plateaux as a mantle of varying depth. To this mantle, as we have seen, were added the sands, clays, marls and gravels of the Tertiary inundations, and as these in turn were subjected to weathering, their residues joined those of the chalk. Where full glacial conditions prevailed, this surface layer must have been over-ridden and eroded by the ice, and incorporated into the great *mélange* of till that made up its burden. In periglacial conditions, the deposits of the plateaux were stirred and mixed by cryoturbation to form the Clay-with-flints, Plateau Drift and Pebbly-clay-and-sand (see Fig. 1.26).

In places, Tertiary deposits (especially Reading Beds) became indurated

Fig. 1.24. The Axmouth–Lyme Regis landslip (now a National Nature Reserve), photographed in 1955. Rabbit-grazed turf intersperses the scrub on Goat Island to the left, while dense ashwood grows in Dowlands Chasm in the centre and foreground of the picture. The cultivated fields no longer extend quite so close to the edge of Bindon Cliff: there is now an intervening strip of tall grassland dominated by *Arrhenatherum elatius*. (Cambridge University Collection, copyright reserved)

to form a solid duricrust of silcrete, sometimes including a conglomerate of pebbles or flints, which subsequently broke up into the familiar sarsens and puddingstones, many of which persist as spectacular monoliths (Davies and Barnes, 1953; Clark *et al.*, 1967). Remarkable assemblages of sarsens are to be found in the rock streams of the Marlborough Downs (Fig. 1.27), while others have been used in the many stone circles of which Stonehenge and Avebury are among the best known.

At times, wind-blown silt (*loess*) was deposited, especially during peaks of glaciation when so much water was locked up in the ice sheet that sea-level fell by as much as 100 m, exposing even the continental shelf to subaerial

Fig. 1.25. Harvesting the wheat from Dowlands Chasm in August 1840. Detail from a print in the Philpot Museum, Lyme Regis. (Roger Mayne)

Fig. 1.26. Solution hollows in chalk, in-filled with more recent deposits, at Kensworth quarry, Bedfordshire. (J. G. Evans)

Fig. 1.27. The numerous sarsens forming the rock stream at Clatford Bottom, Wiltshire. The site, which is owned by the National Trust, is known locally as Piggledene. (C.J.S.)

erosion. The most substantial deposits of loess in the European chalklands are found in northern France (where they are known as *limon*), but significant quantities have been detected throughout the chalk country of southern and eastern England (Perrin, 1956; see also Chapter 3). These deposits are mapped as Head Brickearth in north-east Kent by the Geological Survey, and are well seen there at Pegwell Bay. A coarser periglacial aeolian deposit is seen in the acid and infertile *coversands* of Breckland and elsewhere.

In the bottoms, valley gravels occur, those of the upper dip slopes deriving originally from the plateaux and valley sides, but lower down the dip slopes these grade into the great fans, sheets, lenses and terraces of true glacial gravels of even greater variety left behind by the retreating ice cap, and re-worked and redistributed by more recent river action. The Pleistocene drift deposits are plainly very complex, especially at lower altitudes, and in any one locality their stratigraphy can only be fully appreciated by reference to the appropriate volume of British Regional Geology. But the significance of these drift deposits to soil formation is obvious, and is taken up later on in Chapter 3.

It is very difficult to readily distinguish a "clean" chalk surface from one affected by a few centimetres of drift. The very term chalkland provides a useful blanket to encompass this range of surface lithologies where the exact composition may not be known. On the other hand, where Tertiary strata or Quaternary drift bury the chalk sufficiently deeply, we may legitimately turn the other cheek, and regard these areas as irrelevant to our present study. Nevertheless, we should always be careful that chalk is not closer at hand than we are led to believe. During borings in the valley of the Baughurst Stream near Newbury, Berkshire, for example, Hawkins (1952) encountered chalk much sooner, that is at a shallower depth, than the geological map said he should have done. Further investigation disclosed a truly astonishing subterranean topographical feature:

> . . . a plug of chalk rather like a volcanic neck in which chalk played the part of magma, rising in an almost vertically-sided pinnacle to a height of more than 150 ft [50 m] above its normal level, displacing and contorting the Reading Beds and London Clay as it came up.

A more familiar subterranean feature of chalk where it is overlain by a relatively shallow tertiary or drift cover is the development of solution hollows, or swallow holes, the curse of the civil engineer, with obvious relevance to drainage and water relations and, where subsidence occurs, to local topography (see for example, Wooldridge and Kirkaldy, 1937). Old swallow holes, and other types of solution hollow, refilled with later deposits, may locally greatly increase the depth to the chalk (Figs 1.26 and 3.4).

1.3 The Prehistory of the Chalk Country

1.3.1 Late- and Postglacial times to the Climatic Optimum

1.3.1.1 The general picture

The chronology of this period, which spans some 9000 years, from about 12 000 B.C. when the last major glaciation (the Devensian) began to decline, to the so-called Climatic Optimum, now put at about 3200 B.C. (Evans, 1975), is shown in Fig. 1.28. The ice itself, of course, had wiped the slate clean, all but the hardiest plants and animals retreating southwards to less extreme climes across the bed of the English Channel which, like the southern North Sea, was dry land at that time. An intriguing point raised by Pennington (1974) is that many coniferous species such as the hemlocks (*Tsuga*), firs (*Abies*) and spruces (*Picea*), which were indigenous to Britain before the Pleistocene ice drove them out, were due never to return of their own accord by a quirk of geological history: the east–west orientation of the major mountain ranges of Europe, reflecting the Miocene folding of the Alpine Orogeny. As the ice had crept southwards, so all the zones of vegetation had moved southwards ahead of it, only to encounter the Alps, the Pyrenees or the sea. Many of the less hardy species perished. By contrast, mountain chains in the American continent typically run north–south, so that the less hardy components of the flora (and fauna) could retreat unscathed across the lowlands to lower latitudes, returning when the ice receded again. This has important implications for British forestry (Chapter 9).

Most refugees from glacial Britain, however, spread back northwards as the ice retreated. The hardier (and more mobile) they were, the more closely they followed the periglacial fringe. Thus juniper and birch invaded the tundra, followed in their turn by pine, while the increasingly thermophilous elm, oak and lime came later. A similar pattern can be deduced in the animals. For example, the hedgehog (*Erinaceus europaeus*) with its ability to hibernate, returned well before the mole (*Talpa europaea*) for which hard frozen ground was an insuperable obstacle: the mole never did reach Ireland before the widening and deepening of the Irish Channel blocked its way for good (Matthews, 1952). This general rise in sea-level likewise led to the inundation of the English Channel about 5500 B.C. and many of the more continental plant species, as well as the slower animals, notably snails, which might eventually have re-established themselves in Britain, did not make it in time. Pennington cites the thistle *Cirsium oleraceum*, common in northern France but a rare alien in Britain, as a striking example. Beech (*Fagus sylvatica*) and hornbeam (*Carpinus betulus*)

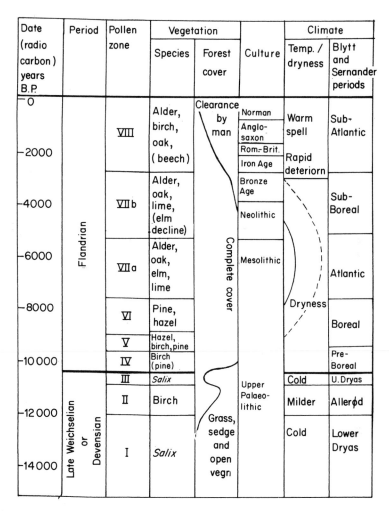

Fig. 1.28. Chronology of Late-glacial and Postglacial climates and vegetation. From Godwin (1976).

apparently only just got back. Many aspects of plant migration are open to argument, however, and we return to this subject briefly in Chapter 5.

Regarding our own remote ancestors, it has long been known that man arrived in Britain early in the Pleistocene Period, but while it was previously thought that these hunter–gatherer cultures can have made no significant impression upon the scrub or forest ecosystems of which, in effect, they were a part, it is becoming increasingly evident that this may not be. true:

apparently anthropogenic forest clearances have been detected which date from about 7500 B.C. and so are attributable to Mesolithic man (Smith, 1970). More substantial inroads began to be made into the climax forests at the beginning of the Neolithic, but here too the calendar continues to be put back: some sites are now considered to date from 4500 B.C., but of this more later.

1.3.1.2 Evidence relating specifically to the chalklands

Of the whole battery of methods which has been used to piece together the information outlined above, pollen analysis features most prominently (Pennington, 1974; Godwin, 1975; Barber, 1976), but although this is useful for reconstructing the general scene, pollen rain is distributed too widely to allow critical work on a more local scale. A second point, which bears specifically on chalk and limestone country, is that, for reasons which are not fully understood, pollen is never as plentiful in calcareous soils and deposits as it is in peat or lake sediments (Dimbleby and Evans, 1974; Evans, 1975). As a consequence, more reliance has had to be placed on physical features of the soil itself (tufa formation, hillwash, buried horizons), on the remains of wood, charcoal and seeds, and on snail-shell fragments (Fig. 1.29), the last having proved particularly informative (Evans, 1972).

Using a combination of these techniques, Kerney et al. (1963, 1964) built up an intriguing picture of the Late- and Postglacial environment of the North and South Downs: the annual accumulation during periglacial conditions of chalky mud at the bottoms of the slopes, containing a network of ramifying rootlet holes suggesting the presence of some sort of grassy vegetation; the establishment around 10 000 B.C. of a primitive rendzina soil supporting an open scrubby woodland of juniper and birch, only to be buried by chalk mud and rubble about 8800 B.C. as freeze–thaw cycles resumed; then the steady improvement again from about 8300 B.C. right through to the warm and humid Atlantic Period (5500–3200 B.C.); and finally the abrupt and unmistakable change both in the mollusc fauna and in adjacent pollen sediments, from woodland to open-ground species, indicative of the first Neolithic forest clearances.

One major drawback of having to rely on mollusc remains is that though these signify types of vegetation, they give no information on species composition. Thus we can only assume, from known soil preferences of the species indicated in the general pollen record, that the woodlands which clothed the Chalk through all those centuries up to the Neolithic— Wordsworth's "primaeval woods, shedding and renewing their leaves", season after season after season—came to consist by Atlantic times mainly of elm, lime, ash and yew (Evans, 1975). Though native in the south-east, beech was to come to prominence later (Section 1.4.1).

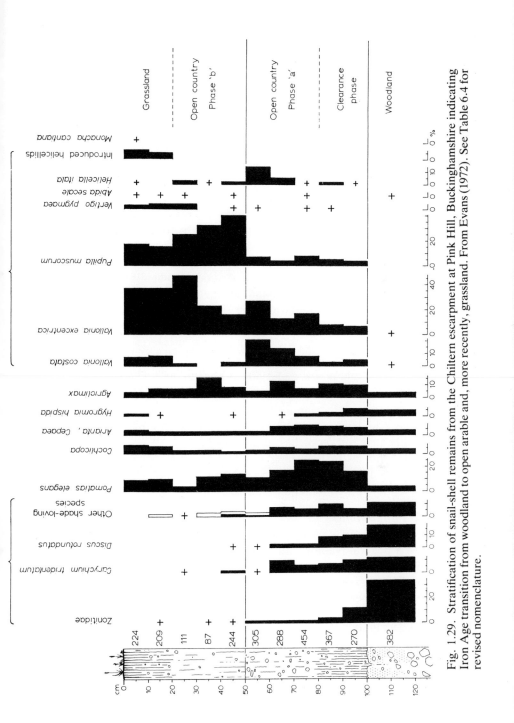

Fig. 1.29. Stratification of snail-shell remains from the Chiltern escarpment at Pink Hill, Buckinghamshire indicating Iron Age transition from woodland to open arable and, more recently, grassland. From Evans (1972). See Table 6.4 for revised nomenclature.

1.3.2 Neolithic to Roman times

1.3.2.1 Forest clearance

There has long been controversy about whether the decline of forest trees and the appearance of species characteristic of more open ground, evident, as we have seen, in pollen and mollusc records alike, was attributable to climatic deterioration or to the hand of man. It is now widely accepted, however, that these changes were indeed the mark of the first Neolithic farmers to reach Britain, and that the chalklands were among the first areas to be used for agriculture once farming penetrated inland (Wooldridge and Linton, 1933).

Another source of argument has been whether, because of their "dryness", the chalk downs were ever as thickly forested, if indeed they supported trees at all, as other lowland areas. Clark (1952), for example, noted the rarity or absence of roe-deer remains from downland sites, and saw this as evidence for a general lack of woodland. This is another concept, however, now regarded as outdated, for the climate appears never to have become sufficiently dry during the Postglacial to inhibit woodland. According to Evans (1975),

> it was the later part of the Sub-boreal which showed a trend towards dryness. Deforestation on the Chalk took place as early as the middle of the fourth millenium B.C. and is now confidently ascribed to the activities of man and his animals.

It is not proposed to examine Neolithic chalkland farming in detail, which has in any case been admirably covered in the review by Evans (1975), to which repeated reference has already been made. Here the intention is simply to emphasize how long the chalklands have been under human influence, and the effects this occupation has had on the landscape, soil, flora and fauna which we tend to regard as natural.

1.3.2.2 Flint mining

The excavation and working of flint, originating far back in the Palaeolithic, reached a peak in Neolithic times. The flint mines at Grimes Graves near Brandon in Breckland are justly famous, though other sites, some older than Grimes Graves, are scattered about the southern chalk country, though not in the Wolds, perhaps because of the difficulty of working the hard chalk there (see Evans, 1975, Fig. 55). In the Brandon mines, only the tough (floorstone) flints, unaffected at a depth of 12 m or more by the periglacial freeze–thaw cycles which weakened those of the upper layers, were sought. As a result, substantial quantities of chalky spoil accumulated around the mines, burying the infertile and unstable cover sands with a medium relatively more favourable for plant growth.

1.3.2.3 Farming practices

Although the first farmers may have made some temporary (Landnam) clearances, growing crops for a few years until the soil was exhausted and then moving on to leave scrub and woodland to grow up again, permanent clearance predominated in the chalklands. Here, domestic livestock supplemented axe and plough in pushing back the forest, preventing regeneration, and encouraging the development of pasture. Ploughs broke up the ground newly claimed from the forest, hoes and spades being used in routine cultivation for the arable crops, which included Emmer wheat, naked barley, beans and flax. Native herbs of open habitats such as plantains (*Plantago* spp.) were able to flourish with renewed scope in the clearances, while other species were brought in as contaminants of the crop seeds; it is possible that the cornflower (*Centaurea cyanus*) was introduced as an arable weed as long ago as this (Salisbury, 1964).

By Early Bronze Age times, arable crops were being grown in small fields about 0·2–0·6 ha in area, typically rectangular to roughly square in shape, the outlines of which can be seen to this day in parts of the more remote and least disturbed downland country, especially from a high vantage point on a clear day with the sun at a low angle. These are the so-called "Celtic" fields (Fig. 1.30), and they were particularly characteristic of the Iron Age chalklands (Bowen, 1961). Some were cultivated as such until as late as A.D. 400 to A.D. 500, but many must have been allowed to revert to grass towards the end of the Iron Age to support the large flocks of sheep known to exist by then. Indeed, by Roman times, wool production was mainly confined to the limestone uplands, including the Chalk, largely through the extension of arable cultivation into virgin land by the sophisticated Belgae who preceded the Romans by about a century (Gilbert, 1969). On Romano-British sites noted by Clark (1952), sheep remains predominate over those of swine (for example on Cranborne Chase) and in one case exceed those of cattle. Hoskins (1955) suggests the latter half of the third century A.D. as the period in which the Wiltshire and Sussex sheepwalks were first established. The association of sheep with the ecology of the chalklands can thus be traced back with certainty for fully 1600 years.

1.3.2.4 Barrows, stone circles, hillforts and hill-figures

Nothing induces a greater feeling for the antiquity of man's association with the chalk landscape than the ancient burial mounds and stone circles, preserved in their greatest numbers on the Wessex downs, where Thomas Hardy captured their atmosphere:

Round about me bulged the barrows
As before, in antique silence—
immemorial funeral piles . . . .

Fig. 1.30. "Celtic" fields on closely grazed downland near Brixton Deverill, Wiltshire. Note also the dewpond in the left foreground. (Cambridge University Collection, copyright reserved)

The mounds range in size from the unique colossus of Silbury Hill (Fig. 1.31) to the much smaller Neolithic long-barrows and Bronze Age disc and bell barrows (Fig. 1.32) Apart from their mystic appeal and great archaeological interest, these structures are also of ecological importance for two major reasons. Firstly, they have preserved contemporary soil profiles, revealing accumulations of loess and other superficial deposits which elsewhere have long since been dissipated by weathering and ploughing, and providing some indication of the rate at which the chalk landscape is eroding away—about 30cm in 4000 years in places, according to Atkinson (1957). Secondly, the earthen barrows themselves, like the Iron Age hillforts which followed them (Fig. 1.33), not only provide distinct microhabitats, but support a truly ancient turf (see Chapter 5). Surprisingly

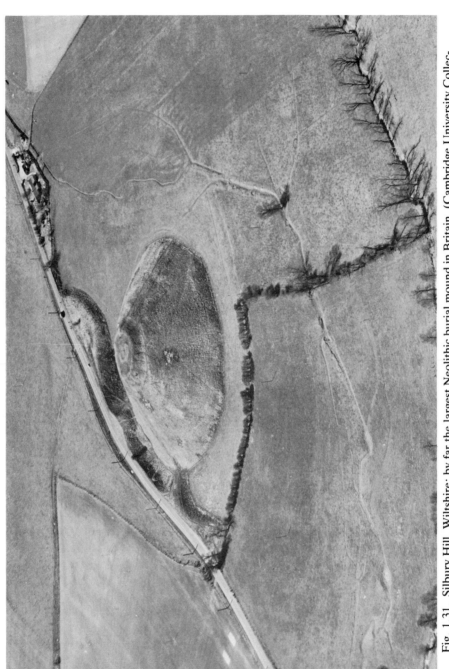

Fig. 1.31. Silbury Hill, Wiltshire: by far the largest Neolithic burial mound in Britain. (Cambridge University Collection,)

Fig. 1.32. Various kinds of Bronze Age round barrows near Winterbourne Stoke, Wiltshire. (Cambridge University Collection, copyright reserved)

few of the figures cut into the chalk hillsides are of archaeological significance: the Cerne Giant (Dorset) and the Long Man of Wilmington (Sussex) are demonstrably ancient, the celebrated Uffington Horse (Fig. 1.34) less certainly so. Most are of much more recent origin (see Marples, 1949), though all provide a bare-chalk habitat for plants and animals (Chapter 4), and are frequently honeypots for countryside visitors (Chapter 10). Signs of early settlement of diverse nature and date may also be seen in soil and crop marks.

1.3.2.5 Ancient tracks and roadways
Throughout the opening up of the British interior, permanent trackways

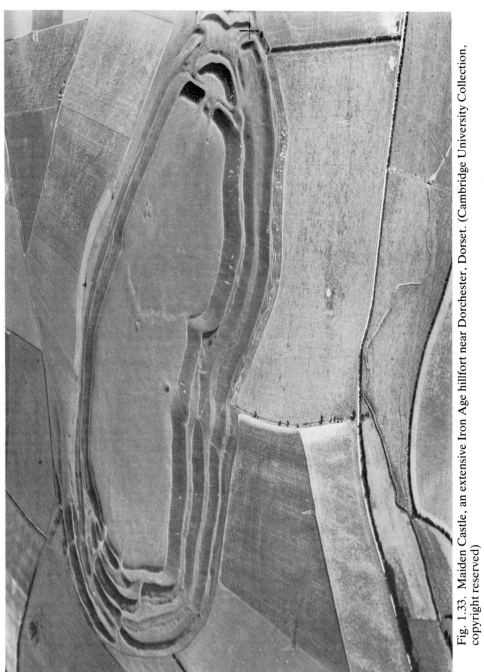

Fig. 1.33. Maiden Castle, an extensive Iron Age hillfort near Dorchester, Dorset. (Cambridge University Collection, copyright reserved)

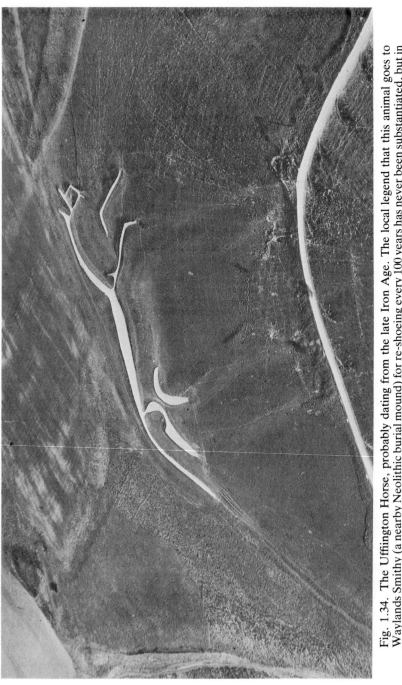

Fig. 1.34. The Uffington Horse, probably dating from the late Iron Age. The local legend that this animal goes to Waylands Smithy (a nearby Neolithic burial mound) for re-shoeing every 100 years has never been substantiated, but in 1974, as a result of the revision of county boundaries, it did move from Berkshire into Oxfordshire! (Cambridge University Collection, copyright reserved)

Fig. 1.35. The Upper Icknield Way between Princes Risborough and Wain Hill, Buckinghamshire. (C.J.S.)

became established, of which many exist to the present day (see Timperley and Brill, 1965; Pyatt, 1973?). The Icknield Way and Ridgeway are familiar examples (Figs 1.35 and 3.1), though apparently narrower now than a century ago, when Richard Jefferies (1879) wrote thus of the broad grassy tracks of his beloved Marlborough Downs:

> Plough and harrow press hard on the ancient track, and yet dare not encroach upon it. With varying width, from twenty to fifty yards, it runs like a green riband through the sea of corn—a width that allows a flock of sheep to travel easily side by side, spread abroad, and snatch a bite as they pass. Dry, shallow trenches full of weeds, and low, narrow mounds, green also, divide it from the arable land. . . .

Hollow-ways (Fig. 1.36) represent in some places trackways of ancient origin which have been worn down by centuries of constant use.

Where these trackways, level or sunken, are hedged, their antiquity can be tested, for the age of a hedge greatly influences its botanical composition and so in turn the fauna it supports. A rough approximation is that in a 30-yard (now, presumably, 30-m) stretch of hedge, one woody species is likely to occur for each century the hedge has stood. Thus a hedge composed only of hawthorn must have been planted within the last 100 years or so, while one in which there are ten species in all is about 1000 years old, and so

Fig. 1.36. An ancient hollow-way descending the Chiltern escarpment above Lewknor, Oxfordshire, within the Aston Rowant National Nature Reserve (NNR). Ironically, the great M40 cutting (Fig. 1.39) lies immediately to the west. (C.J.S.)

on (Hooper, 1970). More recent work has resulted in a refinement of this method (Pollard *et al.*, 1974). The relationship

$$y = 100s + 30$$

where y = hedge age in years and s = the number of woody species in the statutory 30-yard length, holds for the authors' wide-ranging surveys of Devon, Lincolnshire, Cambridgeshire, Huntingdonshire and Northamptonshire, while in a more local assessment of the clay uplands of the Northamptonshire–Huntingdonshire border, where climatic and edaphic variation was reduced to a minimum, the relationship was

$$y = 99s - 16.$$

These newer equations would give a ten-species hedge an age of 1030 and 974 years respectively. The relationship specifically for chalkland hedges has not yet been worked out. The difficulty is not in counting species, but in finding authentic historical records which confirm the genuine age of the hedge against which the field result can be checked. Another problem is that hedges originating from strips of woodland, as distinct from deliberate planting, may be disproportionately species-rich.

Nevertheless, it is plain that the road-verge and hedgerow habitats were established a very long time ago. Moreover in Roman times, regardless of soil type or drainage, the major roads were invariably raised up on a sizeable embankment called an *agger* (Margary, 1967) which, like its modern counterparts in rail and motorway engineering, provided an admirable habitat for indigenous and alien ruderal species. There were many accidental additions to our flora at this time, brought in with the great shipments of troops and grain which accompanied the military occupation; examples are the corncockle (*Agrostemma githago*) and corn marigold (*Chrysanthemum segetum*), both now relatively rare again, and the sow-thistles *Sonchus oleraceus* and *S. asper*. Their spread, together with that of their indigenous counterparts and earlier introductions of Neolithic, Bronze and Iron Age dates, must have been greatly hastened by the roads and traffic of Roman Britain just as, eighteen centuries later, the Oxford ragwort (*Senecio squalidus*) was to spread via the railways (Salisbury, 1964).

1.4 The Historical Period

1.4.1 Anglo-Saxon to mediaeval times

Then came the Dark Ages. Much of the higher chalklands were probably abandoned for many decades before the Anglo-Saxons arrived, who even then tended to go for the valley sites. It was probably this desertion of the high Chalk which encouraged the remarkable spread of the beech about this time, to become the characteristic tree species of the south-eastern Chalk (see Chapter 8). The Danes, of course, settled in the Wolds, place-names ending in –*by* indicating their Danish connections, but this did not result in any significantly different alteration of the landscape (Hoskins, 1955).

The earliest of the narrow cultivation terraces known as strip lynchets, which in places form such a striking feature of the downs (Fig. 1.37), are now thought to date from A.D. 700 to A.D. 800, though many were constructed in the thirteenth and fourteenth centuries up to the time of the population decline which followed the Black Death (H. C. Bowen, personal communication). Like the banks and ramparts of the earlier earthworks, the risers and treads of the strips make important microhabitats for chalk grassland plants and the animals they support.

Undoubtedly the most significant contribution of these times to chalkland ecology was the introduction of the rabbit from Normandy in the eleventh

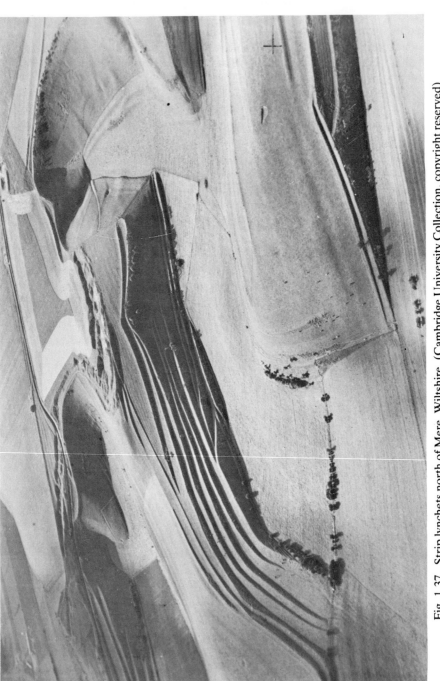

Fig. 1.37. Strip lynchets north of Mere, Wiltshire. (Cambridge University Collection, copyright reserved)

century, the consequences of which are considered in later chapters, particularly Chapter 6. Over this period, too, bear, lynx, wild boar and wolf were exterminated or drastically reduced, at least in the south and east.

1.4.2 The last 400 years

We emerge from these long and still relatively little-known times of the Dark and Middle Ages well into the historical period, when the chalk country was beginning to look very much as it does today, with the obvious exception of modern urban and industrial developments (Figs 1.38 and 1.39). The agricultural history of the chalklands is taken up in Chapter 9, and the effects of both the intensification of farming and the encroachment of urban society on the "natural" chalk landscape and its flora and fauna are discussed in Chapter 10.

Fig. 1.38. Continuing urban development on the chalky slopes of High Wycombe. (C.J.S.)

Here ends, then, a potted account of the 100 million years' history of the chalklands of England, a landscape whose topography has drawn comment from many a pen of repute: comment which has by no means always been favourable. Gilpin (1804) found chalk country "disfiguring", Granville (1841) "dismal", "barren" and "discouraging"; while Johnson, characteristically going the whole hog, regarded the South Downs near "Brighthelmstone" (Brighton) as ". . . so truly desolate, that if one had a mind to hang oneself for desperation at being obliged to live there, it would

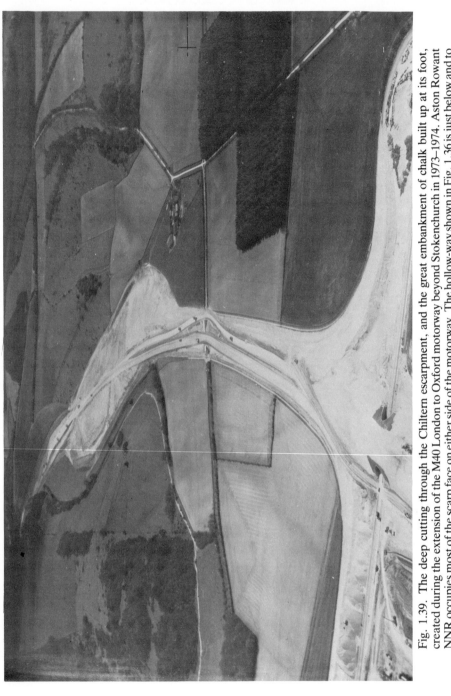

Fig. 1.39. The deep cutting through the Chiltern escarpment, and the great embankment of chalk built up at its foot, created during the extension of the M40 London to Oxford motorway beyond Stokenchurch in 1973–1974. Aston Rowant NNR occupies most of the scarp face on either side of the motorway. The hollow-way shown in Fig. 1.36 is just below and to the left of the cutting, marked by the curving line of scrub and trees. (Cambridge University Collection, copyright reserved)

be difficult to find a tree on which to fasten the rope" (Piozzi, 1786). Happily, these words are countered by the writings of Gilbert White (1788), William Cobbett (1830), Richard Jefferies (e.g. 1879), W. H. Hudson (1900, 1910), and even the censorious Massingham (1936, 1940), all great champions of the chalk country, if at times their ecology was a trifle romantic.

Small (1970) points out that the rounded coombes of the downs are not as unique to the Chalk as many an author or poet would have us believe. Yet, there is no mistaking chalk country. In 1794, William Smith recognised the Yorkshire Wolds to be of Chalk by their topography as seen from York Minster (Avebury, 1902). At closer quarters the chalklands even have a smell of their own. The scent of chalk grassland, for example, on a warm day in the growing season is quite distinct from that of, say, heathland or beechwood. The intricacies of the vegetation are considered in due course, but first we examine the climate and soils of the English chalklands.

2
The Climate of the English Chalklands

2.1 Macroclimate

For the last several thousand years, Britain has enjoyed a maritime, temperate climate. Due to the powerful influence of the Atlantic Ocean, there are normally no great extremes of heat, cold, drought or flood, nor does any one kind of weather last for more than a few days as a rule: a fact which has drawn crisp comment from many a pen. In his well-known scheme still widely used, Köppen (1931) classifies this type of climate as *Cfb*, in which *C* signifies temperance, *f* "adequate" rainfall throughout the year, and *b* that the mean temperature of the warmest month does not exceed 22°C, though the very size and diversity of the area covered by the single description *Cfb* (see, for example, the Appendix in Trewartha, 1968) highlights the limitations of the scheme.

Britain's climate results from the juxtaposition of the Icelandic Low and Azores High pressure systems which, between them, channel in air over a long fetch of Atlantic Ocean, greatly dampening seasonal fluctuations in air temperature over the land (Fig. 2.1). Recent developments in meteorology have pinpointed the importance of jetstreams, zones of very rapidly moving air in the upper levels of the atmosphere, in augmenting this predominantly westerly flow, while the effect is enhanced, particularly in winter, by the warm North Atlantic Drift ocean current (the Gulf Stream) at the surface of the Atlantic itself. The interplay between these systems leads to the characteristically changeable, "westerly" type of weather which is reflected in long-term climatological averages. Periods of cloud, wind and rain attendant upon the strings of North Atlantic depressions and their associated fronts, alternate with warmer, drier spells of weather as the Azores anticyclone exerts its influence (Fig. 2.2a). Of course, other weather patterns intervene, in which polar or continental air may be advected into

(a)

(b)

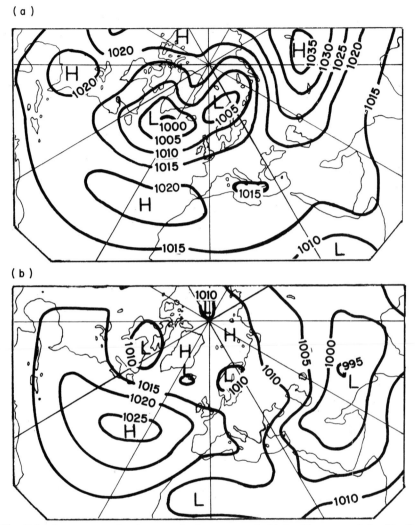

Fig. 2.1. Average sea-level atmospheric pressure (mb) for Europe and the North Atlantic, 1951–1966. (a) January and (b) July. From Lamb *et al.* (1973).

Britain; indeed, recent years have provided evidence of a weakening of the dominant westerly flow. Some of the commoner pressure patterns are shown in Fig. 2.2. More examples can be found in Bilham (1938), Manley (1952) and in particular in Taylor and Yates (1967), who explain the mechanics of these systems in far more detail than can be afforded here. For a topical review of all aspects of British climatology, the reader should consult the collection of papers edited by Chandler and Gregory (1976).

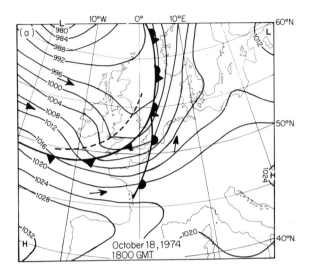

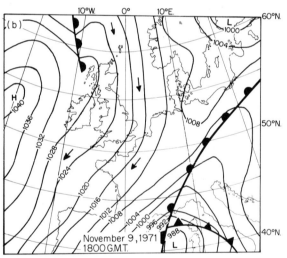

October 18, 1974
1800 G.M.T.

November 9, 1971
1800 G.M.T.

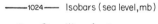

———1024——— Isobars (sea level, mb)

Warm front

Cold front

Occluded front

– – – – – Trough

← Predominant surface wind direction

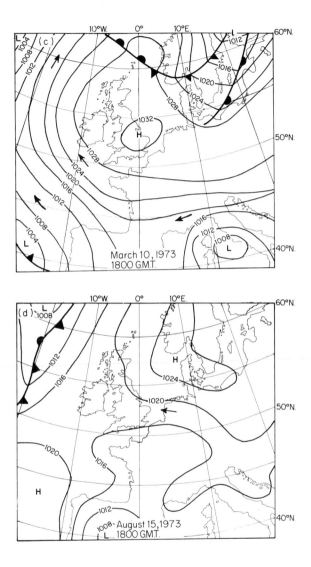

Fig. 2.2. Common pressure patterns over Britain, and their consequences. (a) Classic Atlantic situation, with a changeable westerly airstream maintained by the Icelandic Low and Azores High pressure systems, October 18, 1974. (b) Northerly outbreak, November 9, 1971. (c) Anticyclonic conditions, leading to widespread night-time temperature inversions, March 10, 1973. (d) Continental (summer); slack south-easterlies resulting in very hot weather, August 15, 1973. Simplified from the Daily Weather Report, and reproduced by permission of the Director-General, Meteorological Office, Bracknell.

2.2 Regional Variations in Climate

Plainly, the foregoing is a broad view, and within this overall pattern marked variations in regional climate are to be found. Thus, northern areas of Britain tend to be cooler, wetter, more windy and less sunny than the southern counties. In part this is due directly to latitude, but differences between northern and southern regions result in the main from their respective positions in relation to the dominant centres of low and high pressure. In the longitudinal plane, the hilly western areas draw off much of the rainfall through the effect of orographic lifting, leaving the east in a rain-shadow. These contrasts are augmented by weather patterns which introduce dry easterly or south-easterly winds, as in Fig. 2.2d. When this happens, the south-eastern counties are subjected to a strong continental influence, more so than areas further north and west. The fact that this enhances the likelihood of extremes of summer heat and drought is well known, but the possibility of very low temperatures in winter and early spring should not be overlooked. Altitude complicates these regional patterns, but discussion of this factor is reserved for the section on local climates (Section 2.3.2). A selection of climatological maps is presented in Fig. 2.3, and an example of Walter and Lieth's (1967) mammoth exercise in synthesising data of this kind into regional climate-types is shown for England in Fig. 2.4.

Clearly the bulk of the chalk country coincides with the warmer and drier south-east, although the Dorset downlands experience a somewhat wetter climate and the Yorkshire Wolds a colder one (see also Perring, 1958). Note, too, that Walter and Lieth recognise a warm-temperate category along the southern coastal strip of England which stretches well inland in southern Hampshire. Of course, extreme maritime conditions will prevail along the littoral fringe itself, but this and other more local patterns of climate are considered later in this chapter, and again in Chapter 3.

There are, then, distinct differences in climate across the English chalklands which can help to explain broad-scale patterns of distribution among animal and plant species (Chapters 4–8), but for more detailed investigations such as regional surveys or ecological or agricultural field experiments, it is important to be aware of the limitations of generalised climatological data of this kind. There are two main snags. Firstly, there may be no available data for the particular place in question, necessitating extrapolation or interpolation from information obtained elsewhere. Sometimes information is available but covers only a limited period. Great care should be taken in using such data. Too often, meteorological records

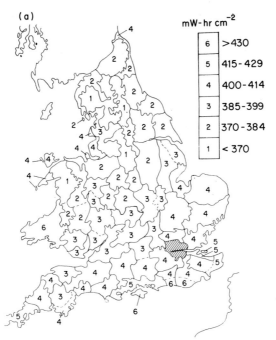

(a)

mW-hr cm^{-2}

6	>430
5	415-429
4	400-414
3	385-399
2	370-384
1	<370

Fig. 2.3. Figures 2.3a–f show climatological data for the agriclimatic regions of England and Wales, 1941–1970. From Smith (1976). (a) Mean hourly short-wave solar radiation receipt, April–September.

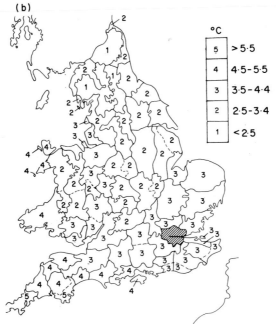

(b)

°C

5	>5·5
4	4·5-5·5
3	3·5-4·4
2	2·5-3·4
1	<2·5

Fig. 2.3b. Thirty-centimetre soil temperature, January.

(c)

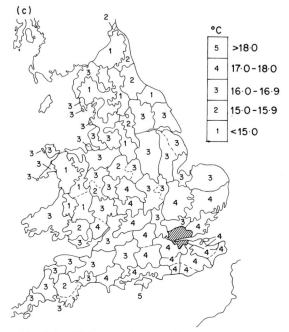

	°C
5	>18·0
4	17·0 - 18·0
3	16·0 - 16·9
2	15·0 - 15·9
1	<15·0

Fig. 2.3c. Thirty-centimetre soil temperature, July.

(d)

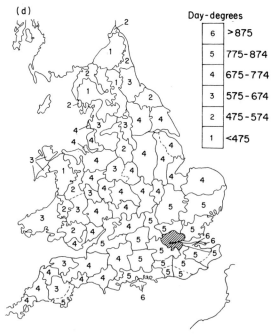

	Day-degrees
6	>875
5	775 - 874
4	675 - 774
3	575 - 674
2	475 - 574
1	<475

Fig. 2.3d. Accumulated screen temperature over 10°C, May–October.

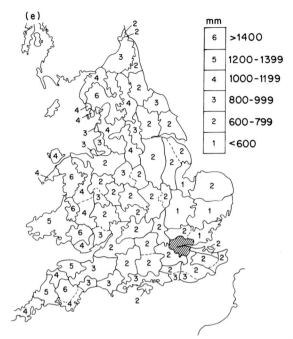

mm	
6	>1400
5	1200-1399
4	1000-1199
3	800-999
2	600-799
1	<600

Fig. 2.3e. Annual rainfall.

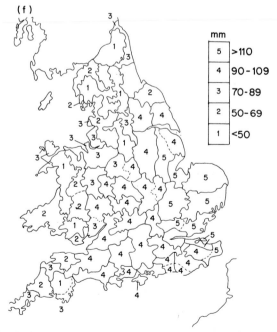

mm	
5	>110
4	90-109
3	70-89
2	50-69
1	<50

Fig. 2.3f. Maximum summer soil moisture deficit (SMD).

are glibly cited which are manifestly inappropriate. In a recent investigation of the botanical composition of a chalk downland site at an altitude of 260 m O.D., the supporting measurements of temperature and humidity were from an airfield 24 km away and a mere 10 m above sea-level.

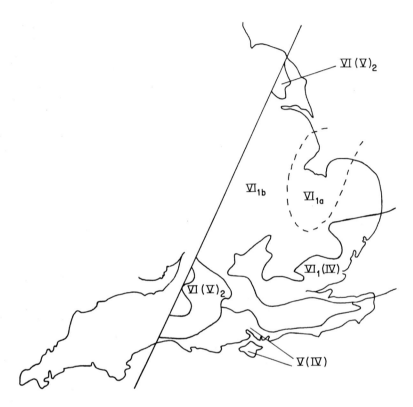

Fig. 2.4. Climate-types of Walter and Lieth within the range of the English Chalk, simplified along the lines of Shimwell (1971b). The main type is VI (typical temperate climate), with V (warm-temperate) and a trend to Mediterranean (IV) in the extreme south and south-east. For a much more detailed analysis see Walter and Lieth (1967), and Walter et al. (1975).

Regarding topicality, it is worth noting that the Climatological Atlas (Meteorological Office, 1952) covers the 30-year period from 1900 to 1930. Yet, this publication is still widely quoted—maps from it have even been included, without comment, in an agricultural atlas published in 1976— despite the availability in published form of much more recent records amassed by the Meteorological Office. It might be argued that the 30-year span of the Atlas is a generous sample. Certainly the data for 1900–1930 can be used with impunity in relation to the classical studies of the vegetation of the English Chalk conducted over that period by Tansley and others (see later chapters), or for that matter in longer-term woodland and forestry studies. But to use this information willy-nilly to back up modern studies on animals, crops and herbaceous plant communities is naive indeed, particularly in view of the changes that have been taking place in recent years. As Lamb (1966) puts it: "Our attitude to climatic 'normals' must clearly change. 1901–30 and still more 1921–50 were highly abnormal periods." The publication by the Meteorological Office of data covering the period 1941–1970, arranged according to agroclimatic regions of England and Wales (Smith, 1976), is a great help in this respect (Fig. 2.3).

A second major drawback of standard climatological data, however, is that they come from instruments sited inside Stevenson screens mounted a metre above the ground, while plants and animals (other than spiders and insects which inhabit Stevenson screens) are exposed to a very different environment. Moreover these data are usually averaged, and critical oscillations of interest to the biologist may be completely missed. To understand more about the finer-scale ecology of the chalklands, or any other biome, local climates and microclimates must be studied more closely than they have been in the past.

2.3 Local Climates, Topoclimates and Microclimates

2.3.1 Some definitions

Expressions such as these tend to overlap between different users, so first a word on terminology adopted in this book. The scale of a *local climate* can be regarded as in the order of 10–200 km², applicable, for example, to Salisbury Plain, Breckland, the mid-Chilterns and to the maritime chalklands. Clearly these grade on the one hand into the regional variations in macroclimate already outlined in the foregoing paragraphs, and on the other into smaller units. To many geographers, anything smaller amounts to a microclimate: indeed, some regrettably apply this term to local climates in the sense defined above. But even on a more conventional basis, microclimate still straddles a considerable range in scale from, say, a cliff-face, a scarp slope or a valley-head perhaps 1 km² in area to a wheel-rut, a rabbit-scrape or an ant-hill, or even individual flowers on a plant, measurable in terms of square centimetres—a range amounting to four orders of magnitude. In this book, the term *microclimate* is reserved for really small-scale variations such as wheel-ruts, etc., while for units of intermediate size *topoclimate* is adopted in view of its strong correlation with topography.

We shall touch only briefly on examples of local climates (for much can be gleaned from what has already been said about regional variations), and examine in more detail the topoclimates of hilly chalk terrain and the microclimates these conceal. With a few notable exceptions, the available information comes from locations comparable with, but not confined to the chalk country. It might be noted in passing that in the introduction to his classic book which until recently was virtually unique in this neglected field of study, Geiger (1965) writes: "Gregor Kraus (1841–1915) . . . the father of microclimatology [first] became aware of the extreme local conditions affecting the chalk countryside of the Main area near Karlstadt . . .". In fact, the geology of this region is Triassic limestone (the *Hauptmuschelkalk*): the distinction of Cretaceous chalk from other limestones is easily blurred in translation.

2.3.2 Local climates

Here is a subject which treads close on the heels of the countryman's weather lore. Rainfall, and especially the more spectacular phenomena such as hail, thunder and lightning, attract most attention, and marked variations in their incidence may be noted from one locality to another in any one year, or even one generation. But unless these contrasts persist over longer periods of time, they must be regarded as chance happenings and not as true climatic differences. The problem of gaps in the national network of meteorological recording stations has already been mentioned, and this applies even more when it is required to identify real differences of this kind in local climate.

P. A. Wells (1971) constructed a map of effective rainfall for eastern Lincolnshire which clearly demonstrates the wetter climate of the Wolds resulting from their relatively high altitude (Fig. 2.5), but unfortunately information of this kind is sparse. Short-term records of rainfall assembled by Adamson (1921) indicated that the wooded western South Downs may receive as much as 1000 mm of rainfall annually in contrast to values of 635–760 mm "for most south-eastern stations". Evidence for a strongly oceanic local climate there was also obtained, with high atmospheric humidity and small annual ranges in air temperature. More recent data (Rodda *et al.*, 1976) show a striking concentration of winter rainfall in this region (Fig. 2.6). Until there is a denser network of recording stations, however, this is a subject which has to remain speculative. As Pedgley (1971) has already recognised in his work in North Wales, there is enormous scope here for meteorological work in schools.

An aspect of local climate sometimes overlooked is the quantity of

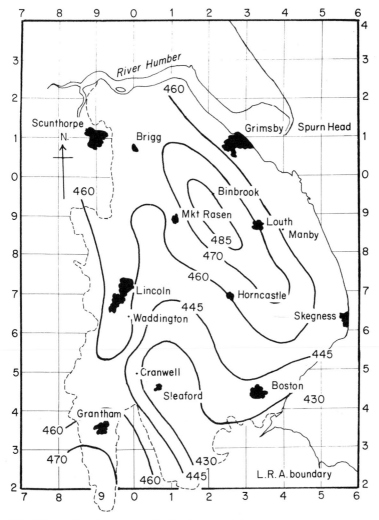

Fig. 2.5. The regional pattern of effective precipitation in eastern Lincolnshire, computed from data collected between September 1952 and October 1968. The influence of altitude on values for the Wolds (as well as the Jurassic Lincoln Edge) is very clear. From P. A. Wells (1971).

impurities in the atmosphere which may be transferred by wind and rain to soil and vegetation. The toxic properties of industrial and domestic aerial pollutants such as soot, sulphur dioxide and lead are well known. Regarding the first two examples, the chalklands are not among the most heavily affected regions, although local pockets of aerial pollution have been

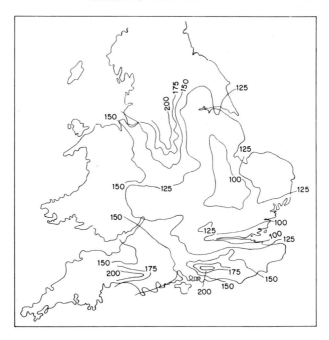

Fig. 2.6. Total winter rainfall (November–March, mm) likely to be exceeded in 98 out of 100 years. Note the remarkably high values in the vicinity of the western South Downs. Data assembled by the Institute of Hydrology (see Rodda *et al.*, 1976).

recorded, especially around London (Bowen, 1970; Cox *et al.*, 1976). Lead levels are likely to be high in the vicinity of roads carrying heavy traffic loads. Rain may leach toxic concentrations of zinc from galvanised fences and electricity pylons to affect the vegetation immediately beneath them. On the other hand, not all aerial contaminants are either man-made or deleterious, and with a parent material as pure as chalk even minute quantities of magnesium, iron, nitrogen, sulphur, boron and other substances carried down in the rain may have ecological significance. The maritime regions are particularly susceptible to these aerial top-dressings (Stevenson, 1968), and it is possible to envisage effects which operate at this intermediate scale (as distinct from topoclimatic patterns considered in the next section). The intriguing paramaritime element of the flora which receives mention in Chapter 4 could, for example, reflect a response to certain nutrients which are carried only a kilometre or two inland by onshore winds.

   Onshore winds affect coastal regions in other ways, too. They reduce the risk of frost by maintaining high humidity and breaking up temperature inversions, and, especially along the cool North Sea coast, they carry sea-frets and low stratus cloud anything from a few hundred metres to several kilometres inland. Along the warmer English Channel coast the effect is to push convection cloud inland, resulting in large annual sunshine totals, a fact well known from the claims of rival holiday resorts. In settled weather, diurnal heating of the land in relation to the sea results in the familiar sea-breeze, and where strong onshore winds are relatively uncommon, sea-breezes, modest as they are, form an important feature of local climate. They can, moreover, spread a considerable distance inland: sea-breeze fronts from the Hampshire coast have been noted as far inland as Reading (Simpson, 1967).

2.3.3 Topoclimates

2.3.3.1 The influence of altitude

The influence of altitude on radiation receipt at the modest heights attained by the Chalk in England is very small: we should not expect to find, for example, the excesses of incoming ultra-violet radiation and large component of back radiation in the total energy budget which, together with relatively high ozone fluxes, are associated with the tenuous air of lofty mountains. However, as was briefly mentioned before, the effect of the decrease with height of mean temperature is sufficient to be felt over more modest topography; on this basis alone, a temperature difference of about 0·6°C could theoretically be expected at any one time between the lowest and highest points of a typical chalk scarp rising 100 m above the plain at its foot. Small as this may seem, the effect can be strongly enhanced by hourly or daily accumulation of the differences.

Moreover, such contrasts in temperature due to altitude are paralleled and even augmented by differences in wind, humidity, cloud cover, rainfall and snow retention. The higher hills tend to be more windy (Fig. 2.7), a feature often betrayed by the shaping of bushes and trees (Fig. 2.8) and at one time by the siting of windmills, though notable wind-gaps sometimes occur at much lower elevations.

In overcast, humid conditions (such as are associated with the situation shown in Fig. 2.2a) when the wind strikes the scarp slope at or nearly at right angles, orographic lifting is often sufficient to cause drizzly rain to fall in a narrow zone perhaps 1 km wide along the length of the ridge, especially in autumn and winter, while the cloud base itself may descend to engulf the higher tops in hill fog (Fig. 2.9). Trees and shrubs, as well as wires, pylons and fences, trap appreciable quantities of water on such days, and this fog-drip may augment orographic precipitation, while the valley slopes and bottoms receive much less, and may even remain dry. When conditions favour convection, hilly terrain can increase the risk of heavy local

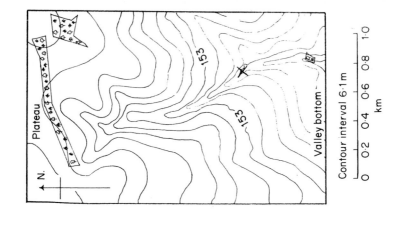

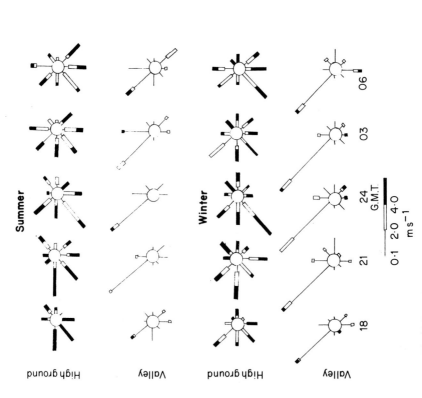

Fig. 2.7. Wind roses for two adjacent topoclimatic stations in the Cotswold Hills, comparing hilltop and valley-bottom situations on clear nights between October 1929 and June 1931. The figures within the circles represent the number of times dead calm was recorded. The position of the valley recorder is marked X. The greater windiness of the hilltop, and the katabatic flow of wind down the valley, are strikingly demonstrated. From Heywood (1933).

Fig. 2.8. A moribund juniper at Coombe Hill, Buckinghamshire, showing marked wind-shaping. (C.J.S.)

precipitation. These and other small but cumulative effects account for many of the regional differences already noted, but they can operate on a surprisingly small scale; for example, the annual rainfall on the crest of the Chiltern scarp at Aston Rowant NNR (244 m) is about 827 mm, compared with only 635 mm immediately below in the Vale of Aylesbury, where the altitude is 122 m (King, 1977b).

Snow accumulates more readily and lies longer on the cooler summits (Fig. 2.10). By reflecting up to 95% of the incoming solar radiation, and adding substantially to the quantity of water held in the surface soil if it melts rather than evaporates, the persistence of even quite a light covering of snow in early spring can seriously delay soil warming, plant growth and tillage operations. Regarding more transient phenomena, parts of East Anglia and the Chilterns are among the most thundery areas of Britain, averaging 18–21 thunderstorm days per year (Roberts, 1968). There is no more devastating environmental factor than a major cloud-to-ground lightning strike, and the chance of this happening is greatly enhanced where relatively high altitude coincides with dry surface soils through which electrical charges, which may themselves have biological effects, are unable to "drain away". The Boer War Monument on the Chiltern escarpment at Coombe Hill, Buckingham-shire, has twice been severely damaged by lightning. In the same area, the scarp is thought to have been a potent factor in the development of the destructive tornadoes of May 1950 (Lamb, 1950).

2.3.3.2 Frost hollows
As long as the thermal lapse rate is negative, that is to say temperature

declines with increasing height, then higher altitudes will tend to be cooler than lower altitudes, and on a general basis this is clearly true. But it is well known that extremes of cold are experienced not on summits but in valleys.

Fig. 2.9. The Chiltern escarpment at Windsor Hill above Princes Risborough, (a, top) on a clear day and (b, bottom) shrouded by low cloud during a spell of cyclonic weather. (C.J.S.)

These extremes are associated with positive lapse rates (temperature inversions) where the air temperature profile shows a temporary decrease towards the ground (Fig. 2.11). Gardeners and farmers know only too well how these fickle valley bottoms, which may in fact get very hot during the day, form frost hollows by night, although, of course, subtle variations in temperature and dewpoint will determine whether dew, frost or fog occur on any one occasion (see Fig. 2.12). During the course of a clear sunny day, the surface of the ground and the air immediately above it warm up as long as the

Fig. 2.10. Snow lingers on the Chiltern scarp at Beacon Hill (in the distance) and Coombe Hill (immediate foreground), having quickly cleared from the Vale of Aylesbury below. (C.J.S.)

gain in energy due to solar heating exceeds losses by conduction, turbulent transfer, evaporation or back radiation. (The multiplicity of factors affecting the microclimate of the soil surface is considered in detail in the next section, and in that part of Chapter 3 dealing with the physics of the soil profile.)

Towards sunset the incoming radiation receipt dwindles, and the energy balance goes into reverse; the soil surface now proceeds to lose heat by long-wave radiation (back radiation) into space. Any factor which abets strong back radiation will intensify the drop in temperature at the soil surface and the subsequent establishment of an inversion. The clear sky, low humidity and lack of wind associated with anticyclonic conditions (as in Fig. 2.2c) are the major factors, enhanced on a seasonal basis by long night-length, and locally by a lack of any overhead shelter or nearby sources of radiation or turbulence such as buildings, trees, strawstacks or the like. Where the soil is covered by vegetation, the zone of minimal temperature is displaced upwards on to the surface of the plants, effectively insulated from the compensating heat of the soil by the rest of the canopy.

Convex tops of isolated hills may lose a significantly greater quantity of heat by back radiation because the solid angle of sky exposure is more than half a hemisphere, but radiative cooling is generally most marked in inland valley bottoms, where direct incoming radiation tends to be cut off relatively early in the day, shelter from wind encourages inversion, and the cooling air accumulates by katabatic drainage (see Fig. 2.7.). Geiger (1965) explains all these processes at length with particular reference to the spectacular Gestettneralm sinkhole in Austria, but he stresses that quite modest relief can lead to striking temperature stratifications.

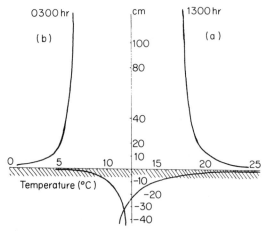

Fig. 2.11. (a) Negative (normal) and (b) positive (temperature inversion) lapse rates in air temperature profiles. From Crowe (1971).

Fig. 2.12. Hughenden Valley, north of High Wycombe, (a, top) on a sunny winter day and (b, bottom) blanketed by thick radiation fog. (C.J.S.)

The classic British example of a frost hollow is in fact a chalkland site: it is a dry valley in the Hertfordshire Chilterns near Rickmansworth, studied for many years by the meteorologist E. L. Hawke, but now largely built-up. The valley is not particularly deep but has a relatively large catchment of katabatic air (Fig. 2.13a), and Hawke (1944) noted that

> . . . excessive cold . . . on radiation nights is sufficiently explained by an environ-mental lie of the land favouring convergence of light katabatic winds from areas of relatively high ground on all sides. Evidence of this effect is particularly noticeable on quiet, clear autumn evenings when garden bonfires of damp leaves are burning on the surrounding slopes and hill-tops. At such times it is common to see, during the hour or so after sun-set, rivers of white smoke slowly winding their way down the northern strip of the valley from points in all quadrants of the compass . . . usually at about 2 miles per hour.

The valley is lined by highly permeable soils, mostly derived from Upper Chalk, but with scattered superficial deposits of mainly Pleistocene gravel, and at the time of Hawke's observations it was entirely down to grassland.

An additional feature is that the valley is dammed by a substantial railway
embankment, pierced by a narrow tunnel used by pedestrians and cyclists.
Hawke was unable to detect any katabatic winds which would blow thistle-
down through the tunnel in a down-valley direction, and assumed that the
air had settled out before it reached the embankment. He was probably

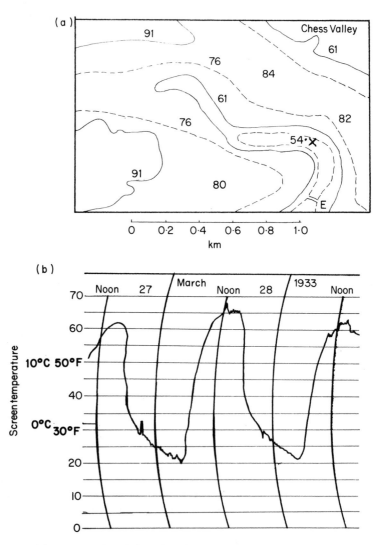

Fig. 2.13. (a) Topography of the Rickmansworth frost hollow and (b) a represen-
tation of the thermograph record for March 27–29, 1933. Altitude is shown in metres;
E = railway embankment. From Hawke (1933).

right, for his bonfire smoke moved plainly enough, but Geiger warns that the popular idea of cold air flowing like water is a misconception, and that because of the very small differences in density between "cold" and "very cold" air, katabatic flow is very sluggish and comparable with syrup rather than water. Unless it has been gaining momentum down a long, smooth, steep slope, the movement of this air may be barely if at all detectable.

Very large ranges of temperature have been recorded at the Rickmansworth site on clear days, such as the spans of 27°C ($-$ 7 to $+$ 20) within eight hours on March 28, 1933 (Fig. 2.13b), and 28°C ($+$ 1 to $+$ 29) within ten hours on August 29, 1936. But it is the night minima which are consistently the most remarkable feature of this topoclimate, and Bilham (1938) has compared the night climate of this valley with that of Braemar at an altitude of 338 m on the Aberdeenshire Plateau. Even in 1952, Manley was still able to describe the climate there as "one of the most exceptionally frosty locations yet known in the British Isles", but there must be many similar localities, which have simply not had the distinction of being the stamping ground of an Honorary Secretary of the Royal Meteorological Society.

2.3.3.3 Other effects of aspect and slope

Here we examine in detail some of the points touched on in the previous section which are not specifically concerned with frost hollows. The relation of the Earth's axis and orbit to the Sun accounts for the familiar diurnal and seasonal fluctuations in both instantaneous and daily radiation receipt. (Rusty readers should consult Crowe (1971) for a lucid refresher course in the geometry of the seasons.) These fluctuations result from changes in the distance traversed by the solar beam as it passes through the Earth's atmosphere (the path-length, insignificant on a topoclimatological scale), and in the angle at which this beam strikes the ground (the angle of incidence). The latter, the effect of which is described by Lambert's Cosine Law (Fig. 2.14), is of enormous importance in topo- and microclimatology.

For a horizontal surface (i.e. one parallel to the curvature of the Earth), maximal values are, of course, received at mid-day (assuming fine weather), while lowest maxima are to be expected in mid-winter when the sun's azimuth angle is lowest. A hillside or bank, however, will concentrate or dilute the sun's rays according to its gradient and aspect. This is illustrated in Fig. 2.15 and Table 2.1. Northerly slopes receive no direct radiation at all in mid-winter and are typically cool and moist. Their vegetation is often dominated by mosses (see Chapter 5). On the other hand, southerly aspects are typically much warmer and drier, of great advantage to the sheep farmer at lambing time, or for starting early potatoes (or even for growing grapes), but tending to run to drought in hot, dry weather in summer. The flora here is more xerophytic. These examples of the ways in which topography can modify radiation receipt are familiar enough to the field agronomist and ecologist, but there are a few minor pitfalls to trap the unwary.

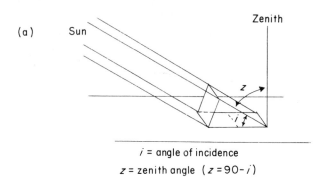

i = angle of incidence

z = zenith angle ($z = 90 - i$)

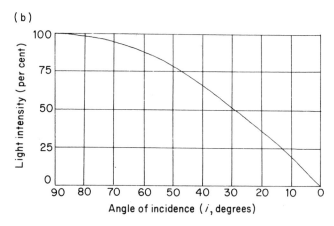

Fig. 2.14. (a) A pencil of rays incident at an angle of 30°. The area illuminated on the horizontal surface is twice that of the cross-section of the beam, so the intensity of illumination at the surface is one half of that normal to the beam. (b) The sine curve which expresses the relationship between light intensity and angle of incidence. Both diagrams from Crowe (1971).

Firstly terminology: remember that the south bank of a valley is north-facing, its north bank south-facing. This may be obvious, but can easily lead to confusion. Secondly, while slopes facing due south might be expected to be warmest, in fact this is not necessarily so. South-westerly slopes have this distinction in fine weather if the morning's radiation is used as latent heat in evaporating water from plant and soil surfaces, but if the gradient is very steep, both southerly and south-westerly slopes are deprived of radiation

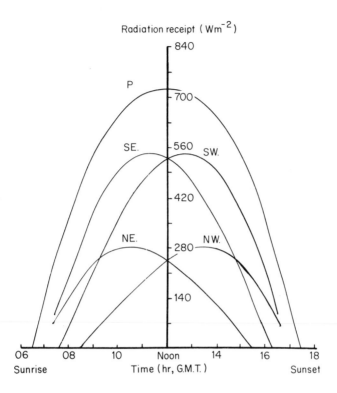

Fig. 2.15. The effect of aspect on insolation, latitude 55°N., March 22. Values for 20° slopes facing NE., SE., SW. and NW. are compared with those received on a surface turning with the sun so as to meet the rays always at right angles (P). From Crowe (1971).

Table 2.1

Daily solar income (MJ m^{-2} day^{-1}) according to aspect, calculated for summer and winter solstices and spring equinox at latitude 45°N.[a]

	June 22	March 21	December 22
Horizontal surface	24·15	13·19	2·85
South-facing slope of 20°	24·70 (102)[b]	17·08 (129)	5·48 (193)
North-facing slope of 20°	20·72 (86)	8·00 (61)	0·08 (3)

[a] From Crowe (1971).
[b] Figures in parentheses indicate the percentage of the corresponding horizontal value.

because direct insolation is obscured for much of the early and latter part of the day. Convection cloud, with a peak around mid-afternoon, may enhance this pattern. Finally, these contrasts in aspect are only significant in relatively clear weather when radiation is mainly direct: on overcast days, diffuse ("sky") radiation predominates. Diffuse radiation is that portion which reaches the Earth's surface "after having been scattered from the direct beam by molecules or suspensoids in the atmosphere" (Rosenberg, 1974). Since this is multidirectional a southerly aspect has no advantage over a northerly one. Under these conditions, the amount of sky "seen" by the site determines its radiation receipt, and a southerly aspect with a restricted outlook would receive less radiation than a northerly one with a broad view (see Pontin, 1962; Wagar, 1964). In chalk quarries, reflection from the banks and cliffs may enhance both diffuse and direct radiation receipt. Some crude measurements I obtained with a Kipp Solarimeter in the Chinnor Cement Works' quarries around noon on a clear September day in 1977 indicated that values of incoming short-wave radiation at ground-level were some 10% higher on the sunlit terraces of the quarry face than on flat, grass-covered ground adjacent to the quarry.

Although radiation receipt is of prime importance in topoclimatology, other environmental factors operate too. Sloping ground collects neither cold air nor excessive rain water—both have a marked tendency to run off—so it is rarely prone to frost or flood, though there is a risk of erosion. Slopes facing south-west are exposed to the full force of the prevailing south-westerly winds (invariably the direction from which the strongest winds blow), and during unsettled "Atlantic" weather there will always be a complicated interaction here between the orographic tendency to collect excessive precipitation, and the loss of water (and heat) by the accentuated evaporating powers of sun and wind. Clearly the lower slopes will tend to be less "exposed" in this respect (see Fig. 2.7. for example), and this is presumably what ecologists usually mean by this rather ill-defined term. Slopes which are not situated in an optimum radiation climate may make up for this by being sheltered from the strongest or coldest winds.

2.3.3.4 Maritime topoclimates

The importance of the maritime influence on the macroclimate of Britain was established early on in this chapter, and we have also touched on certain climatic features of maritime localities. Here we consider the topoclimates of those areas of the English chalklands immediately adjacent to the sea: the spectacular cliffs of the Kent, Sussex and Dorset coasts, the extremities of the Isle of Wight, and the far western and northern outposts of the Chalk at Beer Head in Devon and Flamborough Head in Yorkshire. We have already noted that humidity and windiness are both relatively high near the sea, so frosts are neither as frequent nor as severe as they are further inland, and that the atmosphere is charged with salt (and other chemicals) from the spray and froth of breaking waves. Onshore winds, including powerful vertical currents which sweep up the cliff-faces, pick up this salt and carry it inland, although most of it is deposited within a few metres of the shore.

Most of this information, however, is intuitive, and objective data are remarkably hard to come by. In France, Liger (1956) surveyed the chalk cliffs of the Normandy coast and recognised several distinct plant communities, obviously related to the degree of exposure to the sea; but no measurements of soil salinity or atmospheric humidity were taken. A good source of critical data on maritime topoclimates is Goldsmith's (1973a,b) work at South Stack, Anglesey, admittedly some distance from the Chalk coasts, but obviously applicable to any sufficiently windy sea-cliff topoclimate where the cliffs are of comparable height. Goldsmith measured the volume and conductivity of precipitation collected in beakers over 10 months from a close network of 12 sampling sites, intensified at one stage to 48. A minimal onshore wind speed of 20 knots (35 km hr $^{-1}$) was found to be necessary for salt deposition to occur, but salt was deposited in the largest quantities in gales of more than 30 knots (43 km hr $^{-1}$).

Salt deposition was heaviest at the base and on the face of the 30-m cliff studied, and decreased rapidly inland within a few metres of the cliff edge. But local features of topography, such as bluffs, clefts and gullies, could deflect and concentrate the salt, apparently in much the same way as minor surface features can induce patterns in drifting snow. A particularly significant conclusion of Goldsmith's, however, was that different rates of evaporation from sites receiving similar deposits of salt could result in marked contrasts in the osmotic potential of the soil, so patterns of deposition did not necessarily match up with patterns of vegetation (cf. Liger's conclusions in the previous paragraph).

2.3.3.5 Woodland topoclimates

The woodlands of the Chalk, like the rest of the vegetation, are dealt with in detail in later chapters, but it is convenient to discuss climatic aspects here. The influence of the woodland canopy upon climate is obvious and well known, although we easily forget that the crowns of the trees themselves, which form a distinct, if unfamiliar, habitat, are fully exposed to the elements. But within the wood, the climate differs markedly from that of the open country which surrounds it (Figs 2.16a–e). On a sunny summer day, the wood is shady, cool and humid. On a frosty night the wood retains much of its warmth, for back radiation is greatly hindered by the overlapping crowns of the trees, even when bare of leaf. On a windy day the wood is more sheltered: listen to the roar of a winter gale from inside a wood where the air

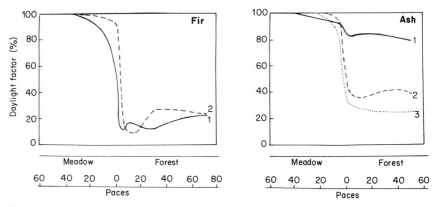

Fig. 2.16. Figures 2.16a–e show woodland topoclimates. (a) The shading effect of fir (1 = full sunlight; 2 = overcast) and ash (1 = winter, overcast; 2 = winter, full sunlight; 3 = summer, full sunlight) canopies (Geiger, 1965). The daylight factor is an expression of woodland light intensity as a proportion of that in the open.

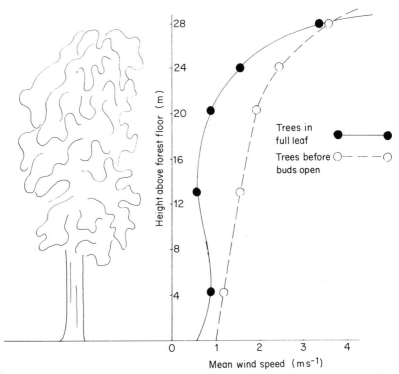

Fig. 2.16b. The influence of winter and summer deciduous canopies on wind speed (Geiger, 1965).

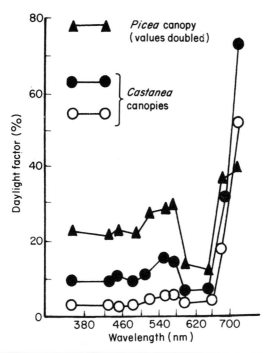

Fig. 2.16c. Spectral composition of shade light under evergreen (*Picea*) and deciduous (*Castanea*) canopies in July (Coombe, 1957).

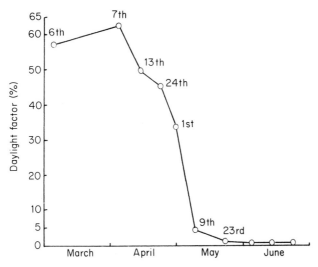

Fig. 2.16d. The decline in relative light intensity under oak–hornbeam woodland in spring (Salisbury, 1916).

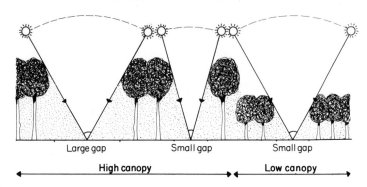

Fig. 2.16e. The effect of gap size in relation to canopy height on the light climate of woodland glades (Cousens, 1974).

is calm enough to light a pipe with ease; or watch young trees in a copse gently swaying in a breeze quite undetectable at ground level. When it rains, the wood offers the dubious shelter of a leaky umbrella, and in foggy conditions the branches may drip with intercepted moisture while the open fields receive far less, if any, precipitation.

The nature of the canopy and its effects on climate vary, of course, from species to species. Evergreens differ from deciduous trees, and among the latter beech, for example, casts a far heavier shade than ash. Many attempts have been made to modify Beer's Law to describe the extinction of light as it passes through, and is absorbed, reflected and transmitted by the leaves and branches. Monsi and Saeki (1953) first proposed the expression

$$\frac{I}{I_0} = e^{-kF}$$

where I_0 is the initial light intensity (more accurately, the flux density), I the flux density after passage through a canopy of leaves of leaf area index F (measured cumulatively, downwards), and k the extinction coefficient of the leaves (varying with their shape, angle, thickness and overlap). This expression, which of course is equally applicable to herbaceous communities, has been superseded by others which take proper account of the variations in texture and geometry of the canopy, and an example is shown in Fig. 2.17. Monteith (1973) and Rosenberg (1974) provide more details than can be given here.

The spectral composition too, is quite different inside the wood from that of the open daylight. It is not always appreciated that the familiar (and to plants all but useless) greenness of the woodland interior is accompanied by a sharp increase in the dark-red and near infra-red components of the spectrum. A hint of this can often be seen in the tints of sunflecks. Work in this field has long been a speciality of the Cambridge plant ecologists, and Coombe's (1957) data from the Heidelberg forests admirably illustrate this important point (Fig. 2.16c).

When it is raining, some of the intercepted rain evaporates again *in situ*, while leaf-drip and stem-flow direct the rest into an uneven patchwork so that while some parts get more than their fair share, others remain quite dry (Blackman and Rutter, 1948). Lee and Monsi's (1963) classic work on the ecology of oriental *Pinus densiflora* woodlands showed that rain water passing through the canopy picked up soluble organic exudates which had a selectively toxic role among the herbaceous plants of the forest floor. No comparable work seems yet to have been done in British woodlands, though on the Chalk many naturalists have long had their suspicions about yew (Chapter 8). Zinke (1962) has demonstrated the effects of trees on various soil properties in their immediate vicinity.

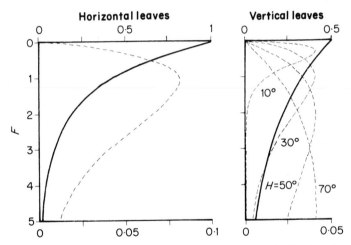

Fig. 2.17. Relative intensity of direct (solid lines, upper abscissa) and scattered (dashed lines, lower abscissa) diffuse light received by foliage, according to leaf orientation, leaf area index (F, measured cumulatively from the top of the canopy downwards) and sun elevation (H). From Kuroiwa (1968).

The topoclimate of deciduous woods exhibits seasonal changes according to whether the trees are bare or in leaf. The spring light phase in deciduous woodlands (Fig. 2.16d) is the result of the lag between increasing solar radiation and leaf expansion, and surprisingly large fluctuations in temperature and humidity can occur when the strengthening sun penetrates to the forest floor for more than a few minutes at a time (Geiger, 1965).

Sharp contrasts in climate can be witnessed in glades and clearings, though much will depend on the size of the gaps in relation to the height of the canopy (Fig. 2.16e). Where these are small enough to let the sun in (and back radiation out) without significantly increasing air movement, hot spots and frost pockets may occur, sometimes severely inhibiting regeneration (Chapters 8 and 9). Larger clearings introduce the risk of wind-throw. The wood-edge experiences unusual combinations of climatic factors. Thus, as in the small gaps, frost may occur at the edge of the wood where there is no overhead shelter to inhibit back radiation, but effective lateral shelter to encourage inversion. If the edge of the wood lies at the bottom of even quite a gentle slope, the effect may be exacerbated by the ponding-up of katabatic air against the wood. Northern and eastern boundaries tend to receive little or no direct sun, but bear the brunt of the coldest winds, while southern and western edges receive strong sun and relatively warm winds.

Not surprisingly, the wood-edge forms a distinct and important habitat for plants and animals. The influence of the wood may be felt a considerable distance away. Shelter extends up to 30 times the height of the wood downwind and 10 times the height upwind (Caborn, 1965), while cooling breezes, similar in cause and effect to coastal sea-breezes, blow out of the wood on warm, settled days well beyond the zone of direct shading (Geiger, 1965).

Hedgerows and even individual trees will show similar but on the whole scaled-down contrasts in climate to those of the wood, and the more obvious effects of shading and shelter are often reflected in the conformation of the trees and bushes themselves, and in the nature of the surrounding herbaceous vegetation or crops. Scrub communities will resemble those of wood, hedge or tree according to their composition, but, in areas where clumps of scrub alternate with open grassland, patterns of topoclimate and microclimate become very complex.

2.3.4 Microclimates

There are no new principles involved here which have not been laid bare in the foregoing pages: merely a reduction in scale. Just as we tend to disregard what happens in the crowns of the woodland trees, our size makes us quite oblivious of the fact that the herbaceous canopy contains no lesser contrasts in light, temperature, humidity and air-movement than the wood. Richard Jefferies (1879) wrote of the "green forest of grass-blade and moss" which confronted the ants he observed, "with no path through . . . the darkness at the roots". In more objective terms, Duffey (1962a) points out that to an erigonid spider 2–3 mm in length, "the vertical height range of 80 cm in *Brachypodium* [grassland] is quite as great as that experienced by a rodent in a deciduous wood".

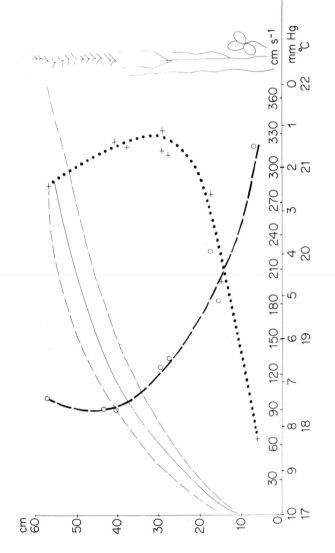

Fig. 2.18. Profiles of wind, humidity and air temperature in a grass-clover canopy about 50 cm high, between 1500 and 1600 G.M.T. on a sunny June day. It had rained the previous day. (———) Mean wind speed; (———) wind fluctuations (cm s⁻¹); (———) saturation deficit (mm Hg); (•••••) air temperature (°C). From Waterhouse (1955).

The thickening up of a turf following the removal of grazing animals has important effects on microclimate (Figs 2.18 and 2.19), the consequences of which are examined in Chapter 6. Also considered later are the importance of flower temperature for seed-set in the stemless thistle (*Cirsium acaulon*) and of nest temperature for brood production in the yellow ant (*Lasius flavus*), as well as various features of the microclimate of the soil surface itself (Chapters 3 and 6).

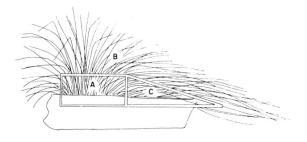

Fig. 2.19. The microclimate of a *Bromus erectus* tussock. During an overcast October day the dry-bulb temperature ranged from 12·8°C to 16·3°C (mean 14·8; ten half-hourly readings between 1100 and 1530) in the open just above the tussock, compared with 12·7–16·0°C (mean 14·6) in region A. Greater differences were found in humidity; measurements of saturation deficit gave 1·7–3·6 g m^{-3} (mean 3·0) in the open and 1·1–2·3 g m^{-3} (mean 1·8) in region A. Much greater contrasts could be expected in warm, sunny weather in summer, or winter cold spells. From Ford (1937); see also Chapter 6.

3
Soils of the Chalk

3.1 Soil Classification

It is quite a revelation, if you have never deliberately done so before, to seek out a vantage point of adequate elevation in a reasonably extensive tract of chalk country given over largely to arable agriculture. Better still, take to the air (Fig. 3.1). Choose a clear day, preferably in late autumn or early winter when the view is of soil rather than crop, and select an area which the geological map shows as "drift-free" chalk, i.e. covered by only a few centimetres of superficial deposits, if any. Note the great kaleidoscope of soils to which the underlying chalk, purely or partially, has given rise, ranging in colour from white to black through every shade of grey and brown, and merging, often quite haphazardly, one into another. Bear in mind, too, that these patterns of colour are paralleled by comparable contrasts in innumerable other features, both physical and chemical. Submit yourself to this simple and very pleasant exercise, and you will not fail to increase your regard, and sympathy, for the soil surveyor.

Of course, this great variety of soils is hardly to be wondered at, taking into account all that has occurred even "recently", since the final retreat of the Devensian ice some 12 000 years ago (Chapter 1). Chalk, marl, flints, clay, sand and gravel have been eroded and mixed, sometimes several times over. Colonisation by plants and concomitant pedogenesis have been repeatedly interrupted, at first by natural processes alone, later increasingly augmented by the activities of farmers. So where do we begin? What exactly is a "chalk soil"? And what is the fundamental unit in soil surveying?

Regarding the first two questions, the shallow soils developed over chalk or chalky drift, as well as exposures of the raw rock itself, obviously justify inclusion, just as the brown earths and podzols derived from Clay-with-flints and other superficial deposits of the plateaux, and the deep alluvial loams

Fig. 3.1. The ancient Ridgeway threads its way south-westwards across the extensive arable chalklands of the Berkshire Downs. The hill fort to the right of the picture is Uffington Castle. (Cambridge University Collection, copyright reserved)

and gleys of the vales, can be disregarded as outside our present terms of reference. The difficulty lies, as always when attempts are made to impose a system of classification onto a continuously variable pattern, in dealing with the transitional soils along the very edge of the clay cappings, or where Lower Chalk and hillwash tail out over Upper Greensand or Gault. For the purpose of this book, a chalk soil is defined as "a calcareous soil overlying chalk or chalky solifluxion or hillwash deposits, and neither covered by a significant thickness of plateau drift or valley alluvium, nor underlain by sufficiently impermeable strata to permanently impede drainage."

The basic unit in soil description is the *pedon*, which can be regarded as a three-dimensional block of the profile typically laid bare in the surveyor's trench, ". . . large enough . . . to include a full set of horizons and permit observations of boundaries between them . . . but small enough to minimise variability within the unit" (Simonson, 1968). The Soil Survey of England and Wales regards 1 m^3 as the volume of soil which "in practice can be adequately described and sampled as a single entity" (Avery, 1973). The *soil series* consists of a collection of pedons with similar (specified) observable or measurable properties, developed in lithologically similar parent material, and is the fundamental unit of soil classification. The soil series name is also used for the *soil mapping unit*, but because of the inevitable heterogeneity already mentioned, this is defined as "a geographical area where the majority of soil profiles (usually at least 60 per cent) conform to the general concept of the dominant soil series" (Curtis *et al.*, 1976). Where soils are markedly heterogeneous, the surveyor resorts to the *soil complex*. A soil series may be further subdivided into *variants* (differing somewhat from the parent series in the relative development of diagnostic horizons) and these into *phases* (according to local contrasts in gradient, stoniness or land-use, for example), although these two terms have been used somewhat interchangeably in the past. As we shall see in a moment, the names for the soil series so far catalogued on the English Chalk have been used in slightly different ways by successive authorities, and this has led to considerable confusion.

3.2 Raw Chalk

The true chalkland rendzina soils are described shortly, but first a word about the status of raw chalk itself as a soil material. Exposed chalk rapidly weathers, as we saw in Chapter 1, to a fine detritus (lithosol), given a multiplicity of names (see Kubiena, 1953), of which simply *raw chalk soil* seems most appropriate. Kubiena describes the raw soil as a young, embryonic, white, earthy mass, with no distinct humus horizon visible to the naked eye. What little humus there is decomposes fairly quickly as long as moist conditions are maintained (wooden fencing posts and stakes rot at the base), but normally the surface dries rapidly after rainfall, and this greatly restricts mineralisation (Section 3.8.3). Indeed, in some parts of northern France, bizarre tales are told of mummified corpses in chalky graveyards which stubbornly resist decomposition! In England, soils of this kind may persist on slopes which are sufficiently steep, as in old chalk-pits and on cuttings, embankments and earthworks (Fig. 4.8; see also Proudfoot, 1965). They are deliberately maintained on some hill-figures (Fig. 1.34). The loose chalk scree at Kimble, Buckinghamshire, has already been noted (Section 1.2.2.3; see also Fig. 8.14).

Where the gradient is less steep, raw chalk soils can only persist in arid climates such as that of Israel (Ravikovitch and Pines, 1967). In Britain and western Europe, organic matter begins to accumulate as a result of the rapid increase in macro- and microbiological activity which accompanies colonisation by plants and animals and the incorporation and conversion into humus (humification) of organic litter. B. W. Avery (personal communication) regards the accumulation of 1% of organic matter in the upper 15 cm as sufficient to distinguish a rendzina from a raw chalk soil. There seems to be very little published information on the rate at which this happens, but it appears that these early stages of pedogenesis occur more rapidly on chalk than on the harder limestones, except, of course, where the chalk itself is of the harder kind (Chapter 1). In his study of an old chalk-pit at Harefield (then in Middlesex), Locket (1946a) first recorded signs of an organic layer accumulating in the top 5 cm of weathered chalk in 1942, 13 years after the pit ceased to be worked.

Reference is made later to a raw chalk soil (created during the construction of a forestry track) which was included in a study of seasonal water and nutrient regimes in the Chilterns by Davy and Taylor (1974a,b; 1975). This soil was given the name "Hobbs Hill" after the wooded knoll of that name on the Hampden Estate where it was studied, but this has not been recognised so far by the Soil Survey.

3.3 The Rendzinas

3.3.1 Types of rendzina

The rendzinas are the characteristic soils of the chalk country, underlying woodland, scrub and grassland alike, and often showing surprisingly little change, other than being truncated, when they are brought into cultivation. We can only guess at the length of time which must elapse between the beginnings of organic enrichment of the surface layers of the chalk as described by Locket, and the stage which can be regarded as mature. But it must be measured in centuries rather than decades (Section 5.4.3.5). The characteristic of grading sharply, within less than half a metre, into the underlying chalk accounts for the alternative titles of lithomorphic, intrazonal or A/C soils, shared with other, rather different, types. As a group, the rendzinas are not restricted to chalk alone; they occur over any suitably calcareous substrate. The term rendzina is derived from the Polish word *rzedcic* or *rzezic*, meaning to roar, an allusion to the noise of the plough scraping through rubble and bedrock, always so near to the surface.

Rendzinas on the English Chalk may be humic or non-humic, colluvial or gleyic. The meaning of these qualifications, and the disposition of these soil types across the chalk country, are considered in the following paragraphs, but unfortunately this exercise is complicated by the fact that only in the last few years has any attempt been made by the Soil Survey of England and Wales to standardise terminology and descriptions. Much of what follows must still be regarded as provisional, if not controversial. The reader is advised to consult Curtis *et al.* (1976), together with Soil Survey publications such as those by Avery (1973), Hodgson (1974) and Cope (1976).

3.3.2 Humic rendzinas

3.3.2.1 Some further definitions

A humic soil is one which contains relatively large quantities of organic matter incorporated into the mineral horizon. Much of this organic matter is in an advanced state of humification, finely divided and strongly bound to the clay fraction in clay–humus complexes, so that in its turn the mineral horizon is said to be *humose*, a situation which, by definition, arises once the organic carbon content reaches $4 \cdot 5$–7% (8–12% organic matter), according to the clay content of the soil (Avery, 1973).

A humic rendzina, then, consists typically of a shallow, humose mineral horizon (signified Ah if uncultivated, Ap if cultivated), virtually black when moist, overlying a strongly calcareous bedrock (C horizon)—in this case chalk. If the chalk is only slightly, or not at all, fragmented, the C horizon is designated Cr; strongly fragmented chalk is given the notation Cu. Sometimes a rubbly A/C horizon may intervene. A combination of physical and chemical weathering with the action of earthworms, which ingest small particles of chalk and deposit them at or near the surface, endow the upper layers with varying quantities of calcium carbonate, which normally results in a relatively high pH (see Salisbury, 1924). Note, however, that humic rendzinas contain less—sometimes much less—calcium carbonate than their non-humic counterparts and may indeed be regarded as only "moderately calcareous" by comparison. The soils have a potentially very active microflora, and strong water-stable aggregate structure and cation exchange properties.

3.3.2.2 The Icknield series

The Icknield soil series, or at least its name, has had a chequered history since Kay (1934) first applied it to the chalk rendzinas of the Berkshire Downs in her survey of the soils of the Vale of the White Horse. Similar soils under cultivation were allocated by Kay to her Upton series, named after the nearby village of that name. In his memoir on the soils of the mid-Chilterns, however, Avery (1964) took a broader view and, discarding the Upton series, included both cultivated and uncultivated chalk rendzinas as mere phases of the Icknield series. In later surveys touching on the South Downs (Hodgson, 1967), the eastern Berkshire Downs and southern Oxfordshire Chilterns (Jarvis, 1968) and, once again, the Vale of the White Horse (Jarvis, 1973), this interpretation of the Icknield series was retained, although Hodgson introduced a brown variant now recognised as the Andover series (Section 3.3.3.2). It ought to be added, however, that the Icknield series was not recognised on the Chalk of the North Downs near Wye, Kent, by Green and Fordham (1973), who proposed instead the reintroduction of the Upton series, but in a sense which included uncultivated as well as cultivated soils (Section 3.3.3.4). Now we have come full circle, as it were, for in his recent survey of the Wiltshire chalklands west of Salisbury, Cope (1976) uses the revised terminology of Avery (1973), and both Icknield and Upton soil series are once again recognised.

Uncultivated Icknield soils are now mostly confined to slopes steeper than about 11°, and are typical of the steeper sides of the characteristically asymmetrical valleys of the chalk downlands (Fig. 3.2). Two sub-phases have been recognised in the Chilterns by Avery (1964), of which one is slightly deeper and more retentive of water, the other shallower and drier. The deeper and moister soils mainly coincide with hollows where litter can accumulate, but they may occur where the clay content is higher, or where the aspect is north. Surface litter is quickly incorporated, mainly by earthworms, into the Ah horizon. Some of this is mineralised, while the rest is humified to a mull-type humus, "saturated with bases and resistant to mineralisation, and . . . strongly bound to the clay fraction in water-stable granular aggregates" (Avery, 1964). The question of mineralisation is returned to later.

Shallower and drier Icknield soils often occur on steep banks subject to strong insolation, and typically coincide with harder or flinty strata in the chalk, relatively free from superficial deposits. In the more windswept woods on these soils, surface litter tends to be swept clean, and it collects in the hollows. In the grasslands, especially where grazing is lax, grassy litter often accumulates as a well-aerated mat which is only very slowly incorporated into the mineral soil. Here, earthworms are largely replaced by other invertebrate species, such as ants and insect larvae, mites (Acarina), woodlice (Isopoda) and millipedes (Myriapoda). These produce a distinct form of humus which Kubiena calls "mull-like rendzina-moder". The organic matter content of this humus is greater than mull humus, and consists of droppings and recognizable plant remains, which are strongly resistant to decomposition. The structural aggregates are finer, softer and more crumb-like, and there is little formation of clay–humus complexes. A particularly fine grade of this soil is strikingly developed in the mounds built by colonies of the yellow ant, *Lasius flavus* (see Wells *et al.*, 1976, p. 612; King, 1977b, p. 271; also Chapter 6 of this book).

Regarding their particle size distribution, Icknield soils (*sensu* Cope, 1976) tend to exhibit two peaks, one in the stone category, the other in the fine silt (see Table 3.1). Stones may consist of flints or pieces of chalk (chalk stones). The silt originates mainly from loess, and, as we have seen, it is now

Table 3.1

Particle size distribution classes and categories of stoniness now accepted by the Soil Survey of England and Wales [a]

	Fraction	Size range
Clay		<0·002 mm
Silt		
	fine	0·002–0·006 mm
	medium	0·006–0·02 mm
	coarse	0·02–0·06 mm
Sand		
	fine	0·06–0·2 mm
	medium	0·2–0·6 mm
	coarse	0·6–2·0 mm
Stones		
	very small	0·2–0·6 cm
	small	0·6–2 cm
	medium	2–6 cm
	large	6–20 cm
	very large	20–60 cm
Boulders		>60 cm

[a] From Hodgson (1974).

apparent that in very few localities is the Chalk entirely free from this deposit. A technicality weights the silt content, however: clay-size particles of calcium carbonate which do not have the electrochemical properties of the true mineralogical clays are, by convention, now assigned to the silt fraction (Hodgson, 1974). The Icknield series is therefore described in full as a "calcareous humose flinty fine silty" soil, which may locally exhibit "chalk-stony fine silty" or "chalky fine silty" phases. A clayey variant grades into the Wallop series (see below). Typical Icknield horizon sequences are shown in Fig. 3.3. Physical, chemical and microbiological properties are discussed later.

3.3.3 Non-humic brown and grey rendzinas

3.3.3.1 Distinguishing features

Non-humic rendzinas contain less organic matter in the A horizon than their humic counterparts so that the colour of the mineral fraction is no longer masked, and various shades of brown and grey predominate. *Brown rendzinas* are brown or reddish, and have an A or thin B horizon (or both) which contains less than 40% $CaCO_3$. Two series within this group, the Andover and Wallop series, have been known for some time on the Chalk, while a third, the Newmarket series, has been proposed in East Anglia by Hodge and Seale (1966). The *grey rendzinas* are found on more chalky substrates, as well as on the marly Lower Chalk. These are typified by the Upton series (*sensu* Cope, 1976) on the Upper and Middle Chalk, and harder beds of the upper zones of the Lower Chalk, and the Wantage series on the Chalk Marl. Chalky Head of appropriate lithology may form the substratum for any of these soils in lieu of the Chalk itself.

It is certain that some, at least, of these brown and grey soils on the Upper and Middle Chalk have developed from Icknield soils within recent historical time as a result of sustained ploughing of the downs, although this should not be taken to mean that the downland soils are necessarily threatened by continuous cereal-growing (see Chapter 9). As far as the Wiltshire chalklands are concerned, Cope (1976) puts the matter in perspective as follows:

> The last major ploughing of downland occurred during the 1940s, and has since proceeded piecemeal. Land ploughed up then is still mostly occupied by Icknield fine silty soils. . . [and] it would therefore appear to need at least 30 years to convert an initially "moderately" calcareous Icknield to an Andover soil due to the abundant humus retained in its stable, calcium-saturated, clay–humus complex.

Cope concludes:

> Moreover, because of their 20–30 cm depth, initially low carbonate content, and

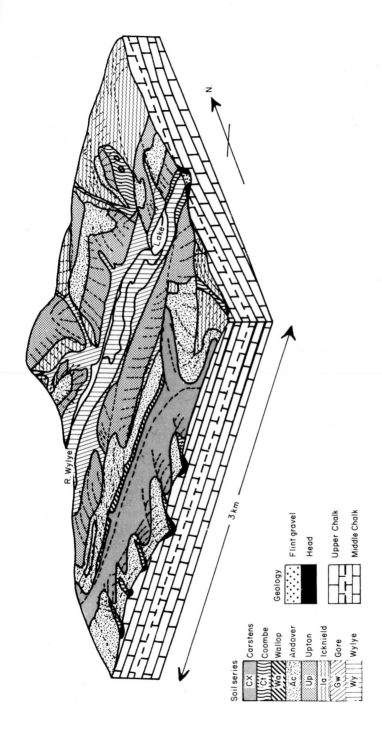

Soil series

CX	Carstens
Ct	Coombe
Wa	Wallop
Ac	Andover
Up	Upton
Ia	Icknield
Gw	Gore
Wy	Wylye

Geology

Flint gravel

Head

Upper Chalk

Middle Chalk

R. Wylye

Lake

N.

3 km

Fig. 3.2. Block diagram of the chalk country in the vicinity of Wylye, Wiltshire, to illustrate the influence of topography on soil distribution (Cope, 1976). Vertical scale exaggerated five-fold. The Carstens series (CX), a brown earth developed on Clay-with-flints, is not described in the text.

flinty or hard chalk substrata, most "recent" cultivation of downland soils has not yet resulted in the formation of chalky Icknield or Upton series soils.

However, deep cultivations to kill couch-grass (*Agropyron repens*), which bring up chalky earth or very small chalk fragments (or both)

are steadily converting the edges of Andover map separates to brownish chalky Upton soils. Local patches of dark greyish brown to greyish brown Upton soils appear to be derived from the recent cultivation of Icknield soils which developed on sites depleted of loessal drift.

Particularly chalky stretches are often seen on the upper margins of fields on steep banks.

3.3.3.2 The Andover series

The Andover is another soil series with a complicated history. The name has been attributed to Robinson (1948) in his essay on the soils of Dorset included in Good's account of the flora of that county (see Chapter 5), but in fact it was first used by Kay in 1940 in her unpublished survey of Carstens and Partridge Farms, Hampshire. However, Kay's Andover series resembled soils now included in the Upton series, while soils now regarded as Andover were placed by Kay in her Ann series! Soils referred to the Andover series as recently as 1973 by Green and Fordham are now classed as Upton flinty phase soils (Section 3.3.3.4). Only Hodgson's (1967) "Icknield series, brown variant" is directly comparable with the Andover series as now recognised, which was, in fact, first proposed and used by Jarvis (1973).

The Andover series is now described as a flinty fine silty brown rendzina. Flints may be locally very numerous: in the topsoil of cultivated fields near Grovely Wood, Wiltshire, Cope records 25–30% by volume occupied by flints. Of the silt, most, as we have seen, comes from loess, though some of the silicate (i.e. non-calcareous) clay probably originates from Clay-with-flints. Here and there a clayey variant referable to the Wallop series occurs (see below). Andover soils are poorly to moderately calcareous (0–40% calcium carbonate) and contain usually less than 10% organic matter. They are usually associated with Upper Chalk and are mainly cultivated, but sometimes occur under woodland. The profile (Fig. 3.3) may include a more distinct B horizon than is the case in the other rendzinas, especially under coniferous woodland, though this tends to be lost in cultivation.

3.3.3.3 The Wallop and Newmarket series

Wallop soils, named by Kay from the series of Hampshire chalkland villages incorporating that name, are, according to current Soil Survey classification, shallow flinty clays overlying chalk. They were formerly (e.g. Avery, 1964) regarded as brown calcareous soils comparable to the Coombe series

Fig. 3.3. Monoliths of (left to right) cultivated Icknield, Andover and Upton rendzinas over white chalk, Wantage rendzina over Chalk Marl, and Coombe brown calcareous soil over chalky head, prepared by the Soil Survey of England and Wales, Rothamsted. The rule is 50 cm long. See text for the significance of the horizon labels. (C.J.S.)

[*Facing p. 101*

(Section 3.4) with which in places they have been combined for mapping (see Hodgson, 1967). The crucial factor is the depth of the intermediate Bw horizon: if this extends below 30 cm, the soil is indeed regarded as a brown earth (such as the Carstens and Winchester series, and others as yet unnamed). Where the Bw horizon is shallower, however, the soil is classified as a brown rendzina and allocated to the Wallop series, differing from Andover soils only in possessing a more clayey lithology, betrayed by a coarse cloddy look after ploughing. Wallop soils may also contain large flints. The Wallop series is nowhere extensive, and is typically found as a narrow band running parallel with the contour along the edge of Clay-with-flints deposits on the crests of upper scarp and valley slopes (see Fig. 3.2). The Newmarket series, named in Cambridgeshire by Hodge and Seale (1966), consists of thin sandy loams over chalk.

3.3.3.4 The Upton series

The Upton series as now recognised appears to include the Andover series of Kay (1940) and of Green and Fordham (1973), as well as the cultivated phase of the Icknield series introduced by Avery (1964) and followed by Hodgson (1967), Jarvis (1968) and Jarvis (1973). Note that it does not correspond exactly to Green and Fordham's "Upton" series. The modern Upton series approximates to Kay's (1934) original description, but in addition it now includes very chalky soils underlying uncultivated downland and woodland, such as where moles, rabbits and badgers are active (see Chapter 6), or where there is a strong tendency for soil creep to occur. Though the soil may contain small quantities of loessal silt, chalkiness is the major feature of the Upton series, which is consequently described as a chalky grey rendzina, sometimes appearing as a flinty or chalk-stony phase. In fact, calcium carbonate may make up more than 60% of the topsoil, and over 30% of the clay fraction (although as noted earlier, this is classified as silt). The mineral horizon of Upton soils contains less organic matter (usually about 2–10%) than either Icknield or Andover soils, and has only a very weakly developed clay–humus complex. Most of these soils have been under cultivation for centuries, but even in uncultivatd profiles the Ah layer is sometimes very thin—less than 15 cm (Fig. 3.3).

3.3.3.5 The Wantage series

The Wantage series was first described by Kay (1934) from the foot of the Berkshire Downs in the Vale of the White Horse but, again, its exact interpretation has varied. Thus, Avery (1964) included in it soils with more or less fragmented bedrock at various depths, whereas Jarvis (1973) defined it as having "bedrock" at 40 cm or less. The soils which have been mapped as Wantage are generally less calcareous and contain more silicate clay than those mapped as Icknield/Upton (see, for example, Kay, 1934; Avery, 1973). The Wantage series has not yet been redefined by intrinsic characteristics in accordance with Avery (1973). Hitherto it has meant little more than stiff greyish arable soil on Lower Chalk (Fig. 3.3). It will probably be set apart from the Upton series on the basis that the A horizon contains more than 18% silicate clay, but the distinction is not an easy one. For example, if the Wantage series is defined as a fine silty (i.e. $> 18\%$ silicate clay and $<$ 50% $CaCO_3$ in the topsoil) grey rendzina over marly chalk or chalky Head, it will include many of the soils identified by Jarvis (1973) as the Gore series, described next (B. W. Avery, personal communication).

3.3.4 Colluvial and gleyic rendzinas

3.3.4.1 The Gore series (colluvial)

Colluvium consists of surface material deposited in the course of subaerial erosion, and is sometimes called "hillwash" or "ploughwash". Colluvium formed from chalk is described by Cope as "pale coloured yellowish or light greyish brown flintless to slightly flinty loose chalky material of irregular depth over compact chalky Head". Head, it will be recalled, encompasses the products of Pleistocene solifluxion such as the Coombe deposits and Coombe Rock (Chapter 1). Rendzinas which form in chalky colluvium are referred to the Gore series, first named in Kent by Brade-Birks and Furneaux (1930) in their survey of the soils of Wye College Farm, and defined by Cope as a "chalky stoneless to slightly stony colluvial rendzina in colluvium or Head with chalk or compact (flinty) chalky Head at more than 40 cm depth". On this basis, the Gore series is recognised widely over the Lower Chalk (Green and Fordham, 1973; Jarvis, 1973), while Cope has mapped it over Middle and Upper Chalk in Wiltshire. As might be expected, Gore soils are typically found at the bottoms of scarps and valley-sides, tonguing-in to Wantage and other soils of the solid chalk and Coombe deposits. Flinty and chalky brown calcareous earth (cf. Coombe series) variants are also recognised.

There is, however, a snag in this interpretation (B. W. Avery, personal communication). Strictly speaking, a rendzina formed in Head cannot be defined as colluvial, although it is often extremely difficult to decide which of the two, Head or colluvium, forms the bulk of the parent material. As was noted in the previous section, soils at present referred to the Gore series which prove to be derived from Head may have to be reclassified as Wantage soils, although from an ecological and agronomic viewpoint, the distinction is of academic rather than practical interest.

3.3.4.2 Gleyic rendzinas

Gleying is not a process normally associated with chalk rendzinas, even in the heavier marls of the Lower Chalk, but incipient gleying, detected by the presence of rusty mottling, may occur where layers of chalky drift fill hollows in much less permeable strata such as the Gault Clay. Such gleyic rendzinas exhibit an Ap–AC–Cg horizon sequence, and are typified by the Ford End series of Avery (1964). Strictly soils of the vales rather than the downs, the gleyic rendzinas are not described further here, but information about them can be found, for example, in Avery (1964), Hodge and Seale (1966), Green and Fordham (1973) and Jarvis (1973).

3.4 Brown Calcareous Earths: the Coombe Series and Others

Wherever the solifluxion deposits and associated colluvium of the chalk scarps, coombes and valley sides are sufficiently deep, soils more reminiscent of the brown earths of the plateau develop, recognisable as such by a distinct brown-coloured B horizon. Where decalcification has taken place, non-calcareous soils of the Charity and related series develop, but where the deposits are moderately calcareous, they give rise to brown calcareous earths (see Avery, 1973).

These chalkland brown calcareous earths are typified by the flinty fine silty soils of the Coombe series (Avery, 1964). This series may display A–Bw–Cu, A–Bw–Cu–2Cr or rarely A–Bw–2Cr profile sequences, depending on the depth and exact nature of the chalky Head or Chalk parent material (Fig. 3.3). Thus, three phases of the series have been recognised: shallow, moderately deep, and deep, according to whether the unmodified (parent) substrate is at 40–60 cm, 60–90 cm or generally more than 90 cm depth. It is the distinct Bw horizon, extending below 30 cm depth, which distinguishes Coombe soils from the rendzinas, although the shallow phase grades into brown rendzinas of Andover or Wallop type, especially on the upper slopes. Indeed, in the Aylesbury and Reading areas, Coombe and Wallop series have been mapped as one unit (Avery, 1964; Jarvis, 1968). The moderately deep and deep phases of the Coombe series typically occur on the lower slopes and valley floors (see Fig. 3.2).

A relatively uncommon clayey variant of the Coombe series is also known, although this has not yet been given a series name. Some soils termed Clatford by Kay (1940) may be of this type, though others are non-calcareous. With increasingly clayey B horizons and lessening calcareousness, Coombe soils are replaced by argillic or palaeoargillic brown earths such as the Charity and Tring series, while in the wetter valley bottoms underlain by impervious clays, chalky–alluvial and calcareous gleys such as the Halton and Wylye series are found (Avery, 1964; Cope, 1976). Brown calcareous earths over chalky loamy drift in East Anglia have been named as the Swaffham Prior and Soham series (Hodge and Seale, 1966), and Soham has since been mapped in the Thames Valley around Marlow and Hurley. In Wiltshire, Kay's Blewbury series has recently been re-surrected to describe brown calcareous soils in drift over Lower Chalk (D. W. Cope, personal communication), while soils of this type over sandy chalky drift in Breckland are referred to the Methwold series (Seale, 1975).

3.5 Concluding Comments on the Soil Survey

These, then, are the main soil series of the chalklands which have been formally described so far. There can be little doubt that they represent the majority of chalk soils, at least in southern and eastern England, but it must be remembered that the areas so far surveyed in detail are still relatively very small, and a great mass of information remains to be collected and pro-cessed. Directly, many of the finer points of soil classification are probably of greater agronomic than ecological significance, but indirectly they can pin-point historical features, highly relevant to chalkland ecology, which might otherwise be overlooked.

In the normal run, mapping units contain subsidiary soil types which are not drastically different from the dominant series. However, dramatic contrasts can be found on a scale too small to map in the normal way where solution hollows in the chalk have been filled with more recent deposits of quite different composition (Fig. 1.27). A particularly striking demon-stration of this phenomenon comes from a unique piece of field work, involving very detailed large-scale mapping, by Cuanalo (1966). He plotted the distribution and recorded detailed profiles of solution pipes containing various combinations of clay, sand and loam in a field near Lambourn on the Berkshire Downs, of which the main soil type was Icknield. His results are shown in Fig. 3.4.

3.6 Physical Properties of Chalk Soils

3.6.1 Water relations and temperature regimes

So far, these soils have been considered from the point of view of their origin, recognition, classification and distribution. Now we look in more detail at their physical, chemical and microbiological features which are of ecological and agronomic importance. Physical properties are considered first. Except for the gleyic soils of the poorly drained "bottoms" (which are atypical of the chalklands proper and so not given further consideration here), all chalk soils are well structured and freely to excessively drained.

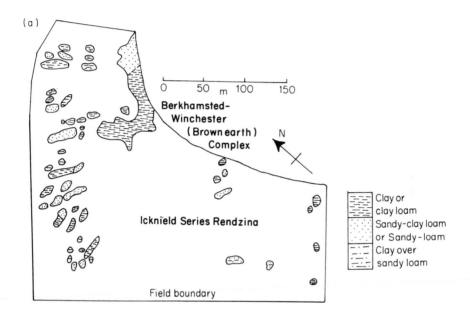

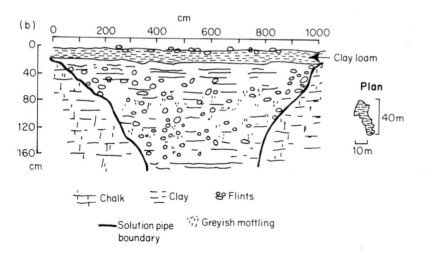

Fig. 3.4. Solution pipes in a field near Lambourn. (a) Distribution and (b) cross-section of an example. From Cuanalo (1966).

Thus they are normally well aerated, so that the major focus of attention is on how warm and dry they can get, and on the nature and significance of their water reserves. Brief mention is made in the following section, however, of the consequences of possible transient anaerobiosis which may affect soils on the Lower Chalk (typically the Wantage series) during wet weather in autumn and spring.

3.6.2 Chalk soil water relations

3.6.2.1 The available water capacity

Few subjects can have led to more starkly opposed viewpoints between naturalists and ecologists on the one hand, and farmers and foresters on the other, than whether the water relations of chalk soils are favourable or unfavourable to plants. Two papers published within three years of each other some 50 years ago graphically illustrate this divergence of opinion. Thus, Miss V. L. Anderson's opening remarks to her classic paper (Anderson, 1927) on the water relations of chalk grassland typify the contemporary attitude of the ecologist:

> The dryness of the chalk has been one of its most widely recognised characteristics . . . . The ecologist has long recognised that the flora is of a xerophytic type and that the water supply is one of the chief determining factors of the environment. . . .

In contrast, representing the point of view of the practical agronomist, the Chiltern forester Sir A. A. Hudson wrote (Hudson, 1930):

> The first thing worth knowing is that a chalk soil is a moist soil. Chalk holds water in suspension and the moisture is shown in many ways. A belt of trees was planted at the side of a carriage drive towards the bottom of the down. The soil in which the trees were planted had been in part denuded of the top soil which was wanted in the garden, and what soil there was left was so chalky that it was nearly white. The belt of trees was carried on further over a fairly deep and light garden soil, far better from a gardener's point of view than the soil on which the trees on the first part were planted. The year following the planting of the trees was exceptionally dry. The result was that all the trees on the poor chalk soil lived, and all the trees on the good soil died.

Why the paradox, and who was right?

It was established in Chapter 1 that solid chalk can hold large quantities of water because it is porous, but that because the pores are of relatively uniform effective radius, this store of water is held over rather a limited range of matric potential. Thus, while a potential of -1 bar must be developed before significant quantities of water begin to be drawn from the chalk, an increase in tension of only a further 4 bar is adequate to remove most of the rest (Fig. 3.5). Since this all happens well before the conventional permanent wilting point (-15 bar) is reached, virtually the whole of the quite substantial reservoir of interstitial water in the chalk is potentially available to plants. This peculiarity of chalk had escaped detection by Anderson, and it was the entomologist G. H. Locket who first reported the phenomenon following his ingenious experiments at Harefield, Middlesex, and later at Whiteleaf, Buckinghamshire (Locket, 1946b). Locket's results have since been confirmed with more sophisticated equipment by Croney and Coleman (1954) (Fig. 3.5), Elston (1963) and Davy and Taylor (1974a). It is important to stress, however, that there are large differences in total water-holding capacity between hard and soft chalks and here is one obvious reason for marked contrasts between one site and another. A steep gradient, strong insolation and a large proportion of chalk stones and flints can all enhance these differences.

The mineral horizon may be able to hold as much water per unit volume against gravity as the solid chalk below—perhaps even more—through comminution of the chalk itself, the incorporation of silt, clay and organic matter, and the formation of aggregates (Table 3.2). Some of this water

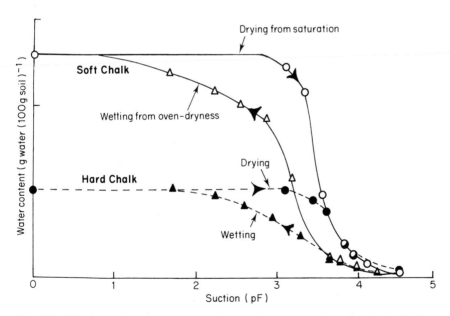

Fig. 3.5. Wetting and drying moisture characteristic curves of hard and soft chalks. The pF scale expresses water potential as the log of the suction in centimetres. Note the tension which must be applied before any water is removed, and the large difference in water content between the two kinds of chalk. After Croney and Coleman (1954).

Table 3.2

Water content and availability in a range of chalk soils[a]

Parent material, soil type, horizon and reference		Depth (cm)	Water content at			Available water	
			Field capacity	Perm't wilting point	%	in/ft	mm/cm
UPPER CHALK							
Icknield type							
Ah horizon, Purley (Anderson, 1927)[b]			45·0	8·5	36·5	5·3	4·4
Ah horizon, Whiteleaf (Locket, 1946b)[b]			40·0	11·0	29·0	4·2	3·5
C horizon, Purley and Whiteleaf (Anderson, 1927, Locket, 1946b)			27·0	4·5	22·5	3·2	2·7
C horizon, origin unknown (Croney and Coleman, 1954)[b]			27·0	2·0	25·0	3·6	3·0
MIDDLE CHALK							
Raw chalk soil							
A horizon, Hobbs Hill (Davy and Taylor, 1974a)			39·7[c]	8·7	31·0	—	—
Icknield type							
Ap horizon, Swyncombe (Elston, 1963)	Sample 1		30·9	8·6	22·3	4·0	3·3
	Sample 2		24·4	6·1	18·3	3·3	2·7
A/C horizon, Swyncombe (Elston, 1963)	Sample 1		22·1	3·9	18·2	3·1	2·6
	Sample 2		21·3	5·7	15·6	2·6	2·2
UPPER/MIDDLE CHALK							
Rolling Upton series[d]							
A horizon, North Downs near Wye (Green and Fordham, 1973)	Sample 1	0–15	37·6	17·5	20·1	2·2	1·8
	Sample 2	0–15	38·4	15·3	23·1	2·8	2·3
COOMBE DEPOSITS							
Moderately Deep/Deep Coombe series							
A/Bw horizons, North Downs near Wye (Green and Fordham, 1973)	Sample 1	0–18	31·6	13·8	17·8	2·5	2·1
		18–36	22·7	9·9	12·8	2·1	1·7
		36–51	20·9	10·7	10·2	1·7	1·4
	Sample 2	0–20	39·8	15·9	23·9	2·9	2·4
		20–40	23·4	10·2	13·2	2·1	1·7

[a] From various sources. Further comparative data can be found in Appendix III of Cope (1976).

[b] Collated by Elston (1963).

[c] Volumetric (ml water (100 ml soil)⁻¹); all others gravimetric (g water (100g soil)⁻¹).

[d] "Rolling" signifies undulating but not particularly steep topography; originally classified as Andover.

may, moreover, be more freely available to plants, in the sense of being held at lower tensions, than is the case in the chalk rock (Fig. 3.6), or in other types of soil (Figs 3.7 and 3.9). Chalk soils are, however, extremely free-draining: they lie within Group B1 of Painter's hydrological classification (Painter, 1971), having a minimum infiltration rate at the surface of 5·8–7·8 mm hr^{-1}, which is second only to regosols and deep sands. The surface layer itself, of course, is strongly susceptible to water loss through evapo-transpiration. Hence, during the dry summers which characterise most of the chalk country, plants must rely on reserves of water deeper in the profile. This applies particularly to the puffy humose horizons of low bulk density, especially when they have been disturbed, as by ploughing or by ant or mole activity. Thus the question returns to the water-supplying power of the chalk itself, and to the ability of plants to tap this source, or endure the conse-quences of not doing so.

3.6.2.2 Seasonal soil water regimes

In southern and eastern England, soils are typically restored to field capacity by December to February, depending on locality and season (Chapter 2), while the chalk itself, especially where it is unfissured or marly, may for a time be saturated because the water with which it is recharged percolates downwards so slowly (Section 1.2.2.3). This low rate of downward move-ment is responsible for the long lag between rainfall and the appearance of water in the ephemeral chalk streams which consequently acquire an appar-ently magical ability to appear overnight even when the weather has for

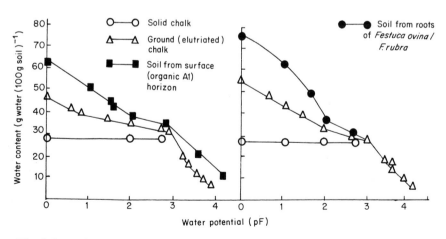

Fig. 3.6. Moisture characteristic curves (drying only) for solid chalk, chalk powder and humose chalk soils (Locket, 1946b).

some time been quite dry again. During spring, increasing solar radiation results in a corresponding increase in evapotranspiration, both directly through the drying and warming of the soil and, indirectly but more substantially, through the rapid increase in growth and water use by plants. The upper layers of the soil begin to dry out and, without rain, plants become increasingly dependent on water held at depth. The deeper and more densely developed the active root system of the vegetation, the more thoroughly the soil water reserve is depleted. The steady fall in soil water content in a relatively rainless summer period was illustrated for an Icknield-type humic rendzina on the North Downs escarpment by Anderson (1927),

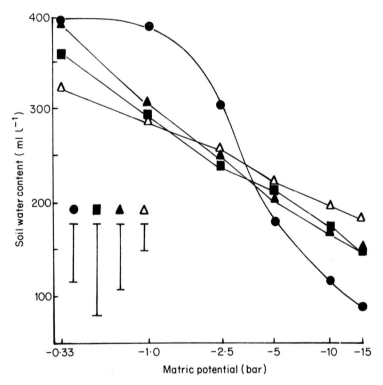

Fig. 3.7. Water release characteristics of four soils from three sites in the Chilterns (Davy and Taylor, 1974a).
▲ F/H + A horizon ⎫
△ Eb horizon   ⎬ Hillock Wood acid mull soil
                    ⎭
■ Coombe series
● Hobbs Hill raw chalk soil
Bars represent Tukey confidence interval for $P=0\cdot05$

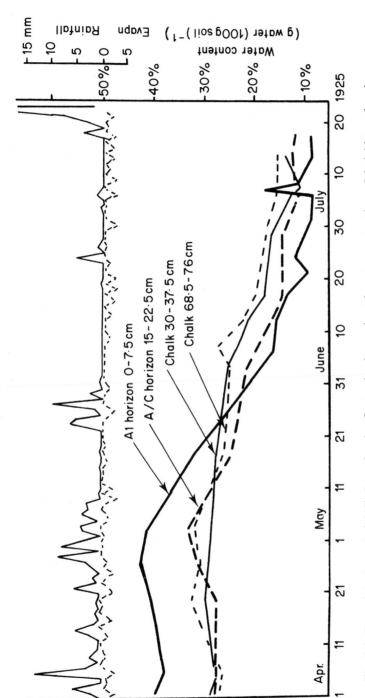

Fig. 3.8. Part of Anderson's (1927) data, showing fluctuation in gravimetric water content in an Icknield-type humic rendzina profile, as affected by depth from the surface, rainfall and evaporation.

who also showed how the susceptibility to both drying and wetting agencies at the soil surface decreased with depth (Fig. 3.8 and Table 3.3). Comparable data, expressed in terms of water potential, were collected in the Chilterns for Hobbs Hill raw-chalk soil and a brown calcareous soil of the Coombe series (as well as for a brown earth) by Davy and Taylor (1974a), and their results are shown in Fig. 3.9.

Table 3.3

The gravimetric water content (g water (100g soil)⁻¹) of an Icknield-type chalk rendzina profile [a,b]

Material	Depth from surface (cm)	Mean	Maximum	Minimum	Range
Mineral soil (A1 horizon)	0·0–7·5	36·78	61·29	8·58	52·71
Soil–chalk rubble transition zone (A/C horizon)	15·0–22·5	26·63	35·59	11·50	24·09
Chalk (C horizon)	30·0–37·5	25·84	34·05	10·76	23·29
	68·5–76·0	27·20	34·93	15·34	19·59

[a] Samples were taken once a week from each depth, and there were 66 sampling dates between January 1924 and August 1925.
[b] After Anderson (1927).

In that part of Anderson's graph shown in Fig. 3.8, there was a small peak in water content during the first week of June (1925) at the 68–76 cm depth which was not paralleled by similar peaks in the upper layers. Disregarding the possibility of experimental error, or that this peak may have been caused by delayed percolation of the falls of the previous fortnight (of which the effects on the upper layers appear to have gone unrecorded), Anderson attributed this kink in the graph to capillary rise from the moist chalk below. In similar vein, Radet (1958) concluded from his long experience of chalk soils in the Champagne district of northern France that "reserves of water at the deeper levels return towards the surface by capillarity in dry periods". Are these ideas still valid? Considerable advances have been made in the past two decades in the understanding of water movement in soils, particularly from the application of Darcy's Law, which, appropriately modified, states that

$$v = -K\left(\frac{\triangle \Phi}{\triangle z}\right)$$

where v = the velocity of flow, expressed as the volume of water passing unit cross-sectional area per unit time; the term inside parentheses represents the change in total water potential (Φ) with depth (z), i.e. the potential gradient; and K is the hydraulic conductivity of the soil, itself a function of the volumetric water content. The whole term is negative since water movement is towards zones of relatively low (increasingly negative) water potential. The reader is referred to any of the many excellent texts on soil water relations (the notation above follows Slatyer, 1967) for either more fundamental or more advanced information than is appropriate here, but the main point is that as the soil dries from field capacity, K, and therefore v, becomes very small.

With regard to capillary rise, it is now recognised "that the upward rise of water by capillary action under conditions permitting evaporation is not so significant as older points of view led one to believe" (Baver et al., 1972). Certainly, there can be no question of water rising through as much as 100 m or even more of unsaturated chalk between the permanent water table and the rooting zone on the high downs. Here, capillary rise could only be from temporarily saturated layers in the chalk sufficiently close to the surface, and this is the phenomenon to which, presumably, both Anderson and Radet refer. Work at present being conducted by the Forestry Commission at a number of chalkland sites in southern and eastern England indicates that water may be expected to rise 10–15 m from saturated chalk, although difficulties have been encountered in fitting experimental data to the Darcy model and other, as yet undetected, processes may influence the redistribution of water in the profile (D. F. Fourt, personal communication). Presumably a layer of flints or hard, rubbly chalk would drastically curtail upward movement.

To take up the question posed at the beginning of this section, of whether chalk soil water relations are "good" or "bad", the answer must be "potentially good". But the key word is "potentially" and, as we shall see in later chapters, much depends on the extent to which plants are able to tolerate or circumvent temporary water and nutrient shortages at critical phases in their establishment, growth and development.

3.6.3 Soil temperature regimes

The microclimate of the soil surface has been covered in Chapter 2, but here it is useful to recall the interaction between water content and temperature. As water in the surface soil is progressively removed by evaporation, so the

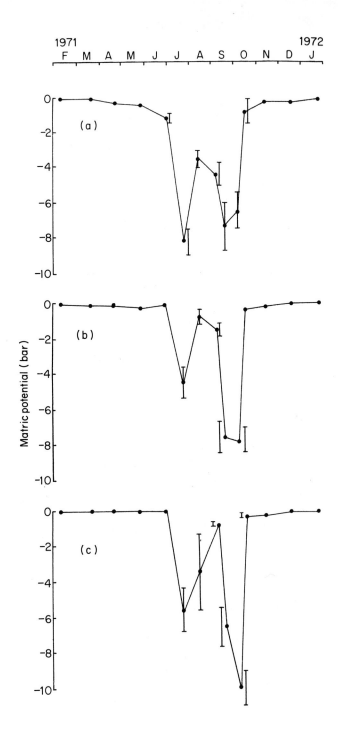

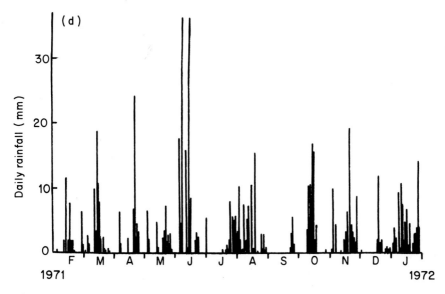

Fig. 3.9. Seasonal variation of matric potential at a depth of 100 mm in (a) raw chalk soil, (b) Coombe series brown calcareous soil and (c) a brown earth over Clay-with-flints, from woodland sites in the Buckinghamshire Chilterns, together with (d) daily rainfall records from a nearby Meteorological Office station. Note how the clay soil reached a considerably lower matric potential (i.e. greater water stress) than either of the chalk soils. After Davy and Taylor (1974a).

cooling effect afforded by the consumption of latent heat of vaporisation $(2470\,\text{J (g water)}^{-1})$ lessens, the heat capacity declines, and the soil begins to warm up, forming a thin surface mulch as it dries. Whether heat continues to accumulate in this surface mulch, or dissipates down the profile, depends on the thermal conductivity and heat capacity of the underlying strata. These two factors are drawn together in the concept of thermal diffusivity ($\propto$) thus:

$$\propto = \frac{K}{C_v}$$

where K is the thermal conductivity and C_v the heat capacity (see Rosenberg, 1974). Both K and C_v decrease as the soil dries, so $\propto$ is not easy to predict, but in one way or another it can be expected to decline rapidly in chalk soils, especially those rich in organic matter, as the water content falls. In other words, the drier the soil the hotter it gets, but the greater is the confinement of the heat to the uppermost layer of the soil, where it may continue to intensify.

Where there is a cover of transpiring vegetation, the soil is likely to dry out to a greater depth than if it is bare of plants (see Section 3.6.2.2). But a cover of vegetation greatly restricts direct insolation of the soil surface, and extremes of water shortage and heat are never normally as marked under the shade of plants as in the topmost layer of bare soil—an important factor in seedling establishment. I have recorded mid-summer extremes of 50°C from the top 2 cm of newly ploughed Icknield soil at Swyncombe, Oxfordshire, when adjacent Stevenson Screen and turf temperatures were 28°C and 36°C respectively, and King (1977a) noted similar contrasts during his studies on the microclimate of ant-hills at the nearby Aston Rowant National Nature Reserve. There, bare-soil temperatures on the south side of the mounds exceeded 40°C at 1 cm depth when the air temperature was more than 21°C, whereas maximum bare-soil temperatures on the north side (and soil temperatures underneath plants on the south side) were at least 10°C lower than this. Under similar conditions, cumulative evaporation in one day (measured with a Piche Evaporimeter) was 2·74 times greater from the south side than from the north side. Dew, frost and snow lay for longer on the northern than the southern sides of the mounds.

Table 3.4

Reflection coefficients of chalk and chalk soils [a]

White chalk	0·70–0·90
Chalk Marl	0·50–0·70
Chalky rendzina (Upton type)	0·10–0·20 [b]
Dark, humic rendzina (Icknield type)	0·10–0·15 [b]
Short turf on chalk	0·20–0·25

[a] Data from J. C. Rodda and J. Elston (personal communications).
[b] Estimates.

The darker the soil, the smaller is its reflection coefficient (r) and so the greater its radiation load (Table 3.4). White and grey soils, whose lightness of colour is emphasised when dry, might be expected to escape the strongest heating. Indeed, in spring, light-coloured, long-cultivated arable chalk soils have been found to be sufficiently slow in warming up to markedly delay the emergence of thermophilous crops such as maize (Section 9.3.4.2). Reflected glare is a significant factor in chalk quarries, however (see Chapter 4).

3.7 The Chemistry of Chalk Soils

3.7.1 The soil atmosphere

In any well-aerated soil, the composition of the soil atmosphere essentially matches that of the open air (Burges, 1967), though minute analysis reveals concentration gradients, particularly around roots and, at least in calcareous soils, previously unsuspected gaseous interactions can occur with important implications for the nitrogen economy (Section 3.8.3.4). Gases diffuse more slowly in moist than in dry soils, and this is especially true of oxygen, of which the rate of diffusion through water is less than one-thirtieth that of carbon dioxide (Russell, 1973). As a consequence the atmosphere of moist soil typically contains less oxygen and more CO_2 than that of dry soil, an effect enhanced by correspondingly greater respiration rates. Concentrations of gaseous CO_2 in calcareous soils range typically between $0 \cdot 2\%$ and 5% (Yaalon, 1957; Woolhouse, 1966a).

In the freely-draining surface horizon of chalk rendzinas, oxygen is never normally likely to be limiting. It is conceivable that the microbiologically highly active humic soils might harbour micro-sites in which temporary shortages of oxygen could occur, but transient anaerobiosis, in which denitrification could be expected to take place, is more likely to develop in Wantage-type soils on the marly Lower Chalk, especially during prolonged, wet spells during autumn and spring (when oxygen demand is high), when the subsoil may be temporarily waterlogged. Even following cultivation, clods may persist which are large enough to harbour anaerobic cores (Greenwood, 1961; Russell, 1973). Recently it has been found that ethylene (C_2H_4) is an important metabolite of waterlogged soils. The gas persists in the soil even after aerobic conditions are restored, and it has been detected in soils over Lower Chalk, as well as in other soil types, in concentrations which in the laboratory have been found to inhibit the growth of barley roots (Smith and Russell, 1969). As yet, the significance of this phenomenon in the field is unknown.

3.7.2 The pH of calcareous soils

Chalk soils are typically alkaline, with a pH between 7 and 8. However, several authors (Yaalon, 1957; Turner, 1958; Woolhouse, 1966a; Russell, 1973; Schinas and Rowell, 1977) have pointed out that the pH of a soil can be markedly influenced by the carbon dioxide concentration of the soil solution. In soils developed over soft calcareous strata, this may, as noted in Section 3.7.1, consist essentially of a solution of calcium bicarbonate in equilibrium with air containing as much as 5% CO_2. The pH of this medium depends on the rather complicated interaction between the solubilities and dissociation constants of calcium carbonate, calcium bicarbonate and CO_2, but Russell shows that the relationship can be expressed thus:

$$pH = 0 \cdot 5\, pCa + 0 \cdot 5\, pCO_2 + K$$

where pCa is the concentration (strictly the activity) of Ca^{2+} ions and pCO_2 the partial pressure of CO_2, both expressed in negative logarithms, while K is defined as "a constant whose value lies between 10 and $10 \cdot 5$", depending on the value used for the solubility product of calcium carbonate.

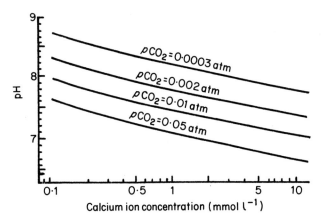

Fig. 3.10. The pH of a saturated $CaCO_3$ suspension as a function of the calcium ion concentration and carbon dioxide partial pressure at 25° C. After Yaalon (1957).

The consequences are that an increase in either CO_2 content or Ca^{2+} ions (or both) will decrease pH. Thus in the rooting zone of plants respiratory carbon dioxide may reduce pH by as much as one unit, compared with that of soil in equilibrium with atmospheric air at the soil surface (Fig. 3.10). This may have caused the marked lateral heterogeneity in soil pH found by Snaydon (1962a) (and quite different from the situation described by Grubb et al. (1969)—Section 5.4.2.5), and augurs strongly against reading too much into unreplicated spot determinations, or for that matter data from large, bulked samples. Conversely, a decrease in Ca^{2+} ions or CO_2 will increase pH, and it has been suggested (Turner, 1958) that during the course of a shower of rain, water percolating down through a calcareous soil may lose enough of both these components for its pH to rise almost to 10— a value which has, in fact, been recorded in the field (see Chapter 4). This could be of significance to plants rooting deeply into solid chalk, though of course respiratory CO_2 from the roots themselves would tend to offset the effect, at least in their immediate vicinity.

3.7.3 Calcium

Apart from the extreme surface layers of certain undisturbed humic rendzinas on level ground, notably under woodland, calcium carbonate normally permeates the entire horizon of chalk soils in the sense defined in Section 3.1, resulting in the familiar effervescence when a drop of dilute hydrochloric acid is applied, and maintaining an alkaline reaction which governs the whole field of chalk soil chemistry. Other salts of calcium occur in chalk soils, including (as we have seen) the bicarbonate, the phosphates, and, to a lesser extent, sulphates and others.

Calcium ions, Ca^{2+}, result from the dissociation of these chemicals, and these too play an important part in the chemistry of the soil medium, as well as being the form in which plants take up the element. Calcium ions flocculate clay and humus particles (prevent them from dispersing), so maintaining favourable soil physical conditions, and this is one of the main reasons for the very high aeration and permeability of chalk soils. The so-called calcium humates are complex molecules derived from the reaction of Ca^{2+} ions with colloidal organic matter, and essential for aggregate synthesis. According to Duchaufour (1970), the divalent Ca^{2+} ions form important bridges between the negative charges of the clay and humus particles to form the clay–humus complexes so vital to the mineral relations of the soil. Moreover, anions such as phosphates and (less strongly) sulphates may be held on to the exchange complex by this bridging action of the calcium ions. Other ionic interactions with calcium are discussed in due course under appropriate headings.

The actual concentration of calcium ions in the soil at any one time depends among other things on the rate of weathering, and on the number of cation exchange sites (on clay and humus colloids) upon which the ions may be retained. Thus raw chalk may contain less available calcium than a medium in which the chalk is associated with other mineral matter and humus (Fig. 3.11; see also Section 4.2.2.1). In a comparison of chalk soils in Rouen, northern France, with those of increasingly oceanic sites in England (see Chapter 5), Perring (1958, 1959, 1960) found lowest values of Ca^{2+} ions on south-facing slopes in the dry, continental conditions of the French locality, and highest values in English sites where rainfall was adequate to ensure weathering yet not sufficiently high to remove the easily leached calcareous materials altogether (Fig. 3.12).

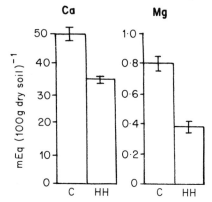

Fig. 3.11. The contrasting exchangeable Ca and Mg status of a Coombe series brown calcareous earth (C) and Hobbs Hill raw chalk soil (HH). Note the 50-fold difference in scale between the two elements. Bars indicate s.e. ($n = 11$). From Davy and Taylor (1975).

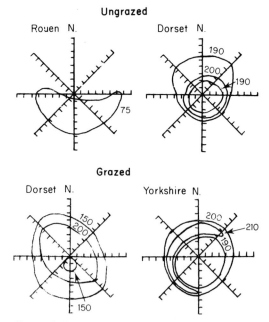

Fig. 3.12. The effects of climate and topography on the distribution of exchangeable calcium in a range of chalk soils. Data express mg exch. CaO $(100g soil)^{-1}$. Major axes indicate aspect, marks on these axes the gradient. The centre corresponds to a level surface, each mark representing progressively increasing steepness at 5° intervals. Thus, the third mark on the "2 o'clock" axis indicates a NE.-facing slope of 15°. From Perring (1960).

3.7.4 Potassium

Except where glauconite occurs, such as towards the bottom of the Lower Chalk and occasionally in other strata, chalk contains very little potassium, and the richness of chalk soils in this element largely depends on the nature and proportion of K-bearing residues with which they have been contaminated. Potassium, moreover, is readily leached. Thus Perring (1958, 1959, 1960) found more exchangeable K in the soils of his drier sites than those with a more oceanic climate. It might be argued that this simply reflects the greater proportion of loess which is known to occur in French soils, but the pattern was repeated at each site on a topoclimatic scale, adding weight to the idea of leaching. It is always essential, however, to allow for biotic influences: Perring separated grazed from ungrazed sites in his analysis, but, as we shall see in Chapter 9, prolonged *former* grazing of chalk downland by sheep folded nightly on arable soils elsewhere could have greatly depleted soils of potassium reserves (Floyd, 1965).

A complication arises here, however, for certain clays (of which even the rendzinas may contain appreciable quantities) are able to trap (adsorb, fix) potassium ions between the layers of their lattice structure as they swell and shrink with fluctuating cycles of wetting and drying, a process apparently encouraged by the presence of calcium ions (Duchaufour, 1970). Strictly this K is non-exchangeable in the sense that it is not immediately available to plants, but where fixation is reversible it may ensure a constant replenishment of exchangeable K as this is removed either by leaching or plant uptake. The potassium status of soils is thus an imprecise and difficult property to pin down, for plant species differ in their ability to take up K, and the more vigorous the rate of uptake, the more fixed K will be released (Arnold, 1960).

Some idea of the availability of K to plants can be gauged either by continuously cropping pots of the soil under test and analysing the herbage, or by extracting the soil with a hydrogen-saturated cation exchange resin (Arnold, 1958). Avery (1964) provides comparative data on, among others, Icknield, Coombe and Wantage soils (Table 3.5). In this admittedly relatively small sample, Icknield soils were low in both exchangeable and reversibly fixed K, Wantage soils eminently better endowed with both, and Coombe soils intermediate. Where Wantage soils impinge on the richly glauconitic seams at the base of the Lower Chalk, their available K content rises accordingly.

Additional information on the available K status of these and other soils comes from more recent surveys by Green and Fordham (1973) and Cope (1976), and tends to confirm that the thinner soils of the Upper and Middle

Chalk are relatively deficient in available K, with those of the Coombe deposits (and for that matter the Clay-with-flints) rather better endowed, though not much (Table 3.6). Note that the availability of K to plants can be modified by the relative proportions of certain other mineral ions. We have already referred to the enhancement of clay adsorption of K by Ca ions; there is in addition a direct antagonism at the site of uptake between potassium and magnesium, particularly applicable to agricultural situations where the application of one may depress the uptake of the other.

Table 3.5

Soil K (mg (100g soil)$^{-1}$) extracted by normal ammonium acetate solution ("exchangeable"), and by H-saturated resin and ryegrass (Lolium perenne) plants ("non-exchangeable") from chalk soils [a]

Soil series		Exchangeable K	Non-exchangeable K	
			H-resin	Ryegrass
Icknield	(1)	—	—	8
Icknield	(2)	8	10	—
Icknield	(3)	7	12	—
Wantage	(1)	—	—	45
Wantage	(2)	44	96	—
Coombe		13	42	—

[a] From Avery (1964).

3.7.5 Phosphorus

We noted in Chapter 1 the occurrence of phosphatic beds in the Upper Chalk, and it has long been known that where these strata outcrop the soils derived from them are accordingly richer in phosphate (Temple, 1929), although they occur only very locally in England. Sites of former habitation sometimes show anomalously high phosphate levels, while in some long-cultivated soils, repeated application of phosphatic fertilisers has contributed substantially to their P content (Cooke, 1967). On the whole, however, the available phosphorus in chalk soils is very low (Table 3.6), partly because what inorganic reserves there are weather only very slowly and partly because of two important effects of pH.

Regarding immobility, Sutton and Gunary (1969) have stressed the particular vulnerability of young seedlings to this phenomenon, in which

Table 3.6

Chemical data on a selection of chalk soils from Wiltshire [a,b]

Map unit	Number of samples	pH	CaCO₃ (%)	Organic matter (%)	Available phosphorus mg l⁻¹ P	Available potassium mg l⁻¹ K	Available magnesium mg l⁻¹ Mg
Icknield	18	7·7 ± 0·2 (2·6)	17 ± 13·1 (77)	12·9 ± 2·59 (20)	22 ± 8·1 (37)	138 ± 41 (30)	81 ± 41 (51)
Andover	35	7·8 ± 0·3 (3·8)	24 ± 14·3 (60)	6·4 ± 1·55 (24)	23 ± 10·7 (46)	177 ± 62 (35)	54 ± 38 (70)
Upton	46	7·9 ± 0·1 (1·3)	55 ± 11·8 (21)	5·5 ± 1·40 (25)	20 ± 9·2 (46)	177 ± 76 (43)	37 ± 16 (43)
Coombe	16	7·7 ± 0·3 (3·9)	19 ± 11·1 (58)	6·2 ± 1·77 (29)	31 ± 17·1 (55)	199 ± 99 (50)	69 ± 31 (45)
Gore	19	7·9 ± 0·1 (1·3)	47 ± 11·5 (24)	5·5 ± 1·72 (31)	22 ± 9·3 (42)	147 ± 44 (30)	39 ± 19 (49)

[a] From Cope (1976).
[b] Each value is followed by its standard deviation (sd) and coefficient of variation (cv), the latter expressing sd as a percentage of the mean. Clearly the variation of some of these attributes is very large.

only a minimal surface of active root tissue is present. Of the two major effects of pH, the first is that as the pH rises above 7, $H_2PO_4^-$ ions (the form in which P is most readily taken up, at least by most crop plants (Hagen and Hopkins, 1955)) give way to the generally less useful HPO_4^{2-} ions (Table 3.7). Secondly, in an alkaline medium, especially above pH 8 and with abundant Ca^{2+} ions, relatively insoluble phosphates are formed, including calcium phosphates and derivatives of apatite, though the chemistry is complicated and not fully understood (Russell, 1973; Duchaufour, 1970).

Perring (1958, 1959, 1960) found available phosphate concentrations to

Table 3.7

The effect of soil pH on the ionic form in which phosphates are most prevalent [a]

pH	Proportion of divalent (HPO_4^{2-}) to monovalent ($H_2PO_4^-$) ions [b]
5	0·6
6	6·0
7	39·0
8	86·0
9	98·4

[a] From Russell (1973).
[b] PO_4^{3-} ions and H_3PO_4 are present in negligible quantities over this range.

increase with the "oceanicity" of his chalkland sites, which he attributed to successive lowering of soil pH. It is possible that the roots of plants are able to effect a decrease in pH in their immediate vicinity by secreting acids which would similarly render phosphates more available. Some of the organic acids thus secreted may in addition act as chelating agents, so by-passing the question of exactly in which form the element is present (Rovira and McDougall, 1967). Some soil P derives, of course, from organic sources through mineralisation and mycorrhizal infection, aspects which are touched on later.

3.7.6 Sulphur

Considering its major nutrient status, sulphur is an element which seems to have been taken very much for granted by soil and plant scientists. There seem to be two main reasons for this. Firstly, attention to the nitrogenous and phosphatic needs of economic plants over the past 130 years or so has automatically taken care of—or at least contributed to—the plant's sulphur requirements, for N and P have commonly been provided as, respectively, sulphate of ammonia ((NH_4)$_2SO_4$, containing 24% S), and superphosphate (in the case of "single supers", a mixture of monocalcium orthophosphate monohydrate and gypsum ($CaSO_4$) containing 11–13% S). Sulphur has also been applied in fungicides.

Secondly, there has been considerable enrichment of the atmosphere with sulphur, especially as sulphur dioxide (SO_2) and sulphuric acid (Whitehead, 1964). Geographically, the English Chalk has never been subjected on the whole to the intensity of industrial or domestic pollution suffered by many parts of Britain, nor are there, except at the coasts, the relatively high levels of airborne sulphates found in the more maritime regions (Chapter 2). Yet, working at Hurley in the mid-Thames Valley, Cowling and Jones (1970) detected a response to atmospheric sulphur in high-yielding grasses growing in pots of arable soil. This suggests that with any further lessening of aerial sulphur levels through pollution control, it may prove fruitful to look at the availability of S in chalk (and other) soils, particularly as farmers turn to fertilisers of much greater purity (the so-called "high analysis" grades). In the Lower Chalk, sulphur is sometimes locally abundant where it is associated with iron in pyrite concretions (see Chapter 1), though its availability is likely to be low. Mineralisation of organic sulphur is briefly mentioned in Section 3.8.4.

3.7.7 Magnesium

Magnesium occurs in soils, like potassium, mainly as silicates in the mineral fraction. Though chemically similar to calcium, magnesium is altogether the less abundant of the two ions throughout the English Chalk (dolomitic chalk occurs only on the Continent). Davy and Taylor (1975) found much less available Mg in raw chalk than Coombe series soil (Fig. 3.11), and the Upton and Gore rendzinas have been found to be relatively deficient (Table 3.6). Magnesium levels may be inadequate on these last two soils for barley (Cope, 1976), while livestock newly turned out to lush grass may develop symptoms of hypomagnesaemia (Chapter 9). A shortage of available Mg in the soil can be envisaged when potassic fertilisers are applied, for there is a direct antagonism between these ions at the site of uptake by plant roots. As with sulphur, areas near the coast may receive appreciable quantities of Mg from the sea.

3.7.8 Micronutrients

3.7.8.1 Some general points

By convention, the elements iron (Fe), manganese (Mn), copper (Cu), zinc (Zn), cobalt (Co), boron (B) and molybdenum (Mo) are grouped under this common heading, even though the amount of each required by plants varies from "almost macro" quantities in the case of iron to infinitesimally small quantities with cobalt. A considerable body of knowledge has now accumulated on the physiological role of these elements in plants, though the exact status of some remains obstinately obscure, and even the form in which they are taken up is not always known.

There are, in addition, mineral ions which, though of no importance to plants as far as we can tell, play a vital role in animal physiology, and so are essential in the diet of grazing and indeed any other animals supported, directly or indirectly, by the vegetation. In addition to most of those essential to plants cited above, these include iodine (I), fluorine (F), selenium (Se) and chromium (Cr). Of the remaining minerals in the soil, some are apparently quite inert to the plants and animals that take them up, but others may (as with the essential ones) be toxic in excess. Mercury (Hg) and lead (Pb) may reach chalk soils from seed dressings and vehicle exhausts, though the chalk country is well out of range of the areas in which these, and others such as cadmium (Cd), are potent industrial pollutants. Aluminium (Al) is not normally found in any quantity in solution in chalk soils, but it occurs widely on acid soils and, as we shall see in Chapter 5, is of great ecological importance in restricting certain plant species to chalk and limestone soils.

The concentration of microelements in chalk soils is, of course, a reflection of the various minerals that their parent rocks contain, and we know by now that this composition can vary widely from the marly materials of the Cenomanian Chalk to the virtually pure calcium carbonate of the white chalks (Chapter 1), and be markedly influenced by the nature and quantity of any additional surface deposits of silts and clays (Chapter 1), and the extent to which aerial enrichment occurs (Chapter 2). But, as with the macronutrients, complications arise through both the alkalinity of the soil medium which can affect availability through solubility and chemical combination, and the large concentration of soil organic matter, with which mineral ions can form complexes. Moreover, the organic surface horizon with which most of the microelements are associated is, as we have seen, markedly prone to drying, and there will be times when low water potentials limit their availability. Green and Fordham (1973) provide a useful summary of their preliminary data on some of these trace elements in Upton and Coombe soils on the North Downs in Kent (Table 3.8), but for obvious reasons these can only be regarded as a guide. As with the major nutrients, ecological and agronomic aspects of trace element chemistry are dealt with in Chapters 5 and 9, respectively.

Table 3.8

Trace element content (mg kg⁻¹) of Upton and Coombe series soils in Kent [a,b]

Trace element	Rolling Upton [c]	Moderately Deep and Deep Coombe
Mn (exchangeable)	12·83 (1·89, 8·0–21·0)	12·17 (2·27, 4·0–20·0)
Zn (0·5N acetic acid)	16·50 (2·60, 5·0–23·0)	12·67 (1·50, 6·0–16·0)
Cu (0·05M EDTA)	2·08 (0·62, 1·0–5·0)	3·03 (0·37, 2·00–4·30)
B (hot water)	1·34 (0·21, 0·70–2·27)	1·07 (0·09, 0·74–1·38)
Co (0·5N acetic acid)	0·51 (0·21, 0·15–1·53)	0·25 (0·05, 0·10–0·39)
Mo (Tamms)	0·08 (0·01, 0·05–0·11)	0·11 (0·04, 0·05–0·28)

[a] From Green and Fordham (1973).
[b] Each value is the mean of five samples of topsoil. Values in parentheses indicate standard deviation and range.
[c] Originally cited as Andover series.

3.7.8.2 Iron

In the freely drained, well-aerated soils of the Chalk, ionic iron is most likely to occur in the ferric (Fe^{3+}) form, although at high pH its availability is jeopardised by combination with other chemicals into less soluble hydroxides and other compounds, and by complexing with phenolic derivatives of the soil organic matter. Certain root exudates, notably $\propto$ - keto-gluconic acid found in the rhizosphere of barley plants (Webley and Duff, 1965), are known to return the ferric ion to solution from such compounds, while Duchaufour (1968) suggests that ferric humates of alkaline soils are less stable than their ferrous counterparts of acid soils, decomposing relatively easily in conditions of high pH, especially when there is an excess of Ca^{2+} ions present. Nevertheless the key to plant–iron relations rests on the extent to which iron–organic complexes can be used directly by the plant, and evidence is accumulating that this is the rule rather than the exception.

A shortage of iron has long been recognised as the reason for the inability of many plant species to thrive on chalk and other alkaline soils, and to develop the symptoms known as lime-induced chlorosis (see Chapters 5 and 9). Such plants are termed calcifuges, as distinct from calcicoles which are unaffected by this condition. Plants able to grow in chalk soils must either have a low requirement for iron, or have evolved a mechanism to ensure that they derive enough from the soil medium despite the high pH, perhaps in an organic form as just suggested. The phenomenon is a complicated one, however, and involves not so much the actual uptake of iron as its mobility and availability within the plant, for chlorotic plants have been found to contain at least as much iron in their tissues as healthy ones. The bicarbonate ion is now known to be an important co-factor in this relationship (Woolhouse, 1966b). Davy and Taylor (1975) report seasonal variation in extractable iron in raw-chalk and Coombe series soils (as well as Batcombe series acid mull) from their Chilterns site near Great Hampden (Fig. 3.13). The acid mull showed marked (and unexplained) fluctuation from zero to a peak in early spring, but during the growing season the two chalk soils, and especially the raw chalk, predictably showed consistently the lowest values.

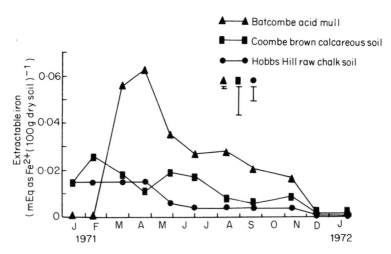

Fig. 3.13. Seasonal variation in ammonium acetate-extractable iron in acid mull, brown calcareous and raw chalk soils. Bars represent Tukey confidence level for $P = 0.05$. From Davy and Taylor (1975).

3.7.8.3 Manganese

Divalent manganese (Mn^{2+}) is the predominant form in the soil solution, and it is in this form that Mn is taken up by plants. There tends to be less divalent Mn in soils of high pH and high oxidation–reduction potentials such as those of the Chalk. Davy and Taylor (1975) found exchangeable Mn "virtually undetectable" in either of their chalk soils, though Green and Fordham found typically 12–13 mg kg⁻¹ exchangeable Mn in Andover and Coombe series (Table 3.8). Manganese complexes strongly with organic matter, and not surprisingly a shortage of available Mn is sometimes found in highly organic soils such as those of newly ploughed downland ("new" corresponding to as much as a decade). This deficiency must either be tolerated or circumvented by plants which grow here naturally, or supplemented for crops to be grown on chalk soils. Conversely, the inability of some calcicolous species to grow on acid soils may be due to their adverse reaction to the excess of Mn^{2+} ions which such soils commonly contain, an effect which could augment that of Al^{3+} ions (see Chapter 5).

3.7.8.4 Other micronutrients

Regarding the remaining micronutrients, soil scientists still know relatively little about their soil chemistry in general, let alone their behaviour in chalk soils. Probably 99% of the copper in the soil solution, for example, is in organic form (E. W. Russell, personal communication). Copper deficiency has been recorded on highly organic, freshly reclaimed downland rendzinas in south-east England, where it expresses itself as a curious blackening syndrome in wheat (Hooper and Davies, 1968; see also Section 9.3.4.6). Boron is probably present in the soil solution as undissociated boric acid, $B(OH)_3$, in which form it seems likely to be taken up by plants, but at high pH it is readily adsorbed by humus colloids. These can act as a reserve in humid conditions, but severely restrict it in times of drought. It is thus not surprising to find reports of B deficiency in well-drained, organic, chalk rendzinas both in south-east England (Hall, 1971) and northern France (Radet, 1958), for even if roots can extend into the moist chalk deeper down, it is in the surface layers that the element is trapped, a phenomenon by no means restricted to boron. Maritime sites might be expected to show a less acute shortage, partly because the climate is usually more humid, but particularly because sea water contains boron which will reach plants near the sea by wind-carried spray droplets and rainfall (Chapter 2).

Finally silicon, always a debatable element in considerations of specific physiological roles in plants. Physically, accumulation of Si endows leaves and shoots with a mechanical resistance to mastication and comminution by the fauna of the soil, litter and leaf canopy, and contributes to the lowering of the feeding value of the mature foliage to wild or domestic grazing mammals. As we saw in Chapter 1, there is no shortage of available Si in chalk, and at least three of the predominant chalk grasses, especially *Brachypodium pinnatum*, accumulate silica to such an extent that a complete siliceous skeleton may persist after decomposition of the rest of the tissues (Elton, 1966). Chemically, the effects of silicates are mainly indirect, increasing the availability of phosphates, for example, and reacting with zinc in a form which protects the latter from irreversible organic complexing (Russell, 1973).

3.8 Microbiology of Chalk Soils

3.8.1 The nutrients involved

Apart from a few minor diversions, the previous section has dealt with soil properties which result essentially from straight chemistry. But one element, of utmost importance to plants and animals, has so far been left aside: nitrogen. The reason for this separate treatment of N is that, apart from traces which reach the soil as oxides of N directly from the air (as well as much larger quantities of manufactured nitrogenous fertilisers which are applied to grassland and arable crops), soil N relations are sufficiently tied to microbiology and biochemistry to warrant review under the present heading, though inevitably some inorganic chemistry is still involved. Although attention is directed mainly at N in this section, a brief return is also made to soil phosphorus and sulphur relations from a microbiological point of view.

3.8.2 Nitrogen: general information

The more complex organic forms of N in soil humus are, of course, unavailable to plants, but these are gradually decomposed to simple compounds of ammonia, from which ammonium ions (NH_4^+) are eventually released. This final stage is called *ammonification.* Ammonium ions may then be oxidised to nitrite (NO_2^-) by bacteria of the genus *Nitrosomonas* and others, and nitrite to nitrate (NO_3^-) by *Nitrobacter*. Both of these oxidative steps constitute *nitrification*, while ammonification and nitrification are both examples of *mineralisation*. Ammonium, nitrite and nitrate ions are collectively termed *mineral N*, but only ammonium and nitrate are used by plants: nitrite is toxic (though see Section 3.8.3.3 and Chapter 5).

At any one time, the mineral N content of a soil, and the exact constitution of this component, will depend on:

(i) the total amount of organic N present,
(ii) the rate at which this reserve is being mineralised,
(iii) the extent to which mineralisation is being supplemented by N-oxides from rainfall (as well as by fertilisers),

(iv) the rate at which all this mineral N is being removed from the system.

Removal may be through (Allison, 1955; Harmsen and van Schreven, 1955; Bremner, 1959; Williams, 1969):

(i) uptake by plants or microorganisms (the latter being referred to as microbial immobilisation),

(ii) chemical immobilisation by adsorption—usually reversible—on to clay lattices,

(iii) various other chemical reactions, mostly irreversible,

(iv) reduction to nitrous oxides and nitrogen gas by anaerobic bacteria (denitrification),

(v) leaching by rainfall to a level beyond the reach of most plant roots in the soil profile.

It has long been known that soils under grassland gradually accumulate organic N. Indeed, this is a fundamental principle of ley farming, and provides a method of estimating the approximate age of a pasture (Wells *et al.*, 1976). The quantity of mineral N present at any one time, however, is always very small by comparison (Richardson, 1938). Under a sward, ammonium N makes up most of this mineral N, nitrate accounting for the rest: nitrite is regarded as transient and negligible, though more often by assumption than measurement. This relative paucity of mineral N must be due mainly to the large "sink" of roots and microbes which quickly mop up free ions, although many soil microbiologists have suspected that secretions by roots might inhibit mineralisation in some way and so augment the gulf between total and mineral N (e.g. Pickering, 1917; Clement, 1958; Gasser, 1969; Moore and Waid, 1971). The ecological reality of such a phenomenon is extremely difficult to demonstrate, however.

On an annual basis, the total N content of a soil remains relatively stable, but mineral N fluctuates, often markedly, from season to season, and even from day to day, so that spot determinations may mean little. Peaks in mineral N under turf are commonly seen in spring when uptake by plants lags behind the first flush of mineralisation. This is particularly impressive after a period of prolonged cold or drought. The effect of the latter was first described by Birch (e.g. 1960) and has consequently come to be named the "Birch effect". However, it is becoming increasingly apparent that mineralisation can proceed at temperatures and water potentials once considered too low for microbial activity. Thus, although mineralisation certainly can be restricted by a shortage of water (Wetselaar, 1968), nitrification has been shown to proceed, albeit slowly, in soils at permanent wilting point, and ammonification in soils containing only half the water present at the wilting point (Clark, 1967). Likewise, both ammonification and nitrification are now known to occur at temperatures at least as low as

2°C (Davy and Taylor, 1974b). Not surprisingly then, significant quantities of mineral N may accumulate in conditions such as these when uptake by plants is at a minimum.

Because of these fluctuations in mineral N, a more useful measure of the N-supplying power of a soil is its *mineralisable N*, determined by aerobic incubation of moist soil samples in controlled conditions for a set period of time, ranging from two to twelve weeks. It hardly needs adding that determinations of mineral N performed on soil samples stored in the laboratory, either moist or air-dry, for many weeks are of no value whatsoever unless the precise conditions of storage are recorded (which is rarely the case, at least in the older ecological literature).

3.8.3 Nitrogen relations of chalk soils

3.8.3.1 Some characteristics of an Icknield series soil

Very little information is available specifically on the nitrogen relations of chalk soils. Attempts were made some years ago, mainly between 1962 and 1971, by A. H. Bunting, J. Elston and myself to unravel the N relations of *Festuca rubra* turf growing in a humic (Icknield) rendzina on the Chiltern escarpment at Swyncombe, Oxfordshire (Elston, 1963; Smith *et al.*, 1971; Smith and Bunting, unpublished), while more recently Davy and Taylor (1974b) have performed similar investigations on tussocky stands of *Deschampsia caespitosa* in woodland on the Hampden Estate, Buckinghamshire, in which two kinds of non-humic chalk soils were compared with an acid mull. Radet and Mantelet (1938) and Radet (1958) provide some interesting and useful data from their work on arable chalk soils of northern France. Supplementary information comes from studies on other kinds of calcareous soils.

At Swyncombe, we established that the humic Icknield rendzina there contained a large quantity of total N (0·64%) which, as expected, greatly exceeded mineral N; in fact, at times there was more than 1000 times as much total as mineral N. Predictably, ammonium exceeded nitrate N in the mineral fraction, but a curious feature was that nitrite was consistently detected, at times in greater concentration than nitrate. Fortnightly measurements of mineral N showed occasional peaks which could be accounted for by weather conditions (Fig. 3.14), but further anomalies cropped up during the incubation of soil samples for mineralisable N determination (as well as in daily monitoring of mineral N in newly ploughed soil in the field), in which both ammonium and nitrate N disappeared unaccountably. Three points of particular interest were thus raised by this work: (i) the restriction of mineralisation of so large a reserve of organic N, (ii) the possible significance of nitrite N, and (iii) the apparent plethora of

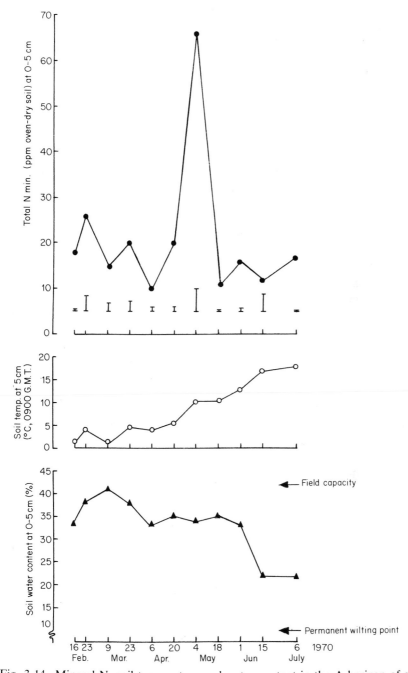

Fig. 3.14. Mineral N, soil temperature and water content in the A horizon of an Icknield rendzina at Swyncombe, Oxfordshire, in the spring of 1970 (Smith and Bunting, unpublished). Bars represent l.s.d. ($P = 0 \cdot 05$); each value is the mean of four replicates.

chemical transformations which could deflect all three forms of mineral N from their "textbook" course of organic $N \rightarrow NH_4^+ \rightarrow NO_2^- \rightarrow NO_3^- \rightarrow$ uptake by plant.

3.8.3.2 The restriction of mineralisation of N

The persistence of so large a quantity of organic N in the soil could stem from several features of the turf–soil system in question, which can here only be touched on briefly.

(i) Because of a general absence of earthworms from this shallow phase of the Icknield series, herbage protected from grazing and mowing accumulated above the soil surface as a mat of tough, grassy litter, well aerated, and with a C:N ratio in excess of 20, and strongly resistant to decomposition as long as it remained uncompacted. On plots of *Festuca rubra* turf at Swyncombe Down which had not been mown for seven years, a staggering quantity of litter had accumulated, amounting to some 30 000 kg ha^{-1} of dry matter. This had a mean N content of 1·5%, so almost 400 kg ha^{-1} of N was being withheld from the soil where at least some degree of decomposition would be likely (Curry, 1969). Interestingly, in the autumn of 1971, the eleventh year of accumulation, leaves shed from invading stinging nettles (*Urtica dioica*) became entangled in the mat which then collapsed and disappeared within a few weeks, presumably through either a lowering of the C:N ratio, or an increase in water content (or both).

(ii) Even when once incorporated into the mineral soil of the Ah horizon, however, *Festuca rubra* litter seems to be particularly resistant to decomposition—notably (Fig. 3.15) more so than other economic herbage species with which it was compared in tests at the Grassland Research Institute, Hurley, using a loamy chalk soil probably referable to the Soham series (Clement, 1958; Clement and Williams, 1964). Litter of *Bromus erectus* (T. C. E. Wells, 1971) and even more so of *Brachypodium pinnatum* (Elton, 1966) appear to have the same effect where they dominate the vegetation. The rendzina-moder humus into which this litter is all eventually transformed is itself more resistant to further decomposition than mull humus. Moreover, phenolic derivatives of the soil organic matter, particularly vanillic acid, have been positively identified as significant components of the humus of the soil from Swyncombe (Burges *et al.*, 1963), and these substances are known to inhibit nitrification in the field (Flaig, 1968). Inhibition by root exudates from the living fescue turf at Swyncombe was suspected but not conclusively demonstrated, although this is an idea which refuses to lie down—see Lee and Greenwood (1976).

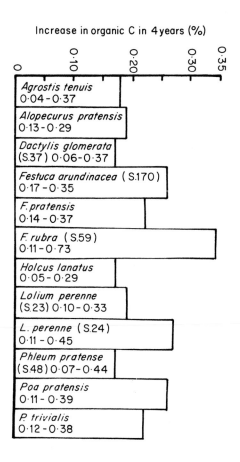

Fig. 3.15. The increase in four years in total organic carbon (%) in the top 15 cm of soil under leys of 12 grass species or cultivars (all sown with S.100 white clover). Each value is the mean of eight samples, and the range is also shown. From Clement (1958).

(iii) The tendency for the Ah horizon of these soils to suffer periodically from a shortage of water has been noted already, though we saw that this may not have as important a direct effect on mineralisation as was formerly thought. Indirectly, of course, the water regime of these well-drained soils accounts for most of the microbiological features—persistent litter, specialised fauna, distinct humus form—described so far, but an intriguing explanation of how sporadic rainfall coupled with high evaporation rates inhibits the mineralisation of organic N specifically in chalk soils was proposed as long ago as 1933 by Radet (see Radet and Mantelet, 1938).

It has long been known that incorporating easily decomposible manures or vegetable litter into a soil triggers the mineralisation of a small quantity of native humus by what has been termed a priming action. Radet had noticed that this feature, though not restricted to chalk soils, was nevertheless markedly characteristic of them, a fact confirmed for samples of Icknield soil from Swyncombe by Elston (1963). In fact, any disturbance of the soil—even the act of soil sampling—results in the decomposition of a small quantity of organic matter. In a recent field trial a significant uptake of N by a grassland sward was detected after a liquid-ammonia applicator had been run through the turf empty (R. S. Tayler, personal communication).

Radet was particularly impressed with the way in which this sharp increase in soil activity, resulting for example from the ploughing-in of a green crop, was so quickly followed by a return to a quiescent state with the humus once more quite resistant to decomposition. Closer investigation revealed, apparently, that as the moist soil crumbs dried, a crust (*pellicule*) of calcium carbonate formed around them, greatly restricting mineralisation. Ploughing in manures and residues boosted respiratory CO_2, and the resulting carbonic acid redissolved the carbonate, allowing mineralisation to proceed, at least for a while. Duchaufour (1970) supports Radet's claim, though soil scientists on this side of the English Channel to whose attention I have drawn this work regard it with some scepticism!

3.8.3.3 The possible significance of nitrite N

Another subject requiring more extensive and critical work is the question of nitrite N. Was this a transient quirk of Swyncombe Down? Or is this something of ecological importance in chalk soils generally? It is known that *Nitrobacter* is more susceptible to high pH than the ammonium oxidisers, and that activity of the former can be inhibited in calcareous soils, especially following heavy dressings of ammonium fertilisers (Fraps and Sterges, 1935). Chapman and Liebig (1952) found that in soils of pH 8 or more, concentrations of 90 mg kg^{-1} nitrite N could persist for several months at temperatures between 10°C and 15°C. We calculated (Smith and Bunting, unpublished) that the concentration of nitrite in the soil solution at Swyncombe could exceed 50 mg kg^{-1} at a water potential at which uptake by plants might still be expected to operate. We found, moreover, that the growth of certain calcifuge grasses was depressed by this concentration of nitrite in sand-culture, while the calcicoles we tested were unaffected by twice as much (see Chapter 5). Nevertheless, these experiments were relatively crude, and the matter requires much more profound investigation.

3.8.3.4 Microbial and chemical transformations

Information in the preceeding paragraphs has come from soils under permanent grassland, undisturbed except by the soil corer, and analysed usually within a few hours (in some cases less than an hour) of collection from the field. The transformations described below have been observed during laboratory incubation of soil samples, and so might be expected to affect arable, rather than grassland soils. Whether or not they occur under the intact turf is not known, although the activities of ants, earthworms, moles and rabbits (see Chapter 6) coupled with soil creep and frost heaving can lead to disturbance of the soil closely comparable with cultivation (and soil sampling).

Firstly, ammonium N: this form usually declines rapidly during aerobic incubation of chalk soils. At times, parallel increases in nitrite and nitrate N suggest nitrificaton as the cause, which is what would be expected. But on many occasions there was no doubt that NH_4^+N disappeared from the mineral pool, mostly within the first day of the incubation—even within an hour. A typical short-term incubation curve is shown in Fig. 3.16. A similar trend in other (non-humic) chalk soils was reported by Davy and Taylor (1974b). There seem to be three possible explanations.

(i) The missing ammonium N had been immobilised by bacteria. The C:N ratio of the Icknield soil ranged between 10 and 15, however, and, except possibly in the vicinity of plant roots, a value in excess of 25 is regarded as more usual in the case of bacterial immobilisation (Gasser, 1969; Whitehead, 1970).

(ii) Ammonium ions had been adsorbed between the layers of the lattices of the soil clays (Bremner, 1959): Icknield soils contain an appreciable quantity of expanding-lattice clays with this property (Avery, 1964).

(iii) Ammonium ions were lost by gaseous transformation. Two such reactions are known to occur in well-drained (even dry) calcareous soils. One results in the release of ammonia gas (NH_3), and is especially characteristic of soils containing large quantities of calcium carbonate (Larsen and Gunary, 1962; Woldendorp, 1968). The other involves the combination of nitrite with ammonium ions, leading to the loss of gaseous nitrogen (Bremner and Nelson, 1968):

$$NH_4^+ + NO_2^- = N_2 + 2H_2O.$$

Declines in nitrite N recorded both in the field and in the incubation flask were assumed to result from the second stage of nitrification since they were invariably accompanied by accumulation of nitrate, but again, it appears that in calcareous soils other processes might have been at work. The ability

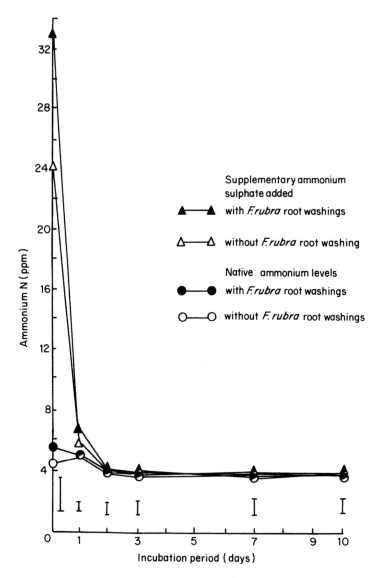

Fig. 3.16. The downward trend in ammonium N content of aerobically incubated Icknield soil from Swyncombe (Smith and Bunting, unpublished). Bars represent l.s.d. ($P = 0.05$); each value is the mean of four replicates.

of nitrite ions to combine with ammonium ions to form nitrogen gas has been mentioned in point (iii) above. In addition, Hauck and Bremner (1969) describe a nitrosation process, which they term chemodenitrification, in which nitrite reacts with phenolic constituents of the soil organic matter to produce a complex organic molecule quite unavailable to plants. This may well take care of any possible toxic effect of nitrite, but in the process prevents an equivalent quantity of nitrate from being formed.

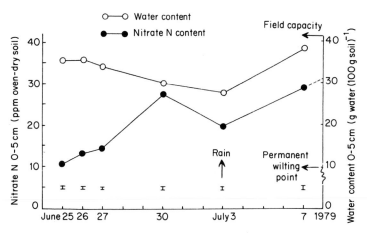

Fig. 3.17. Disappearance of nitrate N from fallowed field plots of Icknield soil during a warm, dry spell at Swyncombe (Smith and Bunting, unpublished). Bars represent s.e. nitrate N; each value is the mean of eight replicates.

A final anomaly encountered repeatedly in our work with soil from Swyncombe Down has been the disappearance of nitrate N. Where this has occurred in the incubation flask it might be argued that despite precautions to keep our soil samples on the dry side to ensure aerobic conditions (and even to feed them with gaseous oxygen), denitrification was nevertheless the cause. This may yet be so, but its sequel cannot be explained away thus. For, during the course of a field experiment in which fluctuations of mineral N were being monitored every few days, nitrate N declined by 25% over a period of three days in bare soil during warm, dry weather (Fig. 3.17). This phenomenon remains unexplained.

3.8.4 Mineralisation of P and S

Relatively little is known about either phosphorus or sulphur in organic combination in soils, nor how they are transformed, although of course such compounds must be mineralised, like those of N, before these elements are available to plants. Conditions favouring mineralisation of S are, in general, the same as those favouring mineralisation of N (Williams, 1968). Soil humus contains 1–1·5 parts of S to each 10 of N by weight, so, as humus is oxidised, nitrates and sulphates are released in this proportion. Like N, S can be immobilised by ploughing in residues of too high a carbon content (in this case too high a C:S ratio). Cowling and Jones (1970) found a greater rate of S mineralisation in planted than in unplanted soils. Mineralisation of P may be stimulated more than N by an increase in pH (Thompson *et al.*, 1954), an observation which has obvious relevance to chalk soils, though which may be masked by chemical immobilisation (Section 3.7.5).

4

The Colonisation of Exposed Chalk

4.1 Exposure of Raw Chalk

We saw in Chapter 1 how, by its physical nature, chalk moulds to a rolling topography in which natural exposures of the bare parent rock are largely confined to the coasts, and are rare inland. Only in the great erosions of the Pleistocene was raw chalk exposed on a large scale, and then, of course, colonisation was by Arctic, not temperate, species. At the present time, most exposures of bare chalk are man-made, and mainly confined to quarries, road construction and other civil engineering operations, among which the present expansion of Basingstoke, on the Hampshire Chalk, is a glaring example. On downland crests exposed to extremes of weathering, where the soil is shallow and the turf sparse, small pockets of chalk may be exposed by a variety of agents ranging from rabbits to mortar bombs. Aerial photographs of closely grazed downland often reveal intriguing patterns of assorted white specks and streaks, some of which defy explanation.

4.2 Pioneer Plants of Inland Chalk Exposures

4.2.1 Lower plants

We have already touched on the colonisation of bare chalk in the discussion of soil formation, where the role of the pioneer plants and animals, and especially the microflora and fauna, was established. The hardest and most freely drained exposures of chalk are sometimes slow to attract plants of any kind, but usually there is a distinct greening of the chalk, particularly striking in mild winter weather, and attributed by Tansley (1939) to green algae of the genus *Chroolepus*. Proudfoot (1965) observed a green layer immediately beneath the surface of the chalk of the experimental barrow at Overton on the Marlborough Downs about three years after it had been built. This layer included species in the Chlorococcales and Ulotrichales. Blue-green algae of the genus *Nostoc* are regarded by Proudfoot as "often the most abundant colonisers of chalk of southern England", and their nitrogen-fixing ability must have some ecological significance. Otherwise, nutrients are presumably derived from rain water (Chapter 2). Occasionally seen on bare or sparsely vegetated chalk slopes in early spring, at least in parts of the south, are the small, flat, bright red fruiting bodies of discomycete fungi of the genera *Scutellinia* or *Peziza*.

Typically appearing simultaneously with the algal colonisers, especially in shade, is the greyer green coloration of the moss *Seligeria calcarea*, which is virtually exclusive to this habitat in southern England, though it colonises other limestones further north. More restricted to the southern chalk is the locally abundant *S. paucifolia*, which grows on loose chalk blocks under trees (Watson, 1968). Another very characteristic moss of chalk quarries is the delicate *Campylium chrysophyllum*, often to be found growing in association with the more familiar *Camptothecium lutescens*. *Thuidium hystricosum* is a fairly early colonist of chalk-pits adjacent to older chalk grassland in the North Downs and Chilterns, and in the South Downs west of Harting (F. Rose, personal communication). *Acrocladium cuspidatum* sometimes invades wetter sites. It seems quite likely that some bryophytes render the ground suitable in some way for invasion by higher plants. Thus, Thomas (1930) noted grasses rooting in patches of moss growing on a waste-lime dump, and Pigott (1955) reports the enhancement of seedling survival in *Thymus* spp. when seeds germinated among moss plants rather than on adjacent bare ground.

4.2.2 Higher plants

4.2.2.1 The strategies of species of open ground

As a general rule, these lower plants are usually followed within the same season (or even preceded) by a plethora of flowering plants which establish readily in the raw-chalk "soil" and in the crevices of the solid rock. The angiosperm flora of disturbed chalk typically contains a fairly wide range of species, of which some are calcicolous (favouring calcareous soils—see Section 5.4.2) but most ubiquitous. (Curiously, some species normally associated with more acid soils, such as creeping buttercup (*Ranunculus repens*) and in the Chilterns even foxglove (*Digitalis purpurea*), seem to crop up on bare chalk more commonly than on the humic rendzinas, perhaps because, as noted in Section 3.7.3, raw chalk is actually poorer by comparison in exchangeable calcium. Tansley and Adamson's (1925, p. 218) records of the calcifuge wood sage (*Teucrium scorodonia*) on bare chalk at Windmill and Butser Hills provide another example.) The flora includes true ruderal species, possessing an opportunist strategy which results from an ability to produce large quantities of viable seed, which disperse readily and establish quickly. Many species possess in addition the ability to withstand, and even to thrive in, extremes of temperature and water stress. Others occur in disturbed sites because they are poor competitors, perhaps on the fringe of their geographical range, and can survive and grow only in relatively open stands.

Grime (1974) has proposed a novel scheme to classify vegetation by a triangular ordination, according to the degree of influence of these three contrasting, but interrelated, strategies: ruderal, stress-tolerant and competitive (Fig. 4.1). The scheme was tested by plotting on the ordination each of 100 species of herbaceous plants from the Sheffield area according to (i) an index of relative competitive ability, and (ii) a measure of relative growth rate. The results are interesting and, by and large, ecologically similar species tend to cluster together within their predicted area of triangle (Fig. 4.2). In Fig. 4.1h the ordination of plant stands of limestone outcrops shows more convincingly the predominance of strategies adapted to disturbance and stress tolerance, and the subordinate role of competition. (A few other types of vegetation are included for comparison.) Of course, these data relate to limestone country contrasting both geologically and geographically with that of the Chalk, and much experimental work remains to be done to test and refine the method. Nevertheless, it does seem a promising line of research in an almost untouched field.

Exactly what plant species appear in any one place will depend on (i) the degree to which the site has been denuded of any former vegetative cover and its reproductive structures, (ii) the availability of migrant seeds and other dispersal units from outside the area, which will vary both geographically and seasonally, and (iii) the nature of the substrate. The way in which this pioneer flora then develops depends in turn on whether the normal course of succession is allowed to proceed undisturbed, or is interrupted or deflected, naturally or otherwise.

4.2.2.2 Establishment from native reproductive structures

Where chalk is exposed simply by skimming off the turf and surface soil, fresh seeds may be scattered, and buried seeds unearthed, which are able to germinate as soon as operations cease. In Clements' (1916) terminology, *ecesis* proceeds forthwith, by-passing the migration phase. The *Reseda* spp. are good examples of pioneer plants which establish in this way, and we return to these in a moment. Some species sprout from pieces of meristematic vegetative tissue, such as decapitated tap roots (e.g. *Galium*

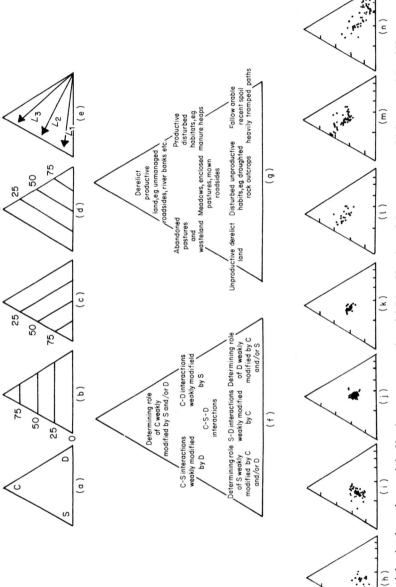

Fig. 4.1. A triangular model of herbaceous vegetation. (a) Identification of the corners at which competition (C), stress (S) and disturbance (D) are exclusive determinants; (b)–(d) relative importance (%—see original paper for computation) of competition, stress and disturbance; (e) course of succession in conditions of low (L_1), moderate (L_2) and high (L_3) productivity; (f) interactions; (g) location on the ordination of selected habitat types; (h)–(n) examples of stand ordinations from (h) limestone outcrops, (i) extensive sheep pastures on limestone, (j) meadows, (k) road-verges mown

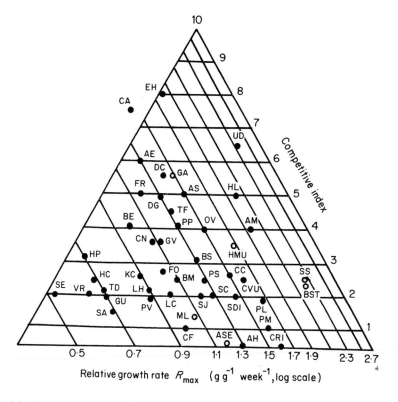

Fig. 4.2. Triangular ordination of herbaceous species of calcareous strata according to a competitive index (*CI*) and maximum relative growth rate (*R* max). After Grime (1974). The computation of *CI*, and the measurement and statistical spread of *R* max can be found in the original paper. The apparent anomaly in which species of open ground and woodland occur together is explained by regarding both desiccation and dense shade as stress factors. In the following list of species, annuals, which are shown as open circles on the triangular ordination, are indicated by an asterisk. AE, *Arrhenatherum elatius*; AH, *Arabis hirsuta*; AM, *Achillea millefolium*; AS, *Agrostis stolonifera*; ASE, *Arenaria serpyllifolia**; BE, *Bromus erectus*; BM, *Briza media*; BS, *Brachypodium sylvaticum*; BST, *Bromus sterilis**; CA, *Chamaenerion angustifolium*; CC, *Cynosurus cristatus*; CF, *Carex flacca*; CN, *Centaurea nigra*; CRI, *Catapodium rigidum**; CVU, *Cirsium vulgare*; DC, *Deschampsia caespitosa*; DG, *Dactylis glomerata*; EH, *Epilobium hirsutum*; FO, *Festuca ovina*; FR, *Festuca rubra*; GA, *Galium aparine**; GU, *Geum urbanum*; GV, *Galium verum*; HC, *Helianthemum chamaecistus*; HL, *Holcus lanatus*; HMU, *Hordeum murinum**; HP, *Helictotrichon pratense*; KC, *Koeleria cristata*; LC, *Lotus corniculatus*; LH, *Leontodon hispidus*; ML, *Medicago lupulina**; OV, *Origanum vulgare*; PL, *Plantago lanceolata*; PM, *Plantago major*; PP, *Poa pratensis*; PS, *Poterium sanguisorba*; PV, *Prunella vulgaris*; SA, *Sedum acre*; SC, *Scabiosa columbaria*; SDI, *Silene dioica*; SE, *Sanicula europaea*; SJ, *Senecio jacobaea*; SS, *Senecio squalidus*; TD, *Thymus drucei*; TF, *Tussilago farfara*; UD, *Urtica dioica*; VR, *Viola riviniana*.

verum), fragmented rhizomes (*Achillea millefolium*), root or stem tubers (*Filipendula vulgaris*), and the like. Certain grasses, such as *Festuca rubra* and *Koeleria cristata*, appear to colonise bare chalk more commonly in this way than from seed. Patches of vegetation composed of single species may thus result, and a "primitive grassland" (Section 4.5.2) frequently develops within a season, telescoping or by-passing altogether the more gradual colonisation which follows the complete removal of topsoil and turf, and deep or extensive excavation of the underlying chalk. In the latter case, the establishment of plants depends to a much greater extent on imported seed.

4.2.2.3 Establishment from migrant reproductive structures

The invasion of an area denuded of vegetation is mainly by seed, but it may also, of course, be vegetative. Daubenmire (1968) conveniently distinguishes *disseminules*, which he defines as "detached units carried through air or water" (and surely through or across soil as well), from *propagules*, which "remain attached to the parent plant while extending through a sub-stratum or over its surface". These terms are often used loosely and interchangeably. Daubenmire goes on:

> Immigration accomplished by disseminules results in the appearance of individuals at many points scattered over the habitat, whereas immigration accomplished by propagules results in a mass invasion from the margins of the bare area.

Clearly the smaller the area, the greater is the relative importance of vegetative encroachment. Even from seed, however, plants may spread only very slowly in from the edges if dispersal is inefficient. Thus, while Tansley and Adamson (1925) record *Thymus "serpyllum"* (probably *T. drucei*) colonising bare chalk adjacent to mature grassland, observations by Pigott (1955)

> suggest that dispersal is inefficient. Plants in garden beds become surrounded by seedlings never more than a metre from the parent . . . . On the railway that crosses the old earthwork of the Devil's Ditch near Dullingham (Cambridgeshire) *T. drucei* has only spread along the cutting to the extent of 0·8 km in over 100 years.

This topic is returned to in Sections 4.5.2 and 10.3.3.3.

The availability of immigrant seed depends upon the nature, state and diversity of the surrounding vegetation, and to some extent the animal population it supports, which may prevent flowering or seeding by grazing, but at the same time may help in seed dispersal. All these factors in turn rest on local geology, soils and land-use. There is bound to be a large difference, for example, between the flora of an exposure of chalk in the midst of 1000 ha of downland on Salisbury Plain, and that of the old chalk-pits of the central Isle of Wight, where the vertically inclined strata range through

Gault Clay, Upper Greensand, Chalk, Eocene sands and clays, and Oligocene marls within a few hundred metres, with patches of superficial gravel thrown in for good measure (Fig. 1.12). Where surrounding land-use is varied too, encompassing perhaps arable land, chalk scrub, acid heath, broadleaf and coniferous woodland, and gardens and allotments, the list of possible species becomes almost limitless. A good illustration comes from Tansley and Adamson's (1925) survey of the grasslands of the western South Downs. They counted 35 species of flowering plants in the quarries between War Down and Butser Hill, "surrounded by great stretches of chalk grassland", and 65 at Buriton limeworks, "much closer to both arable land and woodland". Exactly the same principles apply to the fauna.

Geographical position will also play a part. Exposures of chalk in the southern counties differ in some degree in botanical composition from those in the north; an extreme example can be seen in Thomas's (1930) account of the colonisation of a waste-lime dump at a Glasgow paper mill. Many species which might have been expected further south are absent. A virtually unique chalk flora must grace the old chalk-pits recently rediscoverd in southern Eire (Chapter 1). Near the sea, the maritime influence will be felt more strongly, but this is dealt with separately under the heading of maritime chalk cliffs. More general aspects of plant distribution are taken up in Chapter 5. There are, in addition, always likely to be seasonal differences in seed production and migration, and in any one year the fortuitous seeding of one particular species may be reflected for several to many years thereafter.

4.2.2.4 Some typical herbaceous pioneers

The harder and more freely-drained exposures of chalk weather more gradually, often into slabs and chunks rather than fine talus or spoil. These are colonised more slowly by plants than are the softer chalks and their weathering products, and this is especially true of cliffs and quarry faces, particularly those with a southerly aspect. Here, any plants which do establish either spread vegetatively from more equable spots, or are confined to relatively moist or shaded chinks and clefts in the face. One of the commonest herbaceous species exhibiting the former strategy is the ubiquitous creeping bent (*Agrostis stolonifera*), able not only to root into cracks and fissures as it spreads, but to withstand great extremes of heat and drought. In this situation its stolons become thickened and red with anthocyanin, and its leaves reduced to mere scales. Brambles may similarly clothe the banks with foliage, often spreading vertically downwards from their main point of anchorage, and providing support for the familiar *Clematis vitalba*; but of these woody species, more anon.

Many species can be found colonising the crevices of the chalk face itself from seed (Fig. 4.3), but among the most characteristic are wild mignonette

(*Reseda lutea*), wild carrot (*Daucus carota* ssp. *carota*), greater knapweed (*Centaurea scabiosa*), sometimes parasitised by the tall broomrape (*Orobanche elatior*), the smaller *C. nigra* (usually ssp. *nemoralis* on the southern Chalk), and the grasses *Arrhenatherum elatius* and *Dactylis glomerata*. All of these are able to root deeply into the chalk rock. Anderson (1927) found that *C. scabiosa* roots could penetrate more than a metre into

Fig. 4.3. Colonisation begins: the first flower head of a *Leontodon hispidus* plant in Chinnor quarry attracts a foraging hoverfly. A coltsfoot seedling can also be seen. (C.J.S.)

fissured chalk, and *Arrhenatherum* has a remarkable capacity to send its network of yellow roots into every available crevice of rubbly chalk (Locket, 1946a). Sheep's fescue (*Festuca ovina*), and several of the yellow-flowered ligulate composites, notably *Leontodon hispidus* and *Hieracium pilosella*, are equally able to hold their own on the driest banks, though their more modest root systems suggest that they are able to make do with less water. The water relations of chalk plants are considered in Chapter 5.

As the gradient changes from vertical cliff to steep slope, the lumps of chalk dislodged by weathering accumulate to form a scree. The more stable this scree is, the more closely its flora resembles that of the vertical face, but on loose, unstable banks of coarse chalk rubble, a floristically poorer, though no less interesting, selection of plants occurs. A typical pioneer of chalk scree is the beautiful viper's bugloss (*Echium vulgare*), a bristly-leaved biennial with bright blue flowers, deep pink in the bud, not at all restricted to the Chalk, but very characteristic of open, disturbed ground on light soils. Tansley and Adamson (1925) found it on the coarser quarry talus of their chalk-pit at War Down. Much more locally, steep scress of hard chalk provide ideal conditions for the establishment of box (*Buxus sempervirens*) but, again, more about this later (Chapter 8).

Where the chalk is softer, and the gradient more gradual, a much finer

Fig. 4.4. Chalk spoil at a lower level in Chinnor quarry provides a moist substrate particularly favourable to weld (*Reseda luteola*—here attaining a height of almost 2 m), and coltsfoot (*Tussilago farfara*). (C.J.S.)

grade of detritus accumulates with better water-holding properties, and this supports a wider range of plant species. Absolutely typical of these moister sites, and arriving with uncanny predictability often within weeks of the initial exposure of the substrate, are weld (*Reseda luteola*) and coltsfoot (*Tussilago farfara*). Weld was formerly cultivated for its yellow flavone dye, whence its alternative common name dyer's rocket. Its rosette of dark green strap-like leaves is a striking feature against the stark white of the bare chalk during its first year. In its second, it produces a branched inflorescence which may reach a height of 2 m or more (Fig. 4.4). Weld is a favourite species for demonstrating longevity of buried seed populations. According to Salisbury (1964):

> Excavations of the first century vallum at Cirencester on three occasions resulted in a crop of Dyer's Rocket appearing . . . and the suggestion that these arose from seed over 1800 years old is therefore not incredible.

Whether or not this fantastic claim can be accepted, there is no doubt of the prolonged dormancy of the seed, and it is hardly surprising that it so typically crops up as soon as the chalk is disturbed. The impression of great age given by the occurrence of seeds deep in the subsoil and in chalk quarries is easily explained: the seed has a high density and an extremely smooth testa. As a result, it is one of the most mobile of seeds in soil, and is easily shaken down deeper and deeper as the soil (or quarry) is worked.

Coltsfoot is perhaps the classic coloniser of bared ground, and is by no means confined to chalk. It favours relatively moist sites (one of its many local names is claywort), and it is very tolerant of high pH. Thomas (1930) found it growing on his Glasgow lime-dump in a pH of 10·0. It is quite remarkable how coltsfoot "finds" disturbed ground. Its seeds must be extremely widely and abundantly distributed (Bakker (1960) has recorded dispersal 4 km from source) for, unlike weld, coltsfoot seeds are viable for only 2–3 months (Myerscough and Whitehead, 1966), and so unable to lie dormant for long.

The plant is easily recognised early in spring by its clusters of pink, scaly stems, each bearing a solitary, bright yellow capitulum, its central disc encircled by a large number (up to 300) of ligulate florets in several rows. Coltsfoot is an unusually active plant in flower, and an ideal subject for the time-lapse photographer. In bud and flower, the head is held erect. It closes at night and in cold weather, but in favourable circumstances may open fully within an hour. By flowering so early in the year, coltsfoot provides an important source of nectar and pollen for the first insects to emerge, such as solitary bees (especially *Andrena* spp.) and bee-flies (*Bombylius* spp.) (Proctor and Yeo, 1973). After pollination, the head droops for several days (in a stiff and turgid state, not by wilting) as the stem elongates, and then finally becomes erect again as the white pappus opens out to disseminate the seeds (achenes) in the familiar way of this group of composites. The plant spreads vegetatively by rhizomes, from which the large, rounded–polygonal leaves are produced, usually some time after flowering has finished. These leaves are selected preferentially as resting places by the digger wasp *Mellinus sabulosa* (Step, 1932), and sometimes provide a supplementary diet for caterpillars of the cinnabar moth, *Callimorpha dominula*. A vague pattern of ecological interrelationships centred on coltsfoot begins to emerge, though for a more impressive example see Fig. 4.7.

In the early stages of establishment, vegetative growth and rhizome production take precedence over flowering, but as the network of rhizomes increases the partitioning of assimilates begins to favour flower production, giving the clone the appearance of "deciding to leave" once the bare ground is covered (Ogden, 1974). Accumulation of toxic wastes in the older parts, and the inability of the leaves to maintain an adequate net assimilation rate at low light intensities resulting from the shade of taller vegetation, augment this effect. Coltsfoot thus behaves as a true pioneer. The more fertile the soil, and the greater its stability, the sooner this critical density of rhizomes and leaves is reached, and the sooner the species "leaves" the site. Tansley and Adamson (1925) found coltsfoot growing, predictably, on the finer, more moisture-retentive chalk spoil at their War Down quarry in 1920, but it was still there in 1936 (Hope-Simpson, 1940b), presumably because of either the low nutrient status or the instability of the chalky substrate. The coltsfoot recorded in three sites at the old Harefield chalk-pit described by Locket (1946a) took 7, 9 and 14 years to disappear, but we are not told whether this was due to self-crowding, shading by other vegetation or death from drought (which killed many other species).

Leguminous species form an important group of colonisers of bare chalk, with obvious significance to the accumulation of mineral and organic nitrogen in the raw substrate. *Lotus corniculatus* and *Medicago lupulina* are the most common, though hop trefoil (*Trifolium campestre*) and the melilots (*Melilotus* spp.) (Fig. 4.5a) frequently occur too. The commonest English name applied to *L. corniculatus*, birdsfoot trefoil, describes the resemblance to a bird's foot of the cluster of ripe pods, but "trefoil" is a misnomer, for there are in fact five leaflets. The basal pair are mounted as though they were stipules, but the true stipules are below these, and so minute as to be easily overlooked. *Medicago lupulina* is much more straggly (procumbent, rather than decumbent) and truly trifoliate, or to be precise trifoliolate, for the leaflets are stalked. Its heads of small yellow flowers make it superficially similar to the yellow suckling clover (*Trifolium dubium*), but the flowers of *M. lupulina* do not turn brown after pollination as they do in *T. dubium*, and the fruit of *M. lupulina* develops into an unmistakable coiled black pod, hence its common name, black medick. Among these species, only *Lotus* is truly perennial: the rest are annuals or biennials, or at best short-lived perennials.

Fig. 4.5. (a, left) The common melilot (*Melilotus officinalis*) and (b, right) dark mullein (*Verbascum nigrum*), attractive yellow-flowered plants both typical of the "bare or disturbed chalk" habitat. (C.J.S.)

A little information on the contrasting behaviour of *Lotus*, *Medicago* and *Melilotus* populations can be gleaned from Locket's Harefield data. Thus, two *Lotus* plants recorded in one $0 \cdot 25$ m^2 plot in 1933 had increased to 23 by 1936, and the population maintained itself between 21 and 25 plants over the following seven years with no sign of a decline by the end of the study. Across the floor of the old quarry it became "abundant" and in places "dominant". Plants rooted deeply into the chalk and were "not seen to be attacked by rabbits", though damage was reported from severe frost during the winter of 1939–1940. *Medicago lupulina*, being an annual, fluctuated more in numbers, and it was found to be more susceptible to drought than

Lotus. Locket attributed this to a shallower rooting system, agreeing with Anderson's (1927) measurements of 7·6–17·8 cm (max. 30·4) for *Lotus* and 6·4–14·0 cm (max. 26·7) for *Medicago*. On the scree, a patch of 50 *Medicago* plants per 0·25 m² recorded in 1934 had dwindled to a mere "+" by 1937, and had gone altogether by 1943, but on the more moist floor area, a population ranging from 4 to 13 plants per 0·25 m² persisted throughout the ten-year period of study, and was showing no signs of disappearing when it was concluded.

The melilot (*Melilotus officinalis*) which appeared in the Harefield pit dominated the whole floor area in July 1933 when it grew 60–90 cm tall and flowered and seeded profusely. The population peaked again in 1935 and 1937, presumably as a result of the biennial habit of the species, but declined steadily after that, to virtually disappear by 1942, apparently through susceptibility to competition from other species. It was not grazed by rabbits. Locket notes that *M. officinalis* "flourishes in East Yorkshire on cliffs which are constantly eroded, thus exposing new soil surfaces".

The typically vigorous growth of these species suggests that they have no difficulty in nodulating, and excavation invariably reveals the presence of nodules. This is probably the result of cross-inoculation between species. Thus *Medicago lupulina* is cross-inoculable with *Melilotus* spp. (MacConnell and Bond, 1957). However, the presence of nodules does not necessarily indicate active N fixation, and MacConnell and Bond found that more than half the nodules of the *Medicago* plants they examined contained inefficient strains of *Rhizobium*, a condition which, they thought, was aggravated by the absence of cultivation. Anderson (1927) found nodules of *Medicago* "mostly near the soil surface" and this may have augmented the susceptibility of the species to drought noted by Locket. Even so, the fact that these plants nodulate at all reflects the microbial richness of even raw chalk.

It is a curious fact that almost all these pioneer species have yellow flowers and, as long ago as 1926, Druce commented on this striking feature of newly colonised chalk. Indeed, we might add several more: the beautiful common toadflax (*Linaria vulgaris*), the locally abundant yellow-flowered crucifers *Brassica nigra*, *Sinapis alba* and *Ersymium cheiranthoides*; St John's wort (*Hypericum perforatum*); and yet more Compositae: *Crepis capillaris*, *C. vesicoides* ssp. *taraxacifolia*, *Picris hieracioides*, *Sonchus arvensis*, *Senecio squalidus* and *S. jacobaea*, though this last, ragwort, often spends several years as a rosette before flowering. When ragwort does flower, it frequently supports, and indeed may be totally stripped of its leaves by, the yellow-and-black-striped larvae of the cinnabar moth (*Callimorpha dominula*; see Fig. 4.7). Many of the apomictic hawkweeds (*Hieracium* spp.), notoriously difficult to identify accurately, are to be found in chalk and limestone quarries (see Davis, 1977). Yellow-wort (*Blackstonia perfoliata*) is locally an important colonist of bare chalk.

Another distinct group of pioneers of disturbed chalk which, with the exception of the first, breaks the "yellow flowers" rule includes the robust and showy biennial mulleins, thistles, burdocks and teasels, and the perennial willowherbs, figworts and nightshades. For some reason, most of these tend to be more common in woodland sites than in entirely open country. Perhaps this is because the chalk under former woodland is richer in nutrients than it is under grassland or scrub. Certainly, some at least of the woodland herbs are known to be somewhat eutrophic (Fig. 8.12), and many are often found growing in nutrient-rich soil around rabbit burrows (Section 6.3.3). Indeed, it may be more correct to regard these species as characteristic of disturbed chalk *soils*, rather than raw chalk, and it is a purely arbitrary line dividing the plants reviewed in the following paragraphs from those described later in relation to broken turf and even woodland glades.

The mulleins are, in fact, members of the Scrophulariaceae, but this is not obvious for their flowers are only slightly zygomorphic, and phylogenetically they are close to the Solanaceae. They include the common mullein, or Aaron's rod (*Verbascum thapsus*), dark mullein (*V. nigrum*) and the much more local white mullein (*V. lychnitis*) mainly confined to the Kent Chalk, as well as a few other rare species. The two commoner species are easily distinguished by their very different leaves, as well as by contrasts in floral morphology, though Lousley (1950) reports hybrids between them "showing a perfect mixture of the characters of the two parents".

Verbascum thapsus is the more impressive of the two (Fig. 8.9) with its winter rosette of large woolly leaves, and spectacular summer flowering spikes which in a good season may reach 2 m in height. These spikes often persist, rattling with seeds, for several years after the plant dies. Turrill (1948) recorded plants bearing 700 000 seeds per plant, with 88% viability, but Salisbury (1942) regards 104 000–168 000 seeds per plant as a typical range. The small seeds are easily shaken or blown out by the wind. According to Salisbury, their maximum distance of dispersal is only about 3·7 m, but they can lie dormant for at least 58 years (see Section 8.3.2.1). The dense wool on the leaves of *V. thapsus* presumably serves to check transpiration, but may also insulate the plant from frost. Some solitary bees line their nests with it (Imms, 1971), perhaps aided by the striking larvae of the mullein moth (*Cucullia verbasci*), which tend to leave loose wefts of wool behind as they consume the more substantial tissues of the foliage. The flowering spikes of *Verbascum nigrum* reach barely half the height of *V. thapsus*, but there are usually more of them per plant (Fig. 4.5b). *V. nigrum* produces a deep tap-root and can live for many years, often persisting in dense grassland, for example on neglected road-verges. It is less widespread than *V. thapsus*, and exhibits a curiously patchy distribution: in his survey of the flora of Dorset, Good (1948) found *V. nigrum* only within a few kilometres of the Wiltshire border, and F. Rose (personal communication) notes that it is rare in Kent except in the extreme west near the boundary with Surrey. Good attributed the distribution of his Dorset plants to the fact that they shunned too oceanic a climate, but this conflicts with the pattern in Kent, and in any case dark mullein occurs on the strongly maritime Isle of Wight. The matter deserves investigation.

The larger thistle species, particularly spear thistle (*Cirsium vulgare*) and nodding or musk thistle (*Carduus nutans*), resemble the perennial knapweeds in that they may grow to very large specimen plants and bear large crops of flower heads which attract a wide selection of long-tongued insects, especially Lepidoptera. Of the burdocks, *Arctium minus*, distinguished by its hollow petioles, is the more common chalkland species. The leaves are often reduced almost to skeletons, apparently by slugs. The reddish-purple flower heads differ slightly according to subspecies, but though not particularly showy, all are visited freely by bees, Lepidoptera, and other insects. At maturity, the effectiveness of the involucral bracts as a dispersal mechanism is only too well known.

The teasel most usually found on chalk, *Dipsacus fullonum* (strictly ssp. *fullonum* to distinguish it from the cultivated ssp. *sativus*), is a striking plant of disturbed ground, with a spiky domed head of mauve flowers which attract Lepidoptera and other long-tongued insects. No less worthy of attention, however, are the remarkable water-cups formed by the connate bases of the oppositely paired bristly leaves, and long ago they attracted the attention of Richard Jefferies (1879), who observed:

> . . . of these vessels there are three or four above each other in storeys. When it rains, the drops, instead of falling off as from other leaves, run down these and are collected in the cups, which thus form so many natural raingauges. If it is a large plant, the cup nearest the ground—the biggest—will hold as much as two or three wine-glasses. This water remains there for a considerable time, for several days after a shower, and it is fatal to numbers of insects which climb up the stalk or alight on the leaves and fall in. While the grass and the earth are quite dry, therefore, the teazle often has a supply of water; and when it dries up, the drowned insects remain at the bottom like the dregs of a draught the plant has drained.

The common figwort (*Scrophularia nodosa*) is, like the teasel, more typical of ditches and damp places, but is nevertheless common on chalk. It is the classic example among the flowering plants of adaptation for pollination by short-tongued solitary and social wasps. Its curious, dingy, green-and-reddish-brown flowers, almost globular, resemble open mouths, and their nectar is thus easily accessible. The leaves of this plant are often reduced to a skeleton by the tiny, slimy larvae of the weevil *Cionus scrophulariae*. These then pupate among the seed-crammed capsules of the mature inflorescence, which the cocoons so perfectly resemble as to present an outstanding example of mimicry (Imms, 1971).

The minute seeds of the willowherbs are, like those of the figworts, produced in astronomical numbers, but the willowherbs differ in their possession of a plume of chalazal hairs, by means of which they are carried away on the slightest breeze, ensuring their widespread dispersal. Several species of *Epilobium* may be found, usually *E. montanum*, but most common is the rose-bay (*Chamaenerion angustifolium*). The phenomenal rise of rose-bay from relative obscurity is still recorded as "recent" in many a textbook, but it was an "abundant" and at times "dominant" species of the clearings of the Ditcham Park woodlands as long ago as 1912 (Adamson, 1921). As a result of its ability to spread vegetatively by means of horizontal roots, rose-bay can form quite extensive pure stands which have been known to persist for at least 30 years (van Andel, 1976).

The perennial nightshades include the scrambling *Solanum dulcamara* (woody nightshade) with purple flowers followed by ovoid, red berries, and *Atropa belladonna* (deadly nightshade), whose common name is often misapplied both to *S. dulcamara* and the common annual *S. nigrum* (black nightshade). *S. dulcamara* has an extraordinarily wide tolerance of soil water status, and is apparently equally at home in dry, rocky chalk and half a metre deep in stagnant pond water. True deadly nightshade, *Atropa*, containing the alkaloids atropine and hyoscyamine, is much more poisonous (at least to humans and cattle) than either of the *Solanum* species, and quite distinct from them with its large stature, evil-looking greenish-purple flowers, and spectacular (and sometimes fatally attractive) glossy, round, black berries up to 2 cm in diameter. This species is largely restricted to calcareous soils, especially where rabbits abound, but although spread throughout England and Wales, it is not at all common. A closely related plant found in similar situations, though more frequent near the sea than *Atropa*, is henbane (*Hyoscyamus niger*).

Several campions are characteristic of the vegetation of disturbed chalky ground. The dioecious white campion (*Silene alba*) is readily identified by its large, pure white flowers, those of the female plants developing a much larger capsule than the more shade-loving red campion (*S. dioica*), though hybrids which possess intermediate characters between these two species frequently occur. The smut fungus *Ustilago violacea* commonly infects the female flowers of *S. alba*, inducing the formation of stamens which then give forth violet spores. Infected flowers are conspicuously larger than uninfected ones. The bladder campion (*Silene vulgaris*) has a larger number of much smaller flowers than *S. alba* and is readily identified by its familiar inflated calyces. For some reason, many butterflies single out this species as a place to settle without necessarily feeding from it.

Altogether rarer is the Italian catchfly (*Silene italica*), earning its common name from the sticky secretion which makes its hairy shoots viscid, and which unfortunately attracts dust as effectively as it catches flies or discourages ants from climbing to the flowers. Darlington (1969) describes *S. italica* as "so closely associated with the walls of quarries that on tips it is likely to be found only where these are set up in disused quarries and incorporate exposed talus in their material", though this does not imply that the substrate need necessarily be chalk. The petals of these and other white-flowered campions crumple or roll up during the day, but become conspicuous in the evening, when they also become scented, attracting night-flying moths which pollinate them. Proctor and Yeo (1973) quote observations by Schremmer of the silver-Y moth (*Autographa gamma*) visiting *S. alba* and *S. vulgaris* in the dark.

All the species reviewed so far have been chosen, quite arbitrarily, as probably the most typical of the ruderal chalk habitat. Others which, perhaps without justification, are omitted here receive mention in Chapter 6 in connection with the colonisation of rabbit-scrapes, ant-hills and mole heaps. To round off this section, reference must be made to what are probably the two most significant groups of plants of this biome, deliberately left until now (apart from brief passing comment) to stress their importance. These are the grasses and the woody perennials; it is through these that the pioneer community evolves on the one hand into a grazed turf, and on the other into scrub and woodland.

As with the colonising plants in general, a range of grasses is likely to be encountered in disturbed chalk, and again it is a fairly arbitrary business to narrow the list down to the most typical. Among the annuals, barren brome (*Bromus sterilis*) and wall barley (*Hordeum murinum*) are found in disturbed, dry habitats generally, while *Catapodium rigidum* is more characteristic of calcareous strata. The four commonest perennial grass species of bare chalk, *Agrostis stolonifera*, *Festuca ovina*, *Dactylis glomerata* and *Arrhenatherum elatius*, have already received mention. *Arrhenatherum* soon develops a mass of dark green, leafy foliage, and seeds freely. This species usually succumbs rapidly to repeated defoliation, however, for it has only a small number of basal axillary buds from which new shoots can regenerate, although a chalkland ecotype has been reported which is much more tolerant of grazing (Pfitzenmeyer, 1962). *Dactylis* tends to be nibbled back into compact tussocks, though its flattened shoots and coarse glaucous leaves are apparently less palatable than *Arrhenatherum*. *Agrostis stolonifera* and *Festuca ovina*, on the other hand, tiller freely even under the closest grazing regimes, and between them they can form the first semblance of a turf. Other species, such as *Festuca rubra, Koeleria cristata, Brachypodium sylvaticum* and *Phleum* spp., as well as the glaucous sedge (*Carex flacca*), may also establish fairly early on. It is worth stressing, however, that on the whole the grasses form a relatively unimportant component of pioneer herbaceous vegetation without the presence of the grazing animal or its equivalent (see below).

4.2.2.5 Woody colonisers of disturbed chalk

On disturbed chalk, as on any relatively soft substrate, invading angiosperms are not restricted to herbs, and seedlings of the more mobile shrub and tree species can be seen among the very first colonisers. As with the herbs, almost anything can turn up according to location and season, and a perusal of Chapters 7 and 8 will indicate the range of shrubs and trees which might be expected. On the steeper slopes and banks, wild *Clematis* is usually to be seen, and hawthorn, privet, dogwood, *Viburnum, Rhamnus*, ash, sycamore and yew may similarly establish right from the outset. Beech will do so too, but only in close proximity to older beech trees. A frequent alien in southern districts is the fragrant garden buddleia (*Buddleja davidii*), an unparalleled attraction in flower to the larger Lepidoptera and other insects. But the commonest woody pioneers are the silver birch (mainly *Betula pendula*—Fig. 4.6) and the willows, particularly goat willow (*Salix capraea*) and, in damper sites, sallow (*S. cinerea*).

Clematis vitalba is well known as an indicator of chalk and limestone. Its straggly branches, clambering up anything that will support them by means of a neat twist of their petioles, may reach a height of 30 m, maturing into the familiar lianes with their peeling bark, and more reminiscent of tropical than temperate vegetation. The foliage supports the larvae of several moth species, including the pretty chalk carpet (*Melanthia procellata*). The fragrant, greenish-white flowers appear quite late in summer, and are visited by bees and pollen-eating flies, especially hoverflies (Syrphidae). The long, silvery-grey, plumed styles persist on the fruits to give the plant its attractive winter adornment, whence its common name, old man's beard. The styles are frequently regarded as important agents in seed dispersal, but the very similar structure of the closely related *Anemone pulsatilla* is regarded by Wells and Barling (1971) as of very doubtful significance. In *Anemone*, the heads entangle with each other, the tails get caught up on other plants and there appear to be no hygroscopic movements which might help burial. On the other hand, *Clematis* does have the advantage of its height as an aid to launching.

Locket noted a curious coming and going of young *Clematis* plants over the decade of his Harefield study, and could not explain their inability to establish effectively. Seddon (1971) proposes the intriguing suggestion that

Fig. 4.6. Old chalk workings near Wendover colonised almost exclusively by birches (*Betula*)—an increasingly common sight. (C.J.S.)

this species fails to occupy available sites due to a lack of other woody plants up which it can scramble for support. Certainly, unlike ivy, it seems unable to survive in a completely prostrate position. Hudson (1900) aptly describes the unsupported *Clematis* plant as having a "widowed, forlorn appearance", though he did once find "a perfect wild *Clematis* tree . . . with a round, straight and shapely trunk", looking like a weeping-willow tree. The rare variety *timbali*, with long, narrow lance-shaped leaflets, is reported from Oligocene limestone on the Isle of Wight by Lousley (1950), but apparently not on the Chalk, though clearly this is simply a matter of chance, and can have no ecological significance.

Betula pendula, popularly associated with acid heathland, is in fact widely tolerant of soil pH, provided the soil is well drained. Indeed, Thomas (1930) found *Betula* (and *Salix*) growing in waste lime with a pH of 8 at the surface, 9 at 5 cm, and 10 at 15 cm. It seems distinctly possible that birch is more common on the Chalk now than it was during Tansley's surveys. Even Clapham *et al.* (1962) refer to it as "rare on chalk". Willows were reported by Locket (1946a) to root less deeply into the chalk and to be more susceptible to drought than birch (which itself is fairly shallow-rooted). Nevertheless, the frequency of such distinctly mesophilous species indicates again the large water-supplying potential of solid chalk, provided it can be tapped. The catkins of the willow provide, like the coltsfoot, an important source of pollen and nectar in early spring, while the foliage feeds the larval stage of numerous insects.

4.3 The Animals of Open Chalk Habitats

Rabbits (*Oryctolagus cuniculus*) and badgers (*Meles meles*) are well known for their association with dry chalk banks, although they are not exclusive to this medium and rarely make use of freshly exposed or very hard chalk. In fact, badgers often burrow at the very junction of the chalk with overlying deposits where the latter are of a sandy nature. The effects of these animals, especially rabbits, are considered further in due course. Wherever roots or foliage provide adequate cover, birds are likely to nest in the crevices of the more precipitous and rocky exposures. Likewise the bats, and in particular the little pipistrelle (*Pipistrellus pipistrellus*), which "shows a decided preference for crowding into small crannies" (Matthews, 1952). Larger species may be found in caves, but this depends very much on the extent to which they are likely to be disturbed. Sites of particular interest have been protected by setting grilles across the cave entrance.

Passing reference has been made to some of the better known insects associated with the plants reviewed so far. Obviously, the more specific an insect is to one particular food plant, and the more this in turn is restricted to raw chalk, the more exclusive the insect will be to the chalk-pit habitat. Two small beetles which feed on *Reseda lutea, Phyllotreta nodicornis* and *Baris picicornis*, illustrate very well this kind of restriction (Duffey and Morris, 1966). In fact, most plants are hosts to a whole range of phytophagous invertebrates which feed on leaves, stems, buds, flowers, fruits, seeds and even underground on the roots, and which support, in turn, a great pyramid of consumers of secondary and higher order. Harper and Wood's (1957) example for ragwort (*Senecio jacobaea*) for which almost 200 species of insect visitors and associates have been reported, illustrates beautifully this

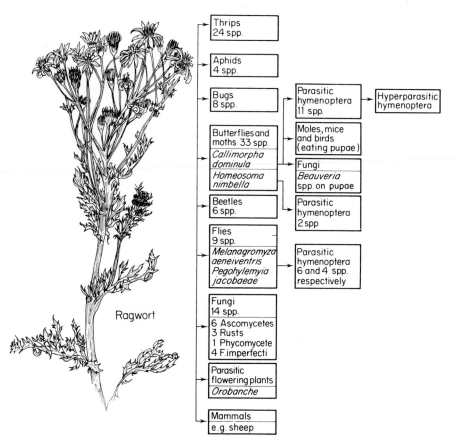

Fig. 4.7. Animals and heterotrophic plants associated with ragwort (*Senecio jacobaea*). After Harper (1957); see Harper and Wood (1957) for more detailed treatment.

intricate web of interrelationships (Fig. 4.7). Note, for example, how, despite their warning colours, even the larvae of the cinnabar moth (*Calli-morpha*) not only provide fare for moles and birds, but are also parasitised by at least nine Ichneumonidae and two Braconidae, of which one of the latter, *Apanteles popularis*, itself has three ichneumonid hyperparasites.

The open ground of chalk-pits, quarries and banks provides an ideal medium for certain invertebrates, notably many of the solitary bees and wasps (Hymenoptera) popularly associated with sandy habitats. Many of the former are solitary only in the sense that they do not share a communal nest, each individual hollowing out its own little tunnel in the ground. In fact, a suitable spot may be peppered with holes and swarming with bees. Sometimes, colonies of two species may be completely intermingled, yet readily distinguishable by the contrasting diameters of their burrows. Many of these diminutive bees, notably members of the genera *Prosopis* and *Andrena*, are voracious nectar and pollen feeders, though others are parasitic on these, in the sense that they rob them of their food, or even take over their nests. The bee-fly *Bombylius major*, unmistakable from its superficial resemblance to a bumble-bee, and its darting flight with its proboscis stuck rigidly out in front of it, can commonly be seen patrolling the bee colonies. Eggs are laid on the soil surface, and the hatching larvae then crawl down the nearest bee-burrow and feed on its contents. The wasps (Vespidae) are mainly predacious, killing and feeding on other invertebrates. All, in their turn, are liable to further hierarchies of parasitism, in which the eggs of the parasites are deposited in the nests, or even within the larvae, of their hosts. Two invaluable volumes, of widely separated vintage and style, contain a fund of information about the ecology of these insects: Step (1932) on the Hymenoptera in general, and Spradbery (1973) on the wasps.

It is undoubtedly the warmth and dryness of the open chalk banks which favour the more thermophilous invertebrates, for many of these are very local even in southern England, and are found more widely in continental Europe and around the Mediterranean. For certain predacious species, such as the spider-hunting wasp *Arachnospila spissa*, the openness of the vegetation is additionally favourable in making the prey more easy to spot and catch (Duffey and Morris, 1966). Ground beetles (Carabidae) and bugs (Heteroptera) likewise have their local representatives in these warm, dry, open habitats, but perhaps the most remarkable group is the spiders, for, according to Duffey and Morris:

> more than fifty species of rare and local spiders have been taken mainly on the chalk, about half of them from sites with open, short vegetation, stones and bare ground. Most of these have a southerly distribution in this country, and may find optimum conditions where there is a marked effect of insolation.

The tiny jumping spider *Europhrys aequipes* is cited as an example, though, like many of the bees and wasps, it is also found on bare, sandy heathland. Some spiders have been found mainly or exclusively in chalk-pits, for example the very local wolf spiders *Lycosa agrestis* and *L. hortensis*. A host of litter and soil-dwelling invertebrates, such as woodlice (Isopoda) and centipedes (Chilopoda), can be found among the crevices of the exposed chalk as the blocks and chunks fall or are prised away from the face.

4.4 The "Wet Chalk" Habitat

In deep chalk quarries there are sometimes places where excavation has reached the water table, and here vegetation characteristic of calcareous fen may develop, including several sedges (*Carex* spp.) and orchids such as *Dactylorchis praetermissa* and *Epipactis palustris*. Surface water may even accumulate, temporarily or permanently, as a lake, startling for its turquoise blue colour imparted by the optical properties of the suspended chalk particles. Waterfowl may often be seen on these strange blue pools, but they are usually transient, for there is little food for them either beneath the surface or at the water's edge. Usually, as spring advances, this water level drops, and on the shoreline a network of cracks develops in the chalk as it dries, augmented in frosty weather by the strangest of patterns on the surface. Seeds, typically of *Reseda luteola* and *Agrostis stolonifera*, lodge in these cracks and germinate there, though growth may be limited in hot weather by the oven-like conditions which result from the combination of shelter from wind with reflected glare from the surrounding chalk (Sections 2.3.3.3 and 3.6.3), and there is always the likelihood of renewed inundation in wet weather, especially in autumn as the water table rises again.

4.5 Succession

4.5.1 The natural process

What happens next depends upon the degree to which the growth of taller herbs and woody plants is held in check by the physical characteristics of the environment, and by animals. Thus on the steep unstable slopes, or on those of southerly or south-westerly aspect which get very hot and dry in summer, a relatively sparse stand of plants may persist for many decades—perhaps much longer. But the natural course of events is for the herbaceous vegetation to give way to scrub and then woodland communities as explained in Chapters 7 and 8.

This, of course, can only occur in the absence of grazing animals, and the way in which livestock, particularly rabbits (and sheep) can deflect this course of succession into a relatively stable grassland community (plagioclimax) is a subject at the very heart of chalkland ecology—almost its *raison d'être!* The later stages of chalk-pit colonisation, particularly where grazing can be controlled in some way, or at least monitored, provide an insight into how this critical deflection comes about.

4.5.2 The development of a grassland sward

From casual observations in the field, it is not difficult to imagine the course of succession on chalk from the first pioneer plants, through open grassland, to the closed, mature turf, as Usher (1973) has recently done from studies of chalk-pits in the Yorkshire Wolds. Yet, amazingly, no one seems ever to have published any detailed and sustained study of so obvious and familiar a course of events. Tansley himself would surely have regretted that in 1980 we still have to fall back on his original and very incomplete data collected in 1920, supplemented by Hope-Simpson's observations of the same spot some 16 years later. It is a great pity, too, that Locket's Harefield site was not monitored for a more prolonged period.

First, a brief digression to define the terms "open" and "closed" vegetation. Shimwell (1971a) provides a useful distinction. *Open* vegetation has "space between individuals which can be colonised". (If the area of space exceeds twice that of the plant cover, so that "substrate, not vegetation, dominates the landscape", the canopy is said to be *sparse*.) In *closed* vegetation, on the other hand, "individual plants in the community are so spaced as to form a continuous lateral contact". Both open and sparse stands can be seen in Fig. 4.8. Of course, these terms relate only to aerial shoots: the roots of an open stand of plants may present quite a different picture from that above ground.

Reporting their 1920 survey of the spoil banks of the old chalk quarries of the Hampshire–Sussex border, Tansley and Adamson (1925) describe three stands of open ("primitive") chalk grassland from War Down, of which two (call them Nos. 1 and 2) were apparently at an earlier stage of development than the third (No. 3). Hope-Simpson later confirmed that the part of the quarry containing stands 1 and 2 had been deserted some 25 years previously (1895), and No. 3 about 40 years previously (1880). Of the two younger stands, No. 1 occupied a south-facing slope, with about 60% of the chalk substrate still bare, while No. 2 occupied the north-facing slope of the same bank, and had only about 40% bare ground. *Festuca ovina* dominated both swards, although Tansley and Adamson tended to overlook *Festuca rubra*, and it is likely that both these fine-leaved fescues occurred, at least in No. 2.

Carex flacca was co-dominant in stand No. 1, and there were 27 species in all, including 23 angiosperms and 4 bryophytes. In stand No. 2, *Tussilago farfara* (in no hurry to "leave"—see Section 4.2.2.4) was co-dominant with the fescues, and the total number of species was 23 (19 angiosperms, 4 bryophytes). The older sward, No. 3 (south-facing, like No. 1), was less open, with only about 10% bare ground. *Festuca ovina* and *Carex flacca* were co-dominant, and there was no *Tussilago* here, but *Echium vulgare* was "prominent".

Fig. 4.8. A sparse, so-called "primitive", community of chalk grassland plants on the steep face of Whiteleaf Cross, Buckinghamshire. The deep gully emphasises the vulnerability of exposed chalk such as this to weathering and erosion, exacerbated here by generations of intrepid climbers. (C.J.S.)

Among the more abundant flowering plant species which had become established in the 25-year-old swards (Nos. 1 and 2) were *Asperula cynanchica*, *Carlina vulgaris*, *Cirsium acaulon*, *Euphrasia nemorosa*, *Gentianella amarella*, *Linum catharticum* and *Helictotrichon pratense*. Bryophytes included (as well as *Camptothecium lutescens* and *Campylium chrysophyllum* persisting from the pioneer phase), *Acrocladium cuspidatum*, *Ctenidium molluscum* and *Neckera complanata*. In turn, the following species were present in the 40-year sward (No. 3) which were not recorded in the younger ones: *Achillea millefolium, Bellis perennis, Campanula rotundifolia, Plantago lanceolata, Ranunculus bulbosus, Ophrys apifera* and *Carex caryophyllea*. Of the younger-phase bryophytes, *Camptothecium, Acrocladium* and *Ctenidium* were "still" present, but *Barbula rigidula, Pseudoscleropodium purum, Fissidens taxifolius* and *Thuidium hystricosum* were "new" arrivals, the last misidentified as *T. abietinum* (F. Rose, personal communication).

Tansley and Adamson described a fourth sward, regarded as still closer to mature chalk grassland, at Buriton lime works. Here the canopy was essentially closed, and it included several species not present at War Down which were regarded as the slowest to appear in the grassland sward. Their very identity, however, suggests that this was a moister, or more neutral site, and that it might be unrealistic to regard the sward here simply as more "advanced" in a temporal sense. The key species here were *Briza media, Cynosurus cristatus, Galium verum, Holcus lanatus, Lolium perenne, Trifolium repens* and *Trisetum flavescens*.

The War Down spoil banks were surveyed again 16 years later, in 1936–1937, by Hope-Simpson (1940b), by which time the swards were over 40 and 55 years old. He found a remarkable increase in the number of species: from 26 to 50 in No. 1, in which the proportion of bare ground had dropped from 60% to 25–30%; 23 to 48 species and 40% to 25–30% bare ground in No. 2; and 33 to 63 species in sward No. 3, which by now was "almost entirely covered" with vegetation. Some of these new species were thought to have come in following the removal of rabbits in 1932; others may have been overlooked in the earlier survey. Others again were not newly recorded, but had greatly increased their contribution to the herbage. Hope-Simpson found that the initial dominance by *Festuca* ("mostly *ovina*") and *Carex flacca* on the two south-facing slopes (Nos. 1 and 3) had declined, but that these two species were still "abundant" or "very abundant". Two major groups of species were recognised, of which the first were "roughly stable . . . or changing very slowly", and the second "showing significant increase or new appearance" (with a few minor conditions). These are listed in Table 4.1.

Hope-Simpson singles out for special mention *Poterium sanguisorba* which had only just begun to appear in the oldest (55-year) stand, and *Scabiosa columbaria* and the common moss *Rhytidiadelphus squarrosus* which were still quite absent from all three stands. This was less surprising in the case of *Poterium* (which Ridley (1930) records as being spread by linnets), for it happened "to be unusually rare on War Down and the adjoining Butser Hill", but the other two may genuinely be slow invaders, since both were "abundant in the vicinity and almost certainly capable of accommodating themselves to the physiographic conditions of at least one of the slopes". *R. squarrosus* does appear to require some depth of humus and litter to become established, however.

Table 4.1

Herbaceous species of possible early and late categories of colonisers of chalk grassland [a]

I. Early species	*II. Late species*
Agrostis stolonifera	*Arrhenatherum elatius* [b]
Dactylis glomerata	*Brachypodium sylvaticum*
Festuca ovina + F. rubra	*Briza media*
Helictotrichon pratense (Avena pratensis)	*Deschampsia caespitosa*
	Holcus lanatus
Carlina vulgaris	*Koeleria cristata*
Cirsium acaulon (C. acaule)	*Poa pratensis*
Echium vulgare	*Trisetum flavescens*
Hieracium pilosella	
Linum catharticum	*Dactylorchis fuchsii (Orchis maculata)* [b]
Lotus corniculatus	*Gentianella amarella (Gentiana axillaris)*
Senecio jacobaea	*Heracleum sphondylium* [b]
Taraxacum spp.	*Leontodon hispidus*
Thymus drucei (T. serpyllum)	*Plantago lanceolata*
Tussilago farfara	*Prunella vulgaris*
	Solanum dulcamara
Acrocladium (Hypnum) cuspidatum	*Sonchus oleraceus*
Camptothecium lutescens	*Trifolium pratense*
Campylium (Hypnum) chrysophyllum	*T. repens*
Ctenidium (Hypnum) molluscum	*Veronica chamaedrys*
	Bryum capillare
	B. inclinatum
	Ditrichum flexicaule
	Pseudoscleropodium (Brachythecium) purum

[a] After Hope-Simpson (1940b); original nomenclature shown in parentheses.
[b] Hope-Simpson considers that these are accounted for by the removal of the rabbits. *Arrhenatherum*, certainly, would not normally be regarded as a late invader. *Solanum* and *Sonchus* are similarly more typical of the earlier phases of colonisation.

Emphasis has been laid in this account on the importance of the ecological richness of the surrounding area in the colonisation of chalk-pits and the like, but of course there is another way of looking at this. Where areas of formerly notable ecological interest have been eaten into by intensive agriculture and forestry, chalk-pits are often the last refuges for chalk grassland animals and plants. Perring *et al.* (1964) describe examples from the Cambridgeshire Chalk, and Ratcliffe (1977) from Breckland, in some detail. The importance of old chalk-pits in the conservation of threatened communities and species in areas such as these is manifestly obvious (see Chapter 10).

4.6 Anthropogenic Modifications

Exposed chalk is often seeded deliberately, with or without a pre-treatment, both to control erosion and to improve its appearance aesthetically. Pre-treatment may include spreading topsoil, adding fertilisers, or, on the steepest banks, spraying on an inert bituminous binding and sealing medium in which seeds may be embedded. Sometimes the surface may be turfed. Seed mixtures are usually composed of standard agricultural grass and legume species and varieties, with red clover (*Trifolium pratense*) a reliable ingredient which rapidly establishes a vigorous sward. It nodulates satisfactorily without the need for inoculation of the seed, suggesting that it finds the appropriate strain of *Rhizobium* even in raw chalk. In a joint project by the Chiltern Society, the Agricultural Development and Advisory Service (ADAS) and the Tunnel Cement Company, several pasture species are currently being tested for their suitability in seeding worked-out chalk quarries (Fig. 4.9).

In certain cases, special "ecological mixtures" have been devised, such as that used to seed the banks of the M40 motorway where it cuts through the Chiltern escarpment on the Buckinghamshire–Oxfordshire border near Lewknor (see Fig. 1.39 and Chapter 10). At the other extreme, I know of one enthusiastic foreman who sows the banks of his chalk quarry with every packet of garden flower seeds he can lay hands on: he "likes to see a bit of colour". He plants herbaceous border throw-outs collected from his neighbours, and transplants quite large trees with a bucket excavator. Lousley (1950) mentions a remarkable collection of 20 alien flowering plant species in an old bomb crater near Box Hill in Surrey, which attracted a no less remarkable assortment of possible explanations. Had the French Resistance movement substituted flower seeds for TNT? Could bombs be added to the list of agencies of seed dispersal? All is now revealed, however, for this was a hoax, a botanical Piltdown, and the seeds were deliberately sown (McClintock, 1977).

Worked-out chalk quarries are often used for dumping, which covers the chalk, impeding any natural colonisation and succession, introduces soil, seeds and even whole plants from elsewhere, and may be accompanied by spontaneous or deliberate burning. This is a far cry from chalkland ecology, although as Darlington (1969) has shown, even refuse tips have educational potential.

Fig. 4.9. In contrast to the management of hill-figures as conspicuous landscape features where the objective is somehow to maintain the bare chalk surface, quarrying and road-building operations frequently demand quick and efficient stabilisation and camouflage. The trials shown here, at Pitstone, Buckinghamshire, are of agricultural grasses and legumes, but this is ideal terrain for the re-establishment of native species of greater ecological interest. (J. Hawkins)

4.7 Coastal Cliffs and Landslips

4.7.1 Coastal cliffs

4.7.1.1 Plant life

We saw in Chapter 1 that, apart from the rare cliffs and screes such as those at Box Hill, Surrey, and Kimble, Buckinghamshire, it is normally only near the coasts that the Chalk is ever exposed naturally. The cliffs of the coasts of eastern Kent, Sussex, Dorset, south-east Devon, the Isle of Wight and east Yorkshire are familiar and spectacular physiographic features. Plants growing on these maritime cliffs face two major environmental challenges: finding a foothold in a substrate which may be quite hard, yet is likely to crumble away fairly regularly; and facing an essentially constant barrage of wind and salt spray from the sea (Section 2.3.3.4). Where the cliff is absolutely sheer, it may be quite bare of plants, especially if it faces the full blast of the open sea, but usually, wherever there is a ledge or cleft, herbaceous plants manage to establish and grow (see Fig. 1.6). The chalk cliffs east of Folkestone in Kent are particularly interesting for their rich calcicolous flora, for here, although there is naturally some instability, protection from the sea has long been afforded by a shingle beach, now supplemented in places by a sea wall.

Most typical of this station is the rock samphire (*Crithmum maritimum*), one of our most strictly halophytic flowering plants, growing within the very splash zone of the sea itself. Although not confined to calcareous rocks, this succulent-leaved umbellifer is very characteristic of the chalk cliffs of the southern and south-eastern coasts. It is perhaps most popularly associated with Shakespeare's Cliff at Dover, from the reference to the samphire gatherer in King Lear. It does not occur much further north on the east coast than Suffolk, so it is absent from the Yorkshire chalk cliffs. In his classic book on the dispersal of plants, Ridley (1930) quotes experiments by H. B. Guppy, who found that the fruits (mericarps) of *Crithmum* could float and maintain their viability in sea water for ten months, and that the same light, corky tissue which renders them thus buoyant, aids in their dispersal up the cliff-face by the sea, Other widespread cliff plants include the sea beet (*Beta vulgaris* ssp. *maritima*), sea campion (*Silene maritima*), thrift (*Armeria maritima*), and the buck's horn plantain (*Plantago coronopus*).

Two very characteristic chalk cliff colonisers are the wild cabbage (*Brassica oleracea*), scraggy ancestor of our domesticated vegetables of that group, and the beautiful yellow-horned poppy (*Glaucium flavum*), whose large yellow flowers give rise in fruit to the remarkable "pods" that give the plant its common name, and which may reach 30 cm in length. Away from the south and south-west coasts, *Brassica* is usually regarded as an escape from cultivation. Both *Brassica* and *Glaucium* are distinctly calcicolous, and are able to spread inland in the south-east if raw chalk is available to them. Thus Scott (1963) reports *Glaucium* growing in fine chalk debris with a pH of 8·1 on the sides of an old anti-tank ditch approximately 1·5 km inland at Folkestone.

Here and there, rather more local species may establish on the cliff-face. From "a ledge of cliff 10 feet above the beach at Cliff End (Cuckmere Haven)", Tansley and Adamson (1926) recorded, as well as *Beta* and *Crithmum*, the saltmarsh species *Puccinellia maritima, Limonium vulgare, Spergularia media* and *Halimione portulacoides. Frankenia laevis* and the grass *Parapholis incurva* occur in abundance at the base of the Folkestone–Dover cliffs, with *Crithmum, Brassica,* and rock sea lavender (*Limonium binervosum*), as well as the much rarer golden samphire (*Inula crithmoides*) and hoary stock (*Matthiola incana*), further up (F. Rose, personal communication). *Inula* was mentioned by Good (1948) as a coloniser of the Dorset chalk cliffs, along with *Crithmum* and the Portland spurge (*Euphorbia portlandica*), while from the westernmost chalk cliff at Beer Head, Devon, Lousley (1950) adds *Limonium binervosum* and the rock spurrey (*Spergularia rupicola*). *Euphorbia* and *Matthiola* are known from the Isle of Wight, to which the latter may be native; definite garden escapes (denizens) there are wallflower (*Cheiranthus cheiri*) and sweet alyssum (*Lobularia maritima*). Although confined mainly to shingle beaches, sea-kale (*Crambe maritima*) has established itself on the chalk cliff of Beachy Head.

Three species which are important components of the maritime grassland vegetation reviewed in Chapter 5 may occur as isolated plants on the cliffs. These are red fescue (*Festuca rubra*), kidney vetch (*Anthyllis vulneraria*) and sea carrot (*Daucus carota* ssp. *gummifer*). Red fescue is widespread throughout Britain, and prominent in such widely different situations as here on the cliffs, in salt marshes, and on chalk downland. Some taxonomists recognise many subspecies and varieties, and it is possible that different ecotypes occupy these contrasting habitats. The grass spreads vegetatively as well as by seed, and where it is ungrazed, particularly on inaccessible cliff ledges, it builds up a great mat of foliage and litter and completely excludes all other plants. This grass is one of the key species in both inland and maritime chalk grasslands, and we shall return to it again in Chapter 5.

Both *Anthyllis* and *Daucus* are more strictly calcicolous than *Festuca rubra*. *Anthyllis* is easily recognised in flower by its paired heads, one maturing before the other, of yellow flowers, their woolly calyces giving them the appearance of being embedded in cotton wool. It is an important plant for insects: its leaves form the food plant for larvae of the small blue butterfly (*Cupido minimus*) for example, and its flowers are visited by many insects including bumble-bees which probably pollinate them. *Anthyllis vulneraria* is a very variable species, and these of the seaside are regarded by some botanists as ssp. *maritima*.

Daucus carota is a widespread umbellifer of chalk grassland, very rough to the touch, and bearing heads of white or pinkish flowers, often with a single deep-red one at the very centre of each umbel. The flowers are freely visited by pollen- and nectar-foraging insects and their predators. The maritime subspecies (ssp. *gummifer*, given species status as *D. gummifer* on the continent) is easily distinguished from ssp. *carota* at maturity, for the inflorescence of *gummifer* remains flat, while in *carota*, the outer peduncles elongate and grow inwards to form the distinct, nest-like structure (Fig. 4.10), often housing a spider such as the well-camouflaged *Araneus redii* (E. Duffey, personal communication). A whole range of variants exists, however, between the two subspecies, and long ago the French botanist Niel

(1886) made a detailed study of these transitional forms. In Britain, *D. carota* ssp. *gummifer* is very local, and occurs mainly along the south coast.

Fig. 4.10. Mature fruiting heads of the common wild carrot (*Daucus carota* ssp. *carota*). Those of the paramaritime *D. carota* ssp. *gummifer* tend to remain flat at maturity.

Among the rest, the plants of fairly general distribution around the coasts may be gleaned from the accounts by Lousley (1950) and Hepburn (1952), but where those of more local distribution are concerned it should be borne in mind that these texts, even in Lousley's second edition of 1969 which in fact was hardly revised at all, relate to events more than 25 years ago, and some may therefore now be more of historical than topical interest. Indeed any statements about cliff plants made in the botanical literature in the present tense should always be read with half an eye on the date of publication, and half on the calendar.

An interesting hypothesis which has been raised several times (Pigott and

Walters, 1954; Rose, 1957, 1973) is that chalk sea cliffs, for example of Sussex and Kent, as well as some steep inland banks including the Mole River cliff at Box Hill, may have formed refugia for relict populations of calcicolous species of open habitats during the Post-glacial period as the vegetation inland developed into woodland. From these refugia, the downland grasses and herbs were able to spread back as the forest was opened up by the first Neolithic farmers to penetrate inland (Chapter 1).

4.7.1.2 Animal life

The outstanding feature of the animal life of the chalk cliffs is, of course, the

Fig. 4.11. The cliffs of hard white northern chalk at Bempton, near Flamborough Head, famous for their wealth of breeding seabirds. The nests of the birds were formerly plundered by teams of "climmers", shown here surrounded by kittiwakes, but the Protection of Birds Act, 1954, put a stop to the practice, and the area now forms part of the RSPB's Bempton Cliffs Reserve. (*Yorkshire Post*).

richness and abundance of the population of sea and coast birds. The ledges and clefts, safe from human interference, make ideal nesting sites, and the commoner species, particularly the herring gull (*Larus argentatus*), cormorant (*Phalacrocorax carbo*) and shag (*P. aristotelis*) among the seabirds, as well as jackdaw (*Corvus monedula*) and rock pipit (*Anthus spinoletta* ssp. *petrosus*) are likely to be found nesting and breeding there. The chalk cliffs of the western Isle of Wight are notable for their puffins (*Fratercula arctica*), which burrow like rabbits, and those of the Sussex Coast west of Beachy Head were formerly famous for their peregrine falcons (*Falco peregrinus*). Many of the more prominent headlands are important navigational features in the flyways of migrating birds. Quite the most outstanding bird cliffs are on the east coast of Yorkshire, however (Fig. 4.11), and the cliffs at Flamborough, Bempton and Speeton contain the largest concentration of nesting seabirds anywhere in mainland Britain (see Sledge, 1971). The gannet (*Sula bassana*) nests on no other mainland site. Cormorant, shag, fulmar (*Fulmarus glacialis*), kittiwake (*Rissa tridactyllis*), guillemot (*Uria aalge*), razorbill (*Alca torda*) and puffin, as well as the rock dove (*Columba livia*), are all notable for their sheer numbers there. Moreover, Gillham's well known studies (e.g. Gillham, 1956) on the Pembrokeshire islands have shown the effects that birds in large numbers can have on the vegetation, particularly by manuring and trampling, and, in the case of the puffin, by undermining.

It is easy to picture these cliffs as the homes of only the hardiest plants and birds, yet the invertebrate fauna is often of great interest. As will be gathered from the previous section, many butterflies and other insects visit the plants in flower on the ledges, secure from all but the most intrepid collectors, and I recall once seeing a mason wasp (*Odynerus* sp.) emerge from a chink in the chalk of the vertical cliff at Beer Head barely a metre above the high-tide mark. The rare spider *Segestria bavarica* occupies a similar habitat in this area (Ratcliffe, 1977). Perhaps one of the most bizarre localities for a British insect is Rosehall Green, where the Freshwater or Isle of Wight wave moth (*Idaea humiliata*) is to be found. Little wonder that the entomologist K. G. Blair had difficulty finding anyone who could locate the place for him, for he was finally shown it from a rowing boat: a precarious slope on the cliff which tapers off to the Needles themselves (O. H. Frazer, personal communication).

4.7.2 Landslips

Raw chalk may also be exposed by landslips, but under modern climatic conditions these occur very rarely inland, and are, like the cliffs, more typically a coastal phenomenon. Indeed, landslips may in their simplest form result merely from the undercutting and slumping of the chalk cliffs already described. But wherever the Chalk overlies an impervious, seaward-dipping clay stratum such as the Cretaceous Gault, whole sections of the coastline are unstable, especially when persistent heavy rain lubricates the surface of the clay and saturates the overburden atop it. Here, large-scale slips occur (see Chapter 1), which result in a jumbled mixture of the geological layer-cake, and a correspondingly complex assortment of soils and vegetation. In a study of the Normandy coast considered in more detail in Chapter 5, Liger (1956) aptly refers to landslips of this kind as *les chaos*.

Some parts slip regularly, and their colonisation by plants rarely gets much beyond a sparse ruderal cover, typically including coltsfoot. Here and there, fresh water may erupt or seep through as a spring, and in these wet flushes plants normally foreign to chalk may be found, such as reeds (*Phragmites communis*), rushes (*Juncus* spp.) and horsetails (*Equisetum* spp.). Elsewhere, the odd knoll or seam of harder rock may shore up the slide for a few years, even for many decades in the drier spots, so that shrubs and even trees may get a hold. But these usually succumb in the end, and present a strange spectacle, whether leaning, dead, at crazy angles, or dumped, still standing vertically, even (briefly) in full leaf, at the water's edge below. Liger makes the interesting point that herbaceous plants deposited thus at the very edge of the sea often show a surprising tolerance of conditions there, provided they arrive with a boll of soil attached—containerised, so to speak.

Where the chalk is not too thoroughly mixed with other kinds of soil, the flora resembles in many respects that of inland chalk-pits, reviewed previously. On the steeper exposures, species of the chalk cliffs are likely to occur, while on the sufficiently stable flatter parts, a maritime chalk grassland sward develops (see Chapter 5). But a characteristic feature of the coastal landslip flora is the addition of that intriguing group of plants which occur within a kilometre or two of the sea, known formerly as the *sub-maritime*, now, less ambiguously, as the *paramaritime* element. Possible factors involved in these distribution patterns are discussed in Chapter 5.

Two easily recognised paramaritime species of disturbed, not necessarily chalky, ground are the introduced Alexanders (*Smyrnium olusatrum*), which produces its umbels of rather sickly-smelling greenish flowers very early in the year, and, less commonly, the slender thistle (*Carduus tenuiflorus*). More restricted to the south and south-west—indeed, classic examples of the Oceanic Southern type of distribution of Matthews (1955)—are wild madder (*Rubia peregrina*), a scrambling prickly evergreen whose foliage often turns a deep purplish-brown on the drier banks, and the wintergreen "rare lords-and-ladies" (*Arum italicum*), both found as far east as Kent, as well as the stinking iris (*Iris foetidissima*), which reaches Essex.

The latter is so named from the unpleasant rubbery smell given off if the leaves are bruised, though this is surely compensated for by the attractive purple (occasionally yellow) flowers, and the bright orange seeds which follow. *Iris* is more strictly calcicolous than *Rubia*, and Good (1948) drew attention to its curious bimodal distribution in Dorset, occurring in two completely disjunct zones. One is along the coast, and the other on the stretch of chalk downland extending from Cranborne Chase towards Dorchester. A similar pattern is seen in Kent: there, *Iris* runs right inland along the North Downs.

Among the woody species, ubiquitous blackthorn, sycamore, ash and ivy (see Chapter 8) are common along the southern chalk coasts, the last of these often parasitised there by the ivy broomrape (*Orobanche hederae*). In places the holm oak (*Quercus ilex*) is very characteristic, perhaps because it is so frequently planted, and regenerates freely from these trees and their subspontaneous offspring. It is particularly abundant on the Isle of Wight landslips, and has spread along the foot of St Boniface Down, where, as a native of the Mediterranean, it is clearly favoured by the warm-temperate climate.

The environs of these landslips are well known to entomologists for some of our rarest butterflies and moths. Thus the Lulworth skipper (*Thymelicus acteon*) and Glanville fritillary (*Melitaea cinxia*) are restricted, respectively, to a small area of the Dorset coast, and the landslips of the southern Isle of Wight. Similarly, Folkestone Warren is one of very few localities in Britain for the rest harrow moth (*Aplasta ononaria*) (Ford, 1957, 1972). It is not at all obvious why these insects should be so local, for their food-plants are common enough: various grasses, including *Brachypodium sylvaticum*, in the case of *Thymelicus, Plantago* spp. in the case of *Melitaea*, and rest harrow (*Ononis* spp.) in the case of *Aplasta*. Coincidentally, the adults of *Thymelicus* appear to favour exclusively the flowers of *Ononis* for feeding (Stokoe, 1944).

5
Chalk Grassland: I. Climatic and Edaphic Influences

5.1. The Origin and Physiognomy of Chalk Grassland

The development of chalk grassland via primary colonisation of the bare rock or disturbed chalk rubble is not of widespread significance. We saw in Chapter 1 how extensive areas of downland turf arose by secondary succession either directly from cleared woodland, or from abandoned arable land, deflected into a relatively stable plagioclimax by grazing animals. This is not to say that all downland swards are necessarily ancient. Indeed recent evidence (Wells *et al.*, 1976) suggests that some areas of chalk grassland may be much younger than was formerly supposed (see Section 5.4.3.5). Nevertheless, commentaries by agriculturists and naturalists over the past two centuries all testify to the extent and renown of these grasslands, and their maintenance by vast sheep flocks. Defoe (1724) wrote of the "fine carpet ground" of the Dorset Downs, "soft as velvet, and the herbage sweet as garden herbs". Praises of chalk downland by Young (e.g. 1769) and Cobbett (1830) sprinkle many a Tour and Rural Ride. Hudson (1900) describes at length the "living garment" of the Sussex Downs, "composed of small grasses and clovers mixed with a great variety of creeping herbs, some exceedingly small".

Small wonder that chalk grassland was such a draw to the early plant ecologists. Indeed, its study on the continent of Europe can be traced back to 1827, when Lestiboudois published his survey of the flora of northern France and Belgium (Stott, 1970). Preliminary field reconnaissance work analagous to Tansley's surveys in the 1920s of the downs of Sussex and Hampshire had been conducted in France in the 1870s and 1880s. Moreover, from a very early stage this information was synthesised into the phytosociological schemes which were evolving at the time (Section 5.2.2), so that by the turn of the century there was a substantial accumulation of organised knowledge of the chalk grasslands of northern France and Belgium. By the early 1920s, Allorge (1921–1922) was able, in his classic account of the grasslands of the *Vexin Français*, to put forward a considerably more sophisticated treatise on European chalk grasslands than the contemporary studies by Tansley and Adamson (1925, 1926) which Tansley (1939) subsequently drew on for his book.

The natural tendency, of course, is for secondary succession to transform these grasslands into scrub and woodland, though the process may be impeded on exposed sea cliffs, in unstable situations such as landslips (Chapter 4), and in very dry localities like Breckland, or more locally on steep and strongly insolated slopes. Occasionally, areas of chalk grassland are encountered which appear to resist invasion for reasons which are not obvious (see Chapter 7). But there is no question that the overriding

Table 5.1
Life-form spectrum of characteristic plants of English chalk grassland:
classification of 182 species recorded in over 1000 1 m² plots [a]

Category	Examples	Number	Per cent
1. Chamaephytes	*Helianthemum chamaecistus* *Thymus drucei*	14	7·6
2. Hemicryptophytes			
(a) Protohemicryptophytes	Many grasses *Asperula cynanchica* *Galium verum*	33	18·1
(b) Semi-rosette hemicryptophytes	*Carex flacca* *C. caryophyllea*	58	31·8
(c) Rosette hemicryptophytes	*Hieracium pilosella* *Cirsium acaulon* *Leontodon hispidus*	22	12·1
(d) Unclassified hemicryptophytes	*Centaurea nigra*	10	5·6
3. Geophytes	Orchids *Filipendula vulgaris* *Cirsium tuberosum*	30	16·5
4. Therophytes	*Linum catharticum* *Euphrasia nemorosa* *Rhinanthus minor*	15	8·2
TOTAL		182	

[a] From Wells (1973).

environmental factor maintaining chalk grassland is the grazing animal, or its equivalents the mower, forage harvester and trampling foot, which effectively top the plants, and favour those whose perennating buds are at the soil surface (hemicryptophytes), or just above (chamaephytes) or below it (geophytes) (Raunkiaer, 1934; see illustrations in Shimwell, 1971a, Fig. 21; Goldsmith and Harrison, 1976, Fig. 3.2). It is the spectrum of these life-forms which gives the turf its distinct appearance (*physiognomy*), and the composition of grazed chalk grassland in this sense is shown in Table 5.1.

There is more to being a successful chalk grassland plant than simply tillering freely under close grazing, however, for even the closest-knit turf contains a proportion of annuals (therophytes) and biennials, as well as shorter-lived perennials, which need to reproduce regularly from seed (see Chapter 6). This may be effected either by flowering and ripening seeds on extremely short stems, or by producing these reproductive structures in flushes so that at least a few escape grazing (see Hudson, 1910, pp. 6–7). Even then, the ingestion of seed heads may not necessarily mean their demise provided they are mature. Seeds of *Plantago lanceolata*, for example, can survive passage through the digestive tract of cattle, for their seedlings have been found in cattle dung (Ridley, 1930), and King (1977c) reports "apparently viable" achenes of *Poterium sanguisorba* in rabbit droppings.

Any tendency to dwarfing may be augmented by extremely close cropping by rabbits, trampling, drought and exposure to strong winds, and the phenomenon is often especially notable on broad footpaths and near the sea. The appearance of the dwarfed plants in flower can be spectacular indeed, for however diminutive their stature (Table 5.2) the actual flowers are usually full-sized (Fig. 5.1). Lousley (1950) tells of the Swedish botanist T. Wulff who published a detailed account of the dwarf plants of Tennyson Down (Isle of Wight), proposing several of them as distinct species, new to science. Another celebrated howler was the "new" gentian, "*Gentianella collina*", which proved to be no more than an extremely stunted clustered bellflower (*Campanula glomerata*), but in a hot, dry summer how easy it is to be thus misled. Most of these dwarfed plants are probably only phenotypic variants, though some may be ecotypes: very little work seems to have been done to confirm this. One notable exception is the tiny dwarf centaury (*Centaurium capitatum*), a mere 3–4 cm tall, and easily overlooked among stunted individuals of the maritime variety of common centaury (*C. erythraea* var. *subcapitatum*). The two species are distinguished by the attachment of the stamens, which in *C. erythraea* are inserted at the top of the corolla tube, but which in *C. capitatum* arise from the base.

Sheep and rabbits are by far the most important grazing animals in the "short chalk grassland" ecosystem. Other animals—including man—can play a part in maintaining the downland turf, though their effects, like those of the rabbit, often border on the destructive (see Chapter 10). Locally, more bizarre agents keep the turf in check, ranging from the increasingly numerous protagonists of grass-skiing (Fig. 10.1) to the wallabies (*Macropus ruficollis*) of the Bedfordshire Chalk escarpment at Whipsnade Zoo.

Table 5.2

Heights of dwarfed chalk grassland plants. These plants, recorded by J. E. Lousley in 1946 from the grassland above Beachy Head, Sussex, and from Freshwater Down, Isle of Wight, were all less than 5 cm high.[a]

Species	Usual stature (cm)
From Beachy Head	
Euphorbia exigua	15
Pimpinella saxifraga	30–45
Carduus tenuiflorus	45–90
Centaurea nemoralis	45–60
Scabiosa columbaria	30–45
Betonica officinalis	45
Phyteuma tenerum	30–45
Gentianella amarella	15–22
From Freshwater Down	
Arabis hirsuta	30
Daucus carota	30–100
Blackstonia perfoliata	15–45

[a] From Lousley (1950).

Enough has been said for the time being to establish the key factors in the maintenance of chalk grassland, which are, of course, biotic. We consider the effects of sheep, rabbits, cattle and other animals in detail in Chapter 6, and return in later chapters to those aspects of management which relate to the conflicting interests of optimal agricultural utilisation on the one hand, and biological conservation on the other. In this chapter, the intention is to review the various types of chalk grassland, the species of which they are composed, and the effects of climate and soil upon their distribution.

Fig. 5.1. Examples of plants with markedly foreshortened inflorescences from closely grazed chalk grassland. (a) *Campanula rotundifolia*, (b) *Succisa pratensis* and (c) *Scabiosa columbaria* from sheep-grazed turf at Coombe Hill, Buckinghamshire; (d) *Hieracium pilosella* and (e) *Knautia arvensis* from rabbit-grazed grassland at Watlington Hill, Oxfordshire.

5.2 Types of Chalk Grassland

5.2.1 Descriptive surveys

The recognition and description of the various types of chalk grassland has long interested ecologists on both sides of the English Channel. In England, most of our knowledge stems from the descriptive field work of the kind traditionally associated with A. G. Tansley. Indeed, examples are repeatedly encountered of the way we still rely on the early work of Tansley and his colleagues, despite some of the shortcomings of their methods: the odd choice of localities which was neither random nor deliberately representative of all degrees of slope and aspect (see Perring, 1958); the variable size of the sampling unit, usually not defined at all; and the differences in the degree of precision with which the plants at each site were recorded.

Yet these were, after all, preliminary surveys. To their credit, these early plant ecologists were aware of the limitations of their findings, and though both Tansley and Adamson themselves (1925, 1926), and later Hope-Simpson (1940a, 1941b), found errors in identification (for example in the fine-leaved fescues), underestimates of the frequency of certain inconspicuous species (such as *Linum catharticum*), and inconsistencies in assigning species to subjective classes of abundance, the overall picture was undoubtedly representative of those grasslands minimally influenced by any other major biotic factors than sheep and rabbit grazing, at least in the South Downs. Indeed, provided this geographical bias is borne in mind, much of the fundamental information on chalk grassland reported in Tansley's "British Islands" is still perfectly sound.

Unfortunately, with the notable exception of Watt's well-known work with permanent quadrats in Breckland, virtually nothing was done to consolidate and expand these early descriptive studies of chalk grassland for almost 20 years, when, in 1954, Margaret Cornish published her valuable review of the grasslands of the North Downs of Kent. Soon afterwards, Thomas et al., (1957) surveyed the Pewsey escarpment of north Wiltshire, and Perring (1958, 1959, 1960) began his careful comparative studies of the chalk grasslands of the Cambridgeshire scarp, the Yorkshire Wolds and the Dorset Downs, as well as localities in France and Germany. From about this time onwards, we have the Nature Conservancy, and its successors the Institute of Terrestrial Ecology (ITE) and the Nature Conservancy Council (NCC), to thank for extending the range of the chalk grassland survey, particularly through the work of A. S. Thomas and T. C. E. Wells. A rich source of new information has recently come to light from the vast and hitherto unrecorded regions of the Wiltshire chalklands under Ministry of Defence ownership to which, happily, officers of ITE and NCC now have access (see Wells, 1975; Wells *et al.*, 1976; Ratcliffe, 1977).

Yet, despite the publication of the "Nature Conservation Review" (Ratcliffe, 1977) with its wealth of invaluable data, there is still no authoritative up-to-date classification of the grasslands of the English Chalk, although this deficiency should be remedied when the National Vegetation Survey begins to publish its findings. Here, in the fervent hope that it will not clash too severely with official schemes to come, a tentative classification is offered of the main types of chalk grassland (Table 5.3). It is, of course, a matter of opinion whether this is an oversimplified—even naïve—list (there are, for example, certainly more than two categories of fine-leaved fescue grasslands), or whether some types noted here are mere variants (facies) of the major units (see Section 5.5.1). Some of these grasslands are illustrated in the photographs, but it is not easy to do justice photographically to many of the distinguishing features which are so manifestly obvious in the field.

Festuca ovina/F. rubra (Type 1) grassland (Figs 5.2 and 6.5a) is by far the most extensive type and is regarded as "typical" chalk grassland in the following pages. It will be recalled (Chapter 4) that *F. ovina* is an early coloniser of bare chalk, and as a rule this species tends to predominate in swards (Type 1A) of relatively warm, dry situations such as banks facing between south-east and south-west, or on other aspects where the soil is shallow and has been disturbed at some time. On the other hand, on deeper soils or on gentle slopes with a northerly component, grasslands in which *F. rubra* is the dominant species occur (Type 1B). When grazing is withheld, this type develops into a hummocky pasture of striking appearance in which only *Helictotrichon pubescens*, *H. pratense* and *Koeleria cristata* are normally able to compete with the dominant *F. rubra* (Fig. 6.17). Under continuous grazing, however, both fescues form the bulk of the turf, although it needs an experienced eye to distinguish them with certainty. Though some patches may be floristically rather poor, others contain a very large number indeed of accompanying species, most of which are mentioned either in this chapter or the next. Some of these may indicate pastures of great age: for example, in those parts of Salisbury Plain surveyed by Wells *et al.* (1976), the following species appear to be restricted to swards probably undisturbed for at least 130 years: *Helictotrichon pratense*, *Carex caryophyllea*, *Asperula cynanchica*, *Filipendula vulgaris*, *Helianthemum chamaecistus*, *Pimpinella saxifraga* and *Polygala vulgaris*. As generally good indicators of ancient grassland, F. Rose (personal communication) adds *Hippocrepis comosa*, *Astragalus danicus*, *Polygala calcarea*, *Thesium humifusum*, *Hypochoeris maculata*, *Senecio integrifolius*, *Orchis ustulata* and *Ophrys sphegodes*.

Table 5.3

A tentative classification of the main types of chalk grassland

Predominant species or community type	Comments

1. Fine-leaved fescues
  1a. *Festuca ovina*
  1b. *Festuca rubra*
	May be very rich in associated species, but sometimes species-poor. Can withstand heavy grazing. Typically associated with Icknield-type soils.

2. *Bromus erectus*	"Brometum."
3. *Brachypodium pinnatum*	"Brachypodietum."

4. *Helictotrichon* grasslands
  4a. *H. pubescens* and *H. pratense* with *Arrhenatherum elatius*
  4b. *Helictotrichon* spp. *with Deschampsia caespitosa* and several mosses
  4c. *H. pratense* and *Koeleria cristata* with *Phleum phleoides* and other Breckland species (Grassland B of Watt, 1940).
	Status uncertain.
	
	Found on steep north-facing slopes receiving little direct insolation, as on the South Downs scarp.
	Confined to Breckland, where it is associated with soils referable to the Methwold and Worlington series.

5. Mixed Gramineae: no single species predominant	Status uncertain. Perhaps merely a variant of Type 1.
6. Maritime chalk grassland	Essentially Type 1 containing maritime/paramaritime species.
7. *Carex humilis*	A distinct, though anomalous feature of the western chalklands.
8. *Arrhenatherum elatius*	"Arrhenatheretum." Characteristic of formerly cultivated, relatively deep soils (e.g. Coombe series), and usually associated with a lack of grazing. A common road-verge community.

9. Herb-rich communities
  9a. Wood-edge communities
 	
  9b. Ex-arable swards on soils of intermediate fertility
	Brachypodium sylvaticum a common component.
	Grasses generally absent. Often heavily grazed. Soils less fertile than in Type 8, but more so than in Type 10.

Table 5.3 continued

Predominant species or community type	Comments
10. Sparse stands with much bare ground; lichens a notable feature at certain times of the year. 10a. Breckland Grassland A of Watt (1940) 10b. Lichen-rich communities of the Isle of Wight, Porton, etc.	Confined to extremely free-draining chalky areas among the more extensive tracts of Type 4c (Grassland B). *Bromus erectus* just able to maintain itself in the absence of grazing. Loose flints often strewn about the surface. These stands appear to be associated with former exhaustive cropping, and probably also rabbit warrening.
11. Chalk heath 11a. Grazed chalk heath 11b. Tall chalk heath (ungrazed)	Communities containing an intimate mixture of calcicole and calcifuge species, occurring where thin loessial deposits overlie chalk.
12. *Sedum–Myosotis*–acrocarpous mosses, with *Iberis amara* and *Crepis capillaris*	A very distinct community of the mid-Chilterns associated with current extreme rabbit pressure. Ragwort may or may not be present.

Fig. 5.2. Extremely species-rich *Festuca ovina* turf at Lodge Hill, Buckinghamshire. The lens cap in this and the following photographs is 5 cm across. (C.J.S.)

Bromus erectus grassland (Type 2), though not as widespread as the first, and showing a curious patchiness of distribution in some regions, is nevertheless very common (Figs 1.21 and 5.3). Tall brome is accompanied by essentially the same species as the fescues (which themselves still form a substantial component of the sward) but this grass too, except on very dry sites, accumulates a thick and smothering litter as soon as grazing is withheld, whereupon floristic diversity is rapidly lost (see Chapter 6). The same applies to *Brachypodium pinnatum* (Type 3), a strongly invasive species confined on the Chalk mainly to the drier south-east, as in Kent and

Fig. 5.3. Moderately (and deceptively) species-rich *Bromus erectus* grassland at Buttlers Hangings, Buckinghamshire. The conspicuous flower head is a carline thistle (*Carlina vulgaris*). (C.J.S.)

the eastern South Downs (though occurring commonly on oolitic soils in the Cotswolds). Brief mention is made of Type 4b in Section 5.5.1. More work is needed to elucidate the factors responsible for the types of grassland in which *Helictotrichon pubescens* and *H. pratense* play a significant role (particularly 4a), and for the "mixed Gramineae" (Type 5).

Maritime (Type 6) and *Carex humilis* (Type 7) communities are discussed under the heading of climatic influences later in this chapter, as is this particular aspect of the Breckland grasslands (4c, 10a). Types 8, 9b and 10b (Figs 6.16, 5.4 and 5.5) are all characteristic of former disturbance and, at least in 10b, of exhaustive cropping and possibly also rabbit warrening. Strictly these are biotic factors, but their effects are clearly edaphic, and it is under this heading that they are given further attention. It might be argued that chalk heath (Type 11, Fig. 5.6) hardly qualifies as a type of chalk grassland, and should be regarded as a class of vegetation in its own right,

Fig. 5.4. This open, herb-dominated sward at Watlington Hill, Oxfordshire, strewn with small flints (as well as some Tertiary pebbles), marks the site of an old field, ploughed and abandoned in the 1930s. There are numerous rosettes of *Cirsium acaulon*, and a small patch of lichen can be seen near the lens cap. (C.J.S.)

but it is convenient to regard it thus here, and it is discussed at some length under the heading of edaphic influences, as well as in later chapters. Further reference is made to the wood-edge community (Type 9a) in Chapter 7, and to rabbit-ground (Type 12) in Chapter 6. Small-scale variations in botanical composition are described in Section 5.5.

There is nothing particularly sophisticated about all this. Any of these types of chalk grassland can be recognised by a competent field botanist from the physiognomically dominant species which give the stand its characteristic appearance, and from certain additional clues provided by such features as topography, soil type and past or present land-use. It is a method of successive approximation which conveniently pigeonholes an immense array of data into reasonably realistic compartments (see Poore, 1962). Few ecologists would dispute the value of such an exercise, at least for a preliminary reconnaissance, though the extent to which any one

Fig. 5.5. A striking example of lichen-rich "grassland" on a remote area of Salisbury Plain. The scale here is a felt-tip pen. (P. D. Goriup, Nature Conservancy Council)

community can be regarded as a discrete unit, or any two as essentially identical to each other, remain controversial topics in ecology.

For general surveys of the kind considered so far, the only feasible alternative to physiognomic classification for communities as species-rich as chalk grassland is phytosociology, but this has been little used in Britain, least of all in chalkland ecology. The subject is briefly touched on in the next section. For more local and detailed work, techniques such as association,

Fig. 5.6. Mown chalk-heath sward at Lullington, Sussex, an intimate mixture of calcicoles (here mainly *Poterium sanguisorba, Filipendula vulgaris* and *Helictotrichon pratense*) and calcifuges (*Erica cinerea*). (P. J. Grubb)

pattern and principal component analysis (see Section 5.5) may have a place, though this is debatable. As will be apparent from the bulk of this chapter, most of what we know about chalk grassland is the result of a combination of semi-quantitative descriptive surveys with *ad hoc* and often rather piecemeal experimental work, well laced with that indispensible ingredient, "ecological intuition"!

5.2.2 Phytosociology

In a phytosociological survey, a series of lists (*relevés*) is drawn up from areas of vegetation deliberately selected for homogeneity. Certain combinations of plants soon become apparent, and reorganisation of the raw data leads to the recognition of associations (used here in a special sense), together with their constant and exclusive species. Associations, which may be subdivided into subassociations and variants, are named by adding -etum to single or paired generic names of their key species. In turn, the associations are grouped into progressively higher taxonomic units: alliances (ending in -ion), orders (-etalia) and classes (-etea). The whole procedure relies heavily on personal experience and judgement, with, to the uninitiated, a touch of mystique thrown in as well. Mainly because of this strongly subjective element, hard phytosociology never really caught on in Britain, and indeed has long been regarded here almost with suspicion (see Poore's series of papers, e.g. 1956). Attitudes seem to have mellowed in recent years, however, partly because of a willingness among British ecologists to acquaint themselves more thoroughly with the techniques, but particularly as allegedly more objective alternatives fail utterly to progress beyond the tail-chasing stage, and to provide anything approaching such a useful pigeonholing system. Shimwell (1971a), who has pioneered the application of phytosociology to the study of British limestone grasslands, provides a useful guide to the intricacies of the subject.

It is not entirely true to say that the English tradition has shunned phytosociology altogether. Tansley and Adamson (e.g. 1926) made selective use of the methods of their continental contemporaries, and some of the terminology filtered through (though not always used strictly correctly—see below). Watt (1940) refers to the grassland associations Xerobrometum and Mesobrometum in Breckland, and the names Festucetum (ovinae, rubrae), Brometum (erecti), Brachypodietum (pinnati), Arrhenatheretum (elatii) and Caricetum (humuli)—the last sometimes illogically called Caricetosum —have been applied to communities dominated by these plants (Table 5.3). In woodland ecology, names such as Fagetum calcicolum, Fagetum rubosum and Fagetum ericetosum are used to describe different kinds of beechwoods (see Chapter 8).

Actually, Tansley and Adamson's data were used on the continent to derive the phytosociological association Xerobrometum Britannicum by Braun-Blanquet and Moor (1938), but this nomenclature is now regarded as inappropriate. Shimwell (1968, 1971b,c, 1973) suggests that the grasslands of the English Chalk are not analagous to the dry calcareous grasslands of Continental Europe (Xerobromion), "since they lack the true character species of such vegetation and contain many mesophilous species indicative of a damper climatic situation"—see Section 5.3.2. Chalk grasslands are referred instead to the Mesobromion. On the basis of a relatively small sample, Shimwell places "typical" chalk grassland in the association Cirsio-Brometum Shimwell 1968, of which the characteristic species are *Cirsium acaulon, Bromus erectus, Brachypodium pinnatum* and *Asperula cynanchica*. Other associations which have been recognised include the rather ambiguously named Helictotricho-Caricetum flaccae and the Helianthemo-Koelerietum. Most of Shimwell's work has, however, centred on the older limestones, which differ from the Chalk in being structurally harder and subjected to a considerably cooler and wetter climate. More work is needed and it is to be hoped that results of surveys of the kind carried out by van der Meulen and Wiegers (1972) on chalk grassland in southern England may be made more accessible, in all senses, to British ecologists. Phytosociological data are being assembled by the National Vegetation Survey.

5.3 Climatic Influences

5.3.1 Typical chalk grassland

It is an axiom of biogeography that climate is the prime factor in the geographical distribution of animals and plants. However suitable in other respects a potential habitat may be, no species will establish itself with any permanence—at least not without human help—if it is outside its normal climatic range. Perusal of any large-scale treatise on plant or animal geography such as that by Good (1974) reveals these major zones and patterns of distribution and their relationship to climate. But note that while adverse climate may ensure the absence of a species, a suitable climate does not necessarily guarantee its presence, even where all other factors are in its favour as well: this is always a matter of chance, though of course the odds are very good that the species will occur in such circumstances.

It is this liklihood that a particular species will turn up which provides the most fundamental method of measuring its occurrence, for this is, in ecological terminology, its *frequency*. In any one survey, a species which is encountered every time a sample is taken is said to have a frequency of 100%. The less widely a species is distributed, the smaller the chance of scoring it, and so the lower is its frequency. (Clearly the result will depend on the intensity of the survey and the size of the sampling unit, but that need not concern us here.) When the results of several surveys are pooled, it is possible to arrive at an average frequency for each species, and this is expressed on a five-point scale as its *constancy*, one of the terms cribbed from phytosociology, as observed in the previous section. The constancy scale is shown in Table 5.4, but note that the one used by Tansley and his colleagues differs very slightly from the continental version of Braun-Blanquet (1932).

Table 5.4

Scales of constancy

Constancy	Mean frequency (%)	
	Braun-Blanquet (1932)	Tansley and Adamson (1926)
I	<21	<20
II	21–40	20–39
III	41–60	40–59
IV	61–80	60–79
V	>80	>79

Most of the plant species of highest constancy in chalk grassland are well within their geographical limits in Britain, and tend to occur throughout the neutral to calcareous grasslands of the lowlands as a whole. Duffey *et al.* (1974) recognise 23 herbaceous species as attaining or approaching Class V constancy throughout the lowland calcareous grasslands, and these are shown in Table 5.5. As an example of a truly ubiquitous species, the distribution of *Lotus corniculatus* in Britain is shown in Fig. 5.7a.

5.3.2 Continental and oceanic influences

Other species, however, are not distributed so widely and uniformly, and occur patchily, relatively abundant in some places, yet virtually or entirely absent in others. This pattern may, of course, be attributable to contrasts in soil type, past land-use, grazing management, and so on, especially if the boundaries are clear-cut and spatial changes in botanical composition are sharp. But a gradual and progressive increase or decline in the frequency of occurrence of a species is normally explained by a parallel (though not always precisely understood) change in some aspect of climate. Again, examples can be readily seen on the 10 km x 10 km grid, and sometimes even on a smaller scale, as in Good's (1948) account of the flora of Dorset (see also Chapter 4).

Table 5.5

Species of high constancy throughout lowland calcareous grassland [a]

Class V constants in all areas	*Briza media* *Festuca ovina* *F. rubra*
	Carex flacca
	Lotus corniculatus *Plantago lanceolata* *Poterium sanguisorba* *Thymus drucei*
Species approaching Class V constancy	*Agrostis stolonifera* *Helictotrichon pratense* *H. pubescens* *Koeleria cristata* *Trisetum flavescens*
	Carex caryophyllea
	Campanula rotundifolia *Helianthemum chamaecistus* *Leontodon hispidus* *Linum catharticum* *Orchis mascula* *Polygala vulgaris* *Ranunculus bulbosus* *Scabiosa columbaria* *Veronica chamaedrys*

[a] From Duffey *et al.* (1974).

We saw in Chapter 2 that the main area of the Chalk coincides, quite fortuitously, with the drier and warmer regions of eastern and southern England. Not surprisingly, many species of strictly northern distribution are relatively poorly represented on the Chalk, if at all, although the Wolds extend far enough north to lose some of the species of more southerly distribution. Writing about the chalkland plants of the Yorkshire Wolds,

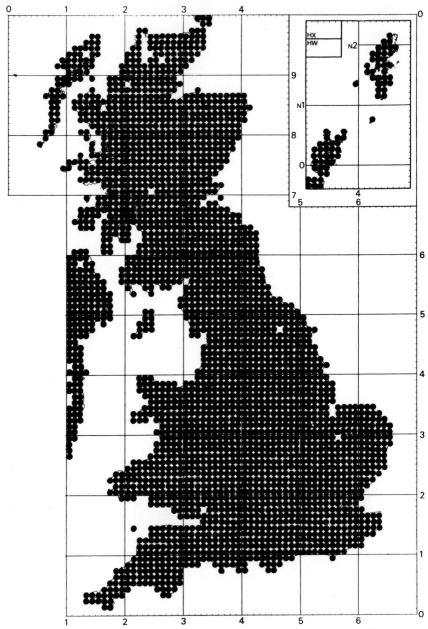

Fig. 5.7a. Distribution of *Lotus corniculatus* in Britain. Each spot indicates an occurrence within a 10 km × 10 km square of the National Grid. Reproduced by permission of the Botanical Society of the British Isles and Thomas Nelson and Sons Ltd, taken from their "Atlas of the British Flora", and revised by the Biological Records Centre, Monks Wood Experimental Station, Abbots Ripton, Huntingdon.

Lousley (1950) comments that "in general, Yorkshire chalk lacks many of the attractive southern plants and has very few additional northerners by way of compensation". It does have some distinctive features, however: the field scabious (*Knautia arvensis*) occurs there as a component of closely grazed turf, as do the ladies' smock (*Cardamine pratensis*), common catsear (*Hypochoeris radicata*) and earthnut (*Conopodium majus*), species which, further south, are normally found on neutral to acid, rather than calcareous soils (J. F. Hope-Simpson, personal communication). Further examples from the Lincolnshire Wolds are noted in Section 5.4.2.5.

A casual flick through the pages of Perring and Walters (1976) shows that there are numerous species with a southerly type of distribution corresponding in the main to Matthews' (1955) Continental and Southern Continental elements of the British flora, whose main centres of distribution lie across the warmer, drier regions of continental Europe. We have already encountered plants of this type in the ruderal flora (Chapter 4) in the *Reseda* and *Verbascum* species, of which *R. lutea* and *V. nigrum* are particularly clear examples.

A good example from chalk grassland is the stemless thistle, *Cirsium acaulon*, whose distribution in Britain, shown in Fig. 5.7b, has been closely studied by Pigott (1968, 1970). In fact, this attractive and familiar downland plant, which possesses the remarkable ability to position itself precisely where hands, knees or bottoms are placed upon the turf, just reaches the Wolds and other calcareous formations of Yorkshire (and Derbyshire), but in these northern stations the species shows a marked restriction to SSW.-facing slopes, in contrast to its indifference to aspect in south-east England and northern France. Pigott writes:

> The geographical shift in response to aspect is correlated with a decrease both in production of fertile fruit and frequency of occurrence of established seedlings. In France and south-east England, *C. acaulon* flowers from late June to September and every year a high proportion of flowers produce fertile fruits, but in Yorkshire and Derbyshire at the northern limit of its distribution, flowering usually begins in August and significant quantities of fertile fruits are only produced in those years when August and early September are exceptionally warm and dry. Thus, in the 15-year period from 1952 to 1967 large amounts of fertile fruit were formed at Tideswell in Derbyshire in 1955, 1959 and 1961 and only following these years were occasional seedlings found at this site.

The key factor is the temperature regime of the flowering heads (capitula), for "the proportion of flowers which form viable fruit is increased by treatments which cause the temperature of the heads to rise frequently above 20–25°C during the period when the embryos are growing . . .". Gradients of fruit set even across individual capitula have been detected at the very margin of the northern boundary of the species. Where persistent cloudiness veils direct insolation, few if any fruits are set, and this effect is

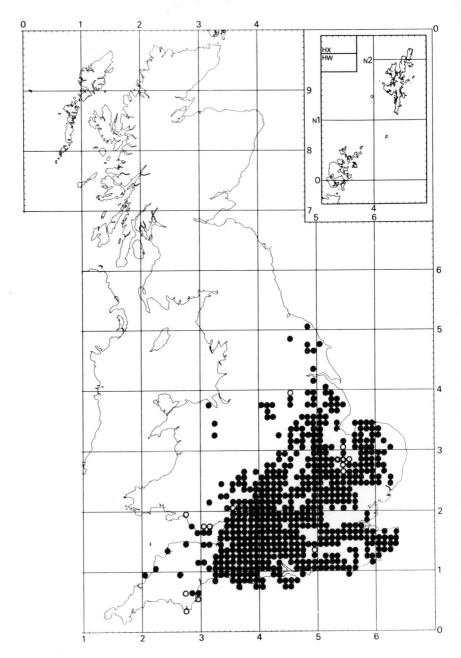

Fig. 5.7b. Distribution of *Cirsium acaulon* in Britain. Open circles represent records made before 1930 but not since. Interpretation and source as for Fig. 5.7a.

enhanced by wetting of the capitula by dew or rain. Not only does the resultant evaporative cooling maintain too low a temperature, but the dampness favours the invasion of the subinvolucral region of the stem by the fungus *Botrytis cinerea*, so that the heads fall off prematurely and decay in the turf. Moreover seedlings may die from damping-off disease, caused by fungi of the genus *Pythium*.

Other species show progressively more restricted patterns of distribution in chalk grassland, and so lower values of constancy. The more locally a plant is found, the more difficult it is to decide whether the cause is climatic or edaphic, or indeed simply a matter of chance. It may, of course, be an interaction between the two, and many of the most strictly continental species, clearly on the very edge of their range in Britain, are unable to maintain themselves anywhere else than in the relatively open swards of warm and freely drained sites, especially with a southerly aspect, which in most localities only the Chalk can provide although, for some, sandy or gravelly strata may do as well provided they are not liable to waterlogging. The most extreme example of this latter situation is Breckland.

With its hot, dry summers (the annual rainfall is only about 530 mm) and cold winters, the climate of Breckland (Chapter 2) resembles more closely than any other British locality the arid steppe country of south-eastern Europe. Its effect is enhanced by the admixture of sand with the underlying chalk, and not surprisingly it supports a flora unique in Britain. Some species

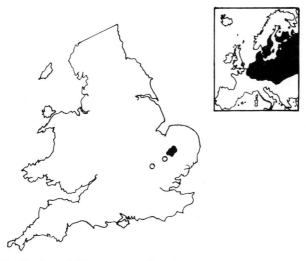

Fig. 5.8. Distribution of *Thymus serpyllum* in Britain, and (inset) Europe. Open circles indicate localities where the species was formerly recorded but is now extinct. From Pigott (1955).

are confined to the sandy facies and so are not relevant to this account, but those of the more chalky areas include *Carex ericetorum*, *Phleum phleoides*, *Medicago falcata* and its hybrid with *M. sativa* (lucerne), *M.* × *varia*, *Veronica spicata*, *Galium parisiense* ssp. *anglicum*, *Silene otites* and *Thymus serpyllum sensu stricto*. (*Thymus drucei* was often reported by its earlier name of *T. serpyllum sensu lato* in the older literature: the distribution of *T. serpyllum* "proper" is shown in Fig. 5.8.) The presence of these species distinguishes Watt's Grasslands A and B (Table 5.3) from otherwise similar grasslands of dry localities. A more detailed review of these Breckland grasslands can be found in Ratcliffe (1977).

Wells (1969) specifies ten flowering plant species which are completely restricted to the Chalk in Britain, and found mainly in eastern or south-eastern localities: a distribution which, he concludes, "strongly suggests that climatic factors are limiting their northern and western range in the British Isles". These are the wild candytuft (*Iberis amara*), ground pine (*Ajuga chamaepitys*), the beautiful round-headed rampion (*Phyteuma tenerum*), the rare umbellifers *Bunium bulbocastanum* and *Seseli libanotis*, and the orchids *Ophrys fuciflora*, *Orchis purpurea*, *O. simia* and *O. militaris* (the late spider, lady, monkey and military, respectively). The very rare clove-scented broomrape (*Orobanche caryophyllacea*), which parasitises *Galium mollugo* (itself of widespread distribution) almost qualifies through its virtual restriction to the extreme east of Kent but, amazingly, this also occurs in Argyll (T. C. E. Wells, personal communication). The lizard orchid (*Himantoglossum hircinum*) occurs on limestone and calcareous dunes as well as on chalk soils in southern England, but merits notice in this section on account of its remarkable increase in numbers during the 1930s which was thought to reflect some subtle change in climate over that period (Good, 1936).

An anomalous distribution is shown by the sedge *Carex humilis*. This is almost entirely restricted to the chalk downs of Dorset and southern Wiltshire (and a few other western locations) in Britain (Fig. 5.9), where it clothes the hillsides with its unmistakable patches of yellowish-green foliage recognisable from a kilometre or more away. Yet, this is a species regarded in Europe as of strongly continental distribution, and its absence from similar terrain in south-east England is difficult to explain. An interesting feature of the upper slopes of the western South Downs, undoubtedly reflecting the high winter-rainfall regime there (Section 2.3.2), is the moss *Rhacomitrium lanuginosum* (Hope-Simpson, 1941a). This is far more characteristic of mountain-top vegetation than of chalk downland, so perhaps Gilbert White's celebrated description of this lofty range was not so fanciful.

It has been customary to compare the flora of the easternmost localities with their counterparts on the opposite side of the English Channel (e.g.

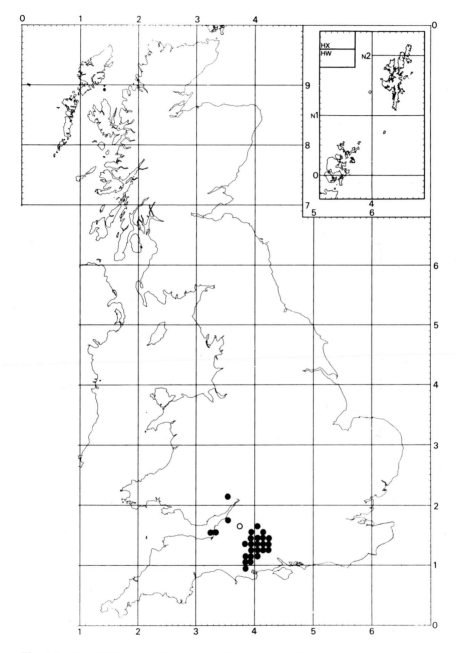

Fig. 5.9. The distribution of *Carex humilis* in Britain. Interpretation and source as for Figs 5.7a,b.

Good, 1928) and to speculate on the extent to which the formation of the Channel was responsible for preventing many more species of the European flora from entering Britain (a point also briefly noted in Chapter 1). Rose (1972) considers that the extremely fine seeds of the orchids and broomrapes could conceivably be blown across the Channel and so have arrived more recently than was previously supposed. On the other hand, he regards the windswept and largely loess- (*limon-*) covered plateau of the Pas de Calais as a far more formidable obstacle than the Channel to the spread of the more continental species into Britain, and he suggests that the Somme Valley, rather than the Channel Coast of northern France, forms a more realistic geographical boundary (see Fig. 5.11). For example, *Seseli libanotis, Teucrium chamaedrys* and several other chalk grassland rarities of south-east England, are not found in France north of the Somme area.

The more strongly oceanic species are, as we saw in Chapter 4 and will note again shortly, most strikingly seen along the chalk cliffs and landslips of the south coasts of the mainland and the Isle of Wight. Some of these plants extend well inland to give the western chalklands a distinct character of their own. A good example is *Iris foetidissima*, even though this does occur in the North Downs as well (Section 4.7.2). Wells (1975) mentions a preponderance of *Sieglingia decumbens* as especially typical of the oceanic chalk grasslands of the central Isle of Wight. This grass also occurs in some quantity on the southern Wiltshire Downs, where two species more typically associated with damp meadows are found: the green-winged orchid (*Orchis morio*) and the meadow saxifrage (*Saxifraga granulata*). Likewise, *Betonica officinalis, Serratula tinctoria* and *Solidago virgaurea* are common in the west, but very rare in, or absent from, chalk turf in the south-east (F. Rose, personal communication).

Perring (1958, 1959, 1960) compared data on plant distribution in four chalkland localities chosen to represent a transition in climate from continental to oceanic. In order to control other environmental variables (soil type, slope, aspect, land-use), he chose sites as similar as possible to each other in all these attributes, and as near in shape to the ideal hemispherical hill as was feasible. His localities were near Rouen in northern France, the Cambridgeshire Chalk scarp, the Yorkshire Wolds and the Dorset Downs. Species were recorded on the basis of percentage cover, and the results plotted, like his soil data (reviewed in Chapter 3), on concentric graphs as a two-dimensional ordination in which the axes were aspect and gradient of slope. Contrasts in geographical range of the species encountered were clearly demonstrated, and these were reflected on a local scale in their micro-distribution. Over the four main geographical areas as a whole, Perring recognised five groups of species ranging from the most continental, typified by the horseshoe vetch (*Hippocrepis comosa*) and

Table 5.6
Grouping of chalk grassland species according to their geographical distribution [a]

GROUP I

Carex humilis

Anemone pulsatilla
Astragalus danicus
Aster linosyris [b]
Bupleurum falcatum [b]
Carlina vulgaris
Centaurea scabiosa [c]
Chrysanthemum leucanthemum
Eryngium campestre [b]
Euphorbia esula ssp. *tristis* [d]
Gentianella amarella agg.
Globularia vulgaris [d]
Helianthemum appeninum [b]
H. canum [b]
Hippocrepis comosa
Hypericum perforatum
Leontodon taraxacoides
   (=*L. leysseri*)

Onobrychis viciifolia
Ononis repens ssp. *procurrens* [d]
Pastinaca sativa
Phalangium ramosum [d]
Phyteuma tenerum
Potentilla tabernaemontani
Senecio integrifolius
Seseli libanotis
Teucrium chamaedrys [b]
T. montanum [d]
Thesium humifusum
Veronica teucrium [d]
Viola hirta

Eurhynchium swartzii
Fissidens cristatus
Neckera complanata
N. crispa
Pleurochaete squarrosa [d]
Weissia microstoma

GROUP II

Carex caryophyllea

Asperula cynanchica
Cerastium arvense
Daucus carota
Euphrasia officinalis agg. (*E. pseudo-kerneri?*)
Filipendula vulgaris
Hieracium pilosella
Hypochoeris maculata
Linum catharticum
Medicago lupulina
Pimpinella saxifraga

Polygala calcarea
Scabiosa columbaria
Taraxacum laevigatum
Thymus pulegioides

Anomodon viticulosus
Barbula convoluta
Camptothecium lutescens
Campylium chrysophyllum
Ctenidium molluscum
Entodon orthocarpus
Hypnum cupressiforme var. *elatum*

GROUP III

Brachypodium pinnatum
Briza media
Bromus erectus
Carex flacca
Festuca ovina
Helictotrichon pratense

Koeleria cristata (=*K. gracilis*)
Phleum bertolonii
   (=*P. nodosum*)
Poa angustifolia
Sesleria caerulea
Campanula glomerata

Table 5.6 continued

Cirsium acaulon
Euphrasia officinalis agg. *(E. nem-*
orosa?)
Galium pumilum
Helianthemum chamaecistus

Lotus corniculatus
Poterium sangiusorba
Thymus drucei

Thuidium abietinum

GROUP IV [e]

Agrostis stolonifera
Cynosurus cristatus
Festuca rubra
Helictotrichon pubescens
Sieglingia decumbens
Trisetum flavescens

Bellis perennis
Betonica officinalis (=Stachys officinalis)
Campanula rotundifolia
Cirsium arvense
C. vulgare
Galium mollugo
G. verum
Leontodon hispidus

Plantago lanceolata
P. media
Polygala vulgaris
Prunella vulgaris
Ranunculus bulbosus
Senecio jacobaea
Succisa pratensis
Taraxacum officinale

Acrocladium cuspidatum
Brachythecium rutabulum
Dicranum scoparium
Pseudoscleropodium purum
Rhytidiadelphus triquetrus

GROUP V

Anthoxanthum odoratum
Arrhenatherum elatius
Dactylis glomerata
Deschampsia caespitosa
Festuca arundinacea
Holcus lanatus
Lolium perenne
Luzula campestris
Poa pratensis
P. trivialis

Achillea millefolium
Alchemilla vestita
Centaurea nigra

Cerastium holosteoides (=C. vulgatum)
Crepis capillaris
Hypochoeris radicata
Lathyrus pratensis
Leontodon autumnalis
Potentilla erecta
Primula veris
Ranunculus acris
Rumex acetosa
Trifolium pratense
T. repens
Veronica chamaedrys
V. officinalis
Viola riviniana

[a] From Perring (1960).
[b] Recorded in this survey only from the Chalk of northern France, but found in certain British localities.
[c] Possible ecotype of chalk grassland.
[d] Recorded in this survey only from the Chalk of northern France, and not found in Britain.
[e] *Crataegus monogyna* (hawthorn) was included in this group.

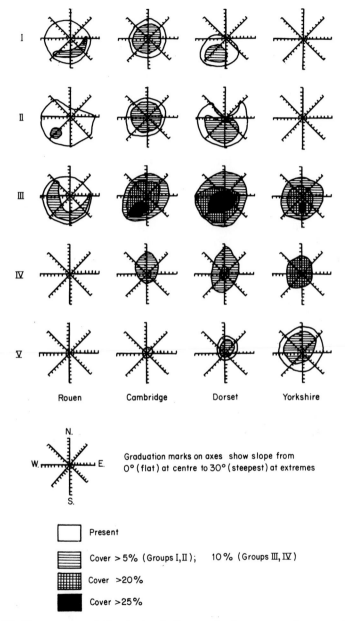

| | Rouen | Cambridge | Dorset | Yorkshire |

N.

W. ———————— E.

S.

Graduation marks on axes show slope from
0° (flat) at centre to 30° (steepest) at extremes

☐ Present

Cover >5% (Groups I,II);   10% (Groups III,IV)

Cover >20%

Cover >25%

Fig. 5.10. Occurrence of Perring's chalk grassland type species according to location, aspect and slope. I, *Hippocrepis comosa*; II, *Asperula cynanchica*; III, *Poterium sanguisorba*; IV, *Plantago lanceolata*; V, *Holcus lanatus*. See text for details.

including the more xerophytic element of the chalk grassland flora, to the most oceanic, of which Yorkshire fog (*Holcus lanatus*) was the type species. The distribution of these type species in Perring's four sites is shown in Fig. 5.10, while the allocation of species to the five groups is given in Table 5.6, though of course the choice of the former and allocation of the latter were done quite arbitrarily.

5.3.3 Maritime chalk grassland

As in the more open vegetation of the cliffs and landslips, proximity to the sea induces distinct floristic changes in chalk grassland, and it is perfectly obvious that these changes arise from the action of constant wind and salt spray. The turf is often as short as any golf green, growth being curtailed not only by grazing but by the osmotic effect of salt water deposited by the wind either on the plants themselves or on the soil surface, in which case the effect, strictly speaking, becomes an edaphic one. In the very last paragraph of his book, Tansley (1939) makes brief mention of the maritime chalk grassland of the Sussex cliff tops. He writes:

> On the flat top of Beachy Head (*c.* 150 m) the turf is very short and is quite a typical sample of chalk grassland without maritime or sub-maritime species. Farther west, where the cliffs are not so high (30–90 m) and may be supposed to receive more spray during strong onshore gales, the following appear in the grassland of the cliff tops: *Agropyron pungens, Armeria maritima, Carduus tenuiflorus, Erodium cicutarium, Glaucium flavum, Plantago coronopus* and *Daucus carota.*

Daucus was "locally very abundant" but in the main the maritime species did not form "a quantitatively important part of the vegetation". F. Rose has recorded a rich maritime chalk grassland flora along the cliff-tops and ledges of the East Kent coast (see also Section 4.7.1.1), which includes in addition to the commoner grasses and herbs several broomrape (*Orobanche*) species, the early spider orchid (*Ophrys sphegodes*), and a surprising abundance of the very local *Silene nutans* var. *smithiana*. No detailed work appears to have been done on these maritime chalk grasslands of the English coasts, however, and in this respect it is interesting to peruse the work of Liger (1956) on the flora of the coast of Normandy.

Liger surveyed the vegetation of the chalk cliffs of the coast of the Pays de Caux, extending all of 120 km from le Havre to Ault, north of the mouth of the River Bresle (Fig. 5.11). He made lists (*relevés*) of species found at each of the 82 sites (not all of which were confined to the Chalk) which he covered. Species were recorded mainly as simply present or absent, but occasionally with a cryptic indication of abundance ranging from 1 to 4, the meaning of which can be surmised, but is not defined. Details of the location, topography, geology, aspect and altitude of each *relevé* are provided, as well as rainfall and temperature data for more extensive regions. He distinguished eight topographic habitats (Fig. 5.12) ranging from the shingle bank at the foot of the cliffs to the edge of the plateau at the top and in addition described several landslips. The features of the cliff-face

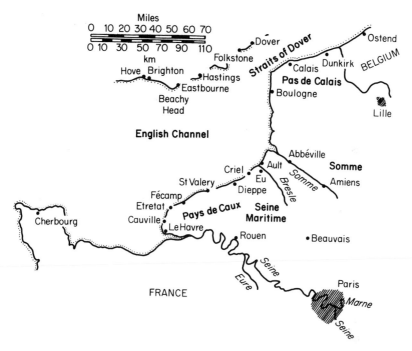

Fig. 5.11. The Normandy coast of France (1:2 000 000). Liger's survey referred to in the text extended from Le Havre to Ault.

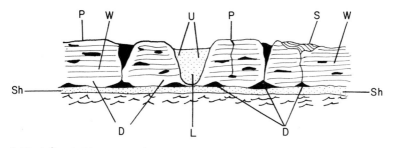

Fig. 5.12. Microhabitats of the chalk cliffs of the Pays de Caux recognised by Liger (1956). Sh, shingle bank; D, deposits at base of cliffs; W, cliff wall; L, lower slopes; U, upper slopes; P, summit plateau; S, debris of summit crests.

itself and the landslips (*les chaos*) were considered in Chapter 4; the shingle bank supported only the extreme halophyte *Atriplex hastata*. Elsewhere, the contrasts in relief were paralleled by botanical heterogeneities assumed to reflect differences in the distribution of salt, though no measurements were made, and we saw in Chapter 2 that patterns of vegetation are more closely related to evaporation than directly to salt deposition.

Nevertheless, Liger's carefully acquired and straightforwardly presented data (with no involved expeditions into phytosociology) are far more extensive than Tansley's on the subject, and give a most useful impression of plant distribution on maritime chalk cliffs. Moreover, a glance through the lists of species shows the close similarity between the floras of this part of the Normandy coast and its English counterpart. Indeed, most differences arise through the greater range of species on the English side as a result of the greater stability and more direct insolation of the English chalk cliffs (Rose, 1972). Out of the total of 146 species recorded by Liger, only 10 are not to be found in the British flora, and of these only "*Ononis procurrens*" (regarded by British botanists as merely a subspecies of *Ononis repens*) is of more than local and limited significance. Table 5.7 lists those species occurring in more than half of the 10–12 *relevés* sampled in each habitat, together with a few others which are clearly restricted to certain habitats only. It would be stretching the terms of reference of this book too much to go more deeply into these results, but it is interesting to note the differences between species in their tolerance of the littoral environment. Particularly clear is the replacement of the one subspecies of *Daucus carota* by the other, and in fact Liger (like Niel 70 years before him) found all shades of transition between the two (see Chapter 4).

The question arises: do the maritime species "need" this particular combination of environmental factors, or are they simply able to tolerate these conditions better than their "inland" counterparts, disappearing in sites away from the coast as a result of competition from the more vigorous inland species on their home ground? On the face of it, the first of these explanations might seem the more reasonable. Can we really believe that robust *Daucus* or *Glaucium* or *Smyrnium* are found at the coast purely for the negative reason that here they meet least competition? Surely they must derive some positive benefit from their coastal station? Cultivated beet responds to added salt. Could not these plants of the cliffs and links have a similar requirement, for salt as such, or Na, Mg, B or whatever else the sea throws at them? Yet, on the other hand, even the smallest differences in local climate are strongly cumulative, and the incidence of both frost and low atmospheric humidity, which increase sharply away from the sea, may be sufficient, if not to kill the maritime species outright, then to render them poor competitors, unable to fend for themselves for any length of time among the plants of the interior.

The fact is that we really have very little idea of causal factors even in distributions as clear-cut as these. Intuitively, and excepting perhaps

Table 5.7

Species of the Pays de Caux exceeding a mean frequency of 50% in at least one habitat, with a few additional species showing clear-cut habitat specificity [a,b]

	Cliff wall	Base deposits	Lower slopes	Upper slopes	Edge of plateau
(a) Species of general distribution along the coastal strip					
Achillea millefolium	8	70	50	66	90
Anthyllis vulneraria	58	80	58	25	50
Centaurea nigra	0	60	83	41	70
Cirsium vulgare	8	60	17	0	20
Dactylis glomerata	50	90	100	83	60
Festuca rubra	100	100	92	92	60
Lotus corniculatus	17	80	92	92	60
Plantago lanceolata	33	80	75	66	60
Senecio jacobaea	8	50	33	8	40
Sonchus oleraceus	33	50	41	17	30
(b) Species tending to increase from shore to cliff-top plateau					
Anthoxanthum odoratum	0	0	0	17	20
Brachypodium pinnatum	8	10	50	58	70
Daucus carota ssp. *carota*	0	0	8	25	40
Galium verum	0	10	25	25	60
Holcus lanatus	0	50	50	33	60
Ononis repens ssp. *procurrens*	0	20	50	41	60
Poa pratensis	0	10	33	25	70
Poterium sanguisorba	0	10	41	66	60
(c) Species tending to decrease from shore to cliff-top plateau					
Armeria maritima	25	10	25	25	0
Beta vulgaris ssp. *maritima*	17	0	0	0	0
Brassica oleracea	66	60	17	33	10
Crithmum maritimum	25	10	8	8	0
Daucus carota ssp. *gummifer*	58	80	75	41	20
Glaucium flavum	8	10	0	0	0
Tussilago farfara	17	50	17	8	0

[a] After Liger (1956).
[b] Numerals refer to the percentage of total *relevés* for each habitat in which each species was recorded. For example, *Brachypodium pinnatum* was recorded from 8% of the cliff *relevés* and 70% of those taken from the edge of the plateau.

Crithmum maritimum, both Tansley aqnd Liger plump for the argument that the maritime species are salt-tolerant, weak competitors, rather than salt-requiring plants. In fact, Goldsmith (1973a,b) offers some convincing objective evidence to support this suggestion from his careful studies in Anglesey. He concludes that

> the species characteristic of inland situations are susceptible to the saline environment (as spray on the foliage or the high osmotic potential of the soil solution or both). Their competitive ability is reduced in saline situations so permitting the growth of the maritime species. This hypothesis indicates that the characteristic sea-cliff species have no salt requirement but are more salt-tolerant than inland species.

Goldsmith found that the growth of even such apparently spray-tolerant species as *Festuca rubra* and *Armeria maritima* (see Table 5.7) was strongly depressed by salt water. He grew these two species in pure and mixed cultures using de Wit's (1960) well-established method, to test their respective competitive abilities in freshwater and saline conditions. These two species were chosen not only because they are typical of the habitat under consideration, but because, conveniently, they have very similar growth characteristics. In mixed stands, *Festuca* outgrew and dominated *Armeria* under freshwater conditions, but when salt water was substituted the situation was reversed and it was *Armeria* which gained the upper hand through its greater salt tolerance. But the crucial point was that in pure stands, both species produced under saltwater conditons only about one-third of the dry matter assimilated in fresh water. Of course, biotic factors can modify this relationship in the field. For example, the effects of rabbits (and even puffins) is to tip the balance even more sharply in *Armeria's* favour by keeping the *Festuca* in check (Gillham, 1956).

5.4 Edaphic Influences

5.4.1 Community boundaries

The reader might query the place of insertion of this subheading, for the discussion so far has certainly strayed more than once into edaphic territory, especially in the case of the maritime species. But in fact it is surprisingly difficult to neatly pigeonhole every environmental variable, and wherever these interact it is a purely arbitrary decision which heading to put them under for closer scrutiny (see Perring, 1958).

We saw that the effect of climate, by and large, is to induce gradual changes in species distribution, and so in botanical composition of communities, from place to place. Edaphic contrasts, on the other hand, typically result in essentially discontinuous variations, in which gradients are much sharper and transitions between communities less diffuse. This is particularly so in lowland Britain, where geological boundaries are sharp, although edaphic gradients may be more extended where boundaries are blurred, as on steep slopes subject to soil creep and similar phenomena (Chapters 1 and 3). An example of a distinct change in species composition across an edaphic boundary—in this case the transition from a brown earth derived from Clay-with-flints to a chalk rendzina—is illustrated in Fig. 2.7 of Kershaw (1973).

5.4.2 Calcicoles and calcifuges

5.4.2.1 Some definitions

We have already seen, both in this and the previous chapter, that many herbaceous (and other) species of chalkland vegetation are not at all confined to the Chalk, nor even necessarily to calcareous strata in general. Others, encompassing the much rarer plants mainly of the southern and eastern Chalk, are restricted to this formation simply by its fortuitous geographical position. There are many plant species of wide distribution, however, which are of great ecological interest, either because they are found only or mainly on chalk and limestone —the *calcicoles*—or because they are normally absent from these strata—the *calcifuges* (or silicicoles). Although it is possible to pick out the more strikingly calcicolous species on the 10 km square grid (for example *Asperula cynanchica*, Fig. 5.13), the very size of these units tends to mask geological and edaphic heterogeneities, and the 2 km × 2 km grid (tetrad) used in certain county floras is often of greater value (see, for example, Dony, 1967, 1976; Knipe and Maycock, in preparation). Nevertheless, even these are large squares in the patchwork quilt of lowland Britain, and many species which might be expected to, show no such clear-cut patterns of distribution on this scale. For these, much closer studies have to be resorted to, and we return to these methods in due course.

Although the term calcicole conjures up for most people the idea of some degree of restriction to calcareous strata, an early definition in the ecological literature (Salisbury, 1921) implies no such limitation. Thus, Salisbury defines calcicoles as plants which find "a suitable home on calcareous soils without necessarily implying any obligatory association with, or even preference for, such soils, apart from that imposed by climatic or biotic factors". Another definition, by de Silva (1934), regards calcicoles as "characteristically luxuriant on highly calcareous soils". It is Hope-Simpson (1938) who narrows down the field: in his well-known and oft-quoted words he defines calcicoles as "species affecting the more important types of calcareous soils and rare or absent from acid soils", with calcifuges defined

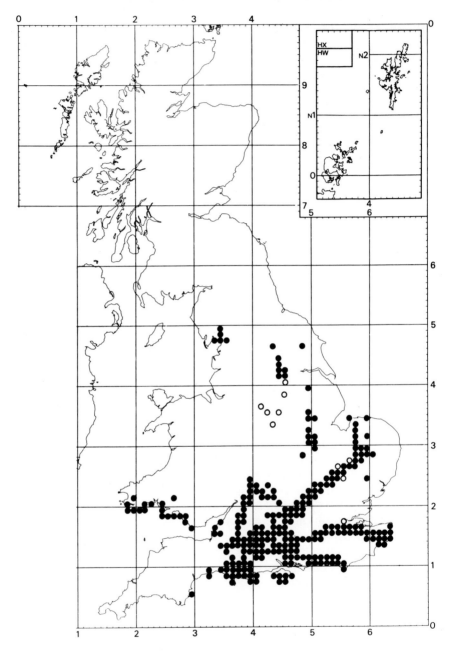

Fig. 5.13. The distribution of *Asperula cynanchica* in Britain. Interpretation and source as for Figs 5.7a,b.

as the reverse of this, though Clymo (1962) warns that the similarity of terms suggests a neat reciprocity of explanations which is not necessarily true. Moreover, plant species exhibit a whole range of responses between extreme calcicoly and extreme calcifugy (Grime and Hodgson, 1969).

Salisbury suggests that calcicoles fall into two groups: physical and chemical. Physical calcicoles favour the relative warmth and free drainage of chalk soils, and may be recognised by the wider edaphic tolerance they display closer to their main centres of distribution. He cites as examples the wall germander (*Teucrium chamaedrys*) as well as the woodland herb *Helleborus foetidus* and several shrubs which, though restricted to calcareous soils in Britain, all occur on siliceous soils in warmer climates. On the same basis, the rare pasque flower (*Anemone pulsatilla*) may be regarded as a physical calcicole (Wells and Barling, 1971). Chemical calcicoles, on the other hand, find either favour in the chemical attributes of calcareous soils or refuge from unfavourable effects of acid ones. In this category, there is an almost infinite number of possible factors and factor complexes responsible for the calcicolous habit.

Some 15 years after Salisbury's review was published, Hope-Simpson (1938) stumbled on a remarkable piece of grassland which, directly and indirectly, was to give a considerable push to the application of experimental techniques (e.g. Rorison, 1960a) to studies of the calcicolous habit. A small field (about three-quarters of a hectare in area), situated on a Lower Greensand soil normally acid in reaction, was found to contain all but 8 of the 45 species of constancy III or over recorded by Tansley and Adamson for chalk grassland (see Section 5.3.1). Enquiries established that the field, about 1 km from the escarpment of the North Downs at Abinger Hammer, Surrey, had been regularly dressed with chalk from a nearby quarry until some 50 years beforehand. In fact, the chalk could still be seen on the surface, intermixed with the sand, testifying to the notable lack of surface leaching in this locality; there was a remarkable similarity to certain Breckland soils. A unique opportunity was afforded, by comparison with adjacent downland and heathland swards, to investigate the extent to which the calcicoles in the field were favoured by physical versus chemical attributes of the calcareous sand. As it happened, the results were rather disappointing, but Hope-Simpson was able to conclude that chemical rather than physical effects accounted for the calcicolous habit in "the vast majority" of chalk grassland species.

Having thus distinguished physical and chemical aspects of calcicoly, these will be looked into in more detail directly. But first, a brief digression is necessary to draw attention to an important biotic aspect of the problem: competition. An early contribution to the subject was the classic study, reported by Tansley (1917), on two bedstraws of contrasting edaphic re-

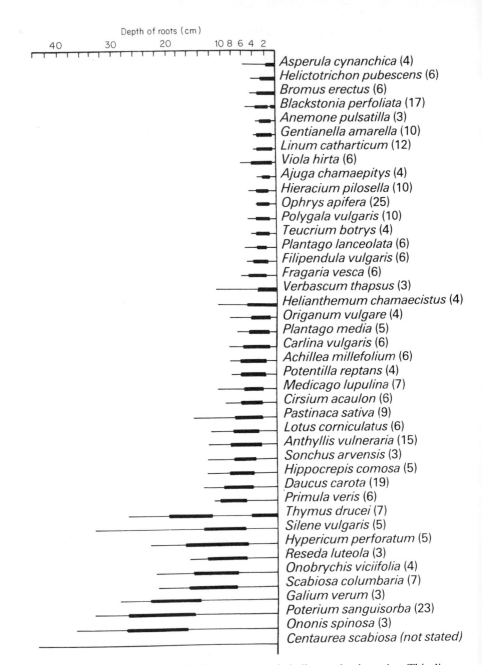

Depth of roots (cm)

Fig. 5.14. The rooting depth of a selection of chalk grassland species. Thin lines indicate the maximum depth of penetration attained, thick lines the average depth at which the maximum development of feeding roots occurred. Figures in parentheses indicate the number of plants of each species examined. From Anderson (1927).

quirements: *Galium pumilum* (referred to as both *G. sylvestre* and *G. asperum* by Tansley), strongly calcicolous (in Britain), and the calcifuge *G. saxatile*. This work, which was curtailed by the outbreak of the 1914–1918 War, was plagued by indifferent seed germination, and the results were by no means as conclusive as the frequent references to the paper would suggest. Nevertheless, important ideas were established on the effect of interspecific competition on edaphic tolerance. Thus, each species grew "best" on the soil most similar to that of its native habitat, and rather poorly on "foreign" soils. In fact, *G. pumilum* fared rather better on the acid soil than *G. saxatile* did on the calcareous soil, the latter species suffering from lime-induced chlorosis from a relatively early stage. But it was competition from the "native" species, rather than adverse soil conditions alone, which tipped the balance and led in the end to seedling death.

5.4.2.2 Water relations and the calcicolous habit

There is no doubt that certain calcicoles grow on chalk because in some way they are favoured by its water relations, but deciding exactly how this operates has led to more contradiction and confusion than probably any other branch of chalkland ecology. We have already established that the chalk bedrock holds a vast potential source of water and that its availability to plants depends on their powers of tapping it. Plants which are less able to reach this supply may nevertheless hold their own either by reducing transpiration and so conserving water, or by being able to draw in water, or redistribute it internally, at very low matric and osmotic potentials. Finally, though succumbing to drought by wilting, plants may possess good powers of recovery. These aspects will now be considered in turn.

Plants able to send roots deep down into the chalk will obviously be at an advantage during times when the topsoil dries out. Anderson's (1927) pioneer work shows the marked range in rooting depth among chalk grassland species (Fig. 5.14). Usually the deepest roots explore natural fissures in the chalk (see for example *Poterium sanguisorba*, Fig. 5.15a), and sometimes a plate of ramifying rootlets may develop in these fissures. Other investigators (e.g. Locket, 1946a; Wood and Nimmo, 1962; Litav and Orshan, 1963; Grubb *et al.*, 1969) have found similar patterns of root distribution and they, too, report the deeper roots of herbaceous and woody plants penetrating fissured and even solid chalk. Anderson went as far as to suggest that respiratory carbon dioxide from these roots might be an agent in increasing the volume of the fissures and cavities in the chalk. The idea is plausible, and implicit in the mechanisms of pedogenesis, but seems never to have been confirmed. The ability to bore through solid chalk is not specific to, nor even necessarily characteristic of calcicoles. Anderson found *Convolvulus arvensis*, and Locket the ubiquitous *Cirsium arvense*, capable of so doing. Grubb reports even gorse roots penetrating chalk (see Section 5.4.2.5), though this seemed to be exceptional. Conversely, Anderson found that roots of *Pastinaca sativa* tended to avoid contact with the solid chalk, in one case even by growing vertically upwards for 10 cm.

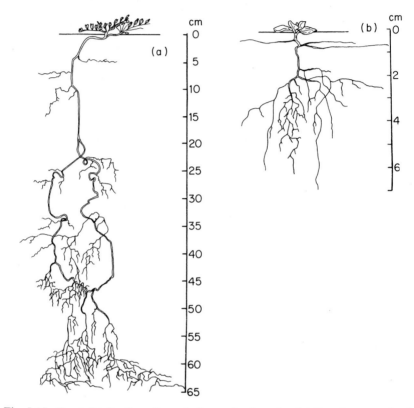

Fig. 5.15. Root distribution of two chalk grassland species, (a) *Poterium sanguisorba* and (b) *Blackstonia perfoliata*. Note the different scales. From Anderson (1927).

Observations such as these, especially on the rarer species, are infrequent for both ethical and technical reasons. Even when Anderson conducted her study, she was obliged to draw conclusions from samples which for some species were very small (see Fig. 5.14). The excavation of statistically viable samples of bee orchids, complete with roots, would hardly be advocated nowadays with a clear conscience, except perhaps from a site earmarked for ploughing or development and impossible or impractical to conserve. But in any case, from a purely technical viewpoint, it is extremely difficult to be sure that the whole root system has been won intact. Moreover, there is no way of telling purely from looking at them which roots are active and which are not.

The most reliable information on root activity comes from following the depletion of water from different depths in the soil, or from monitoring the uptake of radioactive tracers. One of the most striking demonstrations of the exploitation of water in solid chalk comes from Cohen and Sharabani's (1964) work on vines in Israel. They had been impressed with reports that before 1948 grapes were widely grown on the very shallow chalk soils of the Judaean Hills, without irrigation. In an investigation employing neutron tubes, they found that vine roots penetrated and withdrew water from chalk down to 120 cm. They make the important point that roots are able to penetrate the chalk only in spring when the rock is still wet and soft from the winter's rains; later in the season, when it is dry, little root development occurs.

Elston (1963) used a radioactive tracer technique to observe rooting depth, or more accurately the vertical distribution of root activity, in a chalk grassland sward at Swyncombe, Oxfordshire. Rubidium-86 (as $^{86}RbCl$ in neutral solution) was applied via polythene tubing to different depths in the soil profile, and the degree of radioactivity recovered from herbage cut one week later indicated the activity of roots at different levels, though strictly in terms of rubidium rather than water uptake. Elston found that *Filipendula vulgaris* and *Hieracium pilosella* extended roots much more deeply into the chalk than was suggested by Anderson. There was good agreement for *Poterium sanguisorba*, however. These comparisons are shown in Table 5.8, which includes additional data on species not studied by Anderson.

Table 5.8

Maximum depth from which roots took up ^{86}Rb for six herbaceous species from Swyncombe Down (Elston, 1963), showing comparisons for three of them with Anderson's (1927) root excavation data

Species	Maximum depth from which uptake took place (cm)	Maximum depth from which roots were excavated (cm)
Festuca rubra	30	—
Koeleria cristata	30	—
Filipendula vulgaris	30	12·5
Taraxacum officinale	30	—
Hieracium pilosella	60	11
Poterium sanguisorba	91	86

Yet, it is not only the deep-rooted herbs which remain green during prolonged dry spells. In the face of the strong ecological dogma of her day stressing the dryness of the chalk, Anderson had to admit:

> For 7 weeks the upper 3 inches of the soil had been the driest part, and there are a number of plants, including the bulk of the grasses, which are chiefly dependent on this zone for their water supply, yet none of these appeared to be seriously affected.

Cobbett (1830) makes frequent reference to the verdure of the chalkland sheepwalks even in very dry summers, while a telling point made by Hudson (1900) was that this greenness prevailed in chalk pastures *"provided that they had not at sometime been cultivated"* (see Section 5.4.3.2). All these observations strongly support the involvement of capillary rise (Section 3.6.2.2). A large volume of roots, as produced by *Bromus erectus* and other grasses, may make up for a lack of depth to some degree, as may the production of surface roots to catch dew and otherwise insignificant rainfall (see Fig. 5.15b).

Many chalk grassland plants possess classic xerophytic adaptations. Rosette hemicryptophytes such as the plantains (*Plantago*) and hawkbits (*Leontodon*) press their leaves close to the surface of the ground, a habit which, despite the extremes of temperature there, is alleged to protect them from excessive evaporation. Leaves may be hairy or roll up, or their stomates may be deeply sunken. As a member of the Crassulaceae, the plant family renowned for its ability to fix CO_2 at night, *Sedum acre* may be presumed to open its stomates only at night and close them during the day (Nishida, 1963). This last species displays, in addition, the capacity to store water in its succulent leaves, while the root tubers of *Filipendula vulgaris* appear to serve, in part, a similar purpose (Fig. 5.21c). Although the importance of stomatal modifications is now well established (Meidner and Mansfield, 1968), many morphological adaptations are only assumed to have ecological significance, and it should not be taken for granted that all species follow the classic examples of the physiology textbooks. As long ago as 1912, Stapledon concluded from his extensive investigations into the effects of prolonged drought on grassland that "drought resistance is not correlated with any one set of morphological characters".

In fact, it is extremely difficult to demonstrate the role of these features in plant–water relations. One line of approach has been to observe the response of potted plants brought to a state of full turgor and then subjected to artificial drought. Anderson conducted numerous tests along these lines with plants carefully transplanted from the field, and more recently Fenner (1975) has been doing the same thing with a range of chalk grassland species raised from seed. Fenner found, for example, that *Iberis amara* is particularly susceptible to drought, at least in the seedling stage, compared with *Senecio integrifolius*. Fenner and Anderson both confirm that the more gently conditions of water stress are imposed, the less drastic is the effect: the familiar idea of hardening-off.

Fenner also measured the rate of water loss by cut shoots, and how this was influenced by the rapidity and effectiveness of stomatal closure. The results were most interesting. *Blackstonia perfoliata, Gentianella germanica* and *Polygala vulgaris*, for example, had relatively high rates of transpiration (measured as water loss from freshly cut shoots) while *Anacamptis pyramidalis* and *Iberis amara* had notably low values. *Thesium humifusum* lost nearly two-thirds of its total water content (reaching a relative turgidity of 36%) before its stomates closed, while in *Clinopodium vulgare* closure occurred when its relative turgidity had fallen only to 87%. In this last species, however, significant cuticular transpiration continued after stomatal closure, a phenomenon which was even more marked in *Arabis hirsuta*.

Recovery from wilting was also investigated by both Anderson and Fenner. Again quoting from the more recent work, Fenner exposed cut shoots to increasingly prolonged drying, and measured the minimum water content from which 90% relative turgidity could be regained. (Ninety per cent was chosen for statistical reasons, and no attempt was made to monitor any possible long-term damage to stomatal function.) Here, *Scabiosa columbaria* and *Clinopodium vulgare* were found to recover from remarkably large deficits (70% and 68%, respectively), with *Arabis hirsuta* showing much poorer powers of recovery, able to withstand deficits of no more than 28%. *Iberis amara* (47%) and *Senecio integrifolius* (46%) were among the intermediate species in this respect. Fenner derived two useful terms from his data, an index of water loss, and an index of shoot drought resistance. These are illustrated and explained in Fig. 5.16.

What conclusions can be drawn from this sort of information? Above all, the fact that no single property can necessarily explain the success of a species in a particular situation. Of the seven species which were included in all Fenner's experiments, four (*Ajuga chamaepitys, Clinopodium vulgare, Scabiosa columbaria* and *Teucrium botrys*) behaved "consistently". In the other three, however, Fenner found

> little relationship between the behaviour of the cut shoots and that of whole plants when subjected to water stress. For instance, the shoots of *Iberis amara* have the lowest rate of cuticular loss of the species tested; its stomata close early, and it recovers well when dried out. Yet, *I. amara* was unquestionably the most sensitive species in whole-plant experiments. Contrariwise, excised shoots of *Senecio integrifolius* lose water very rapidly, but the whole plants were among the most resistant of the species tested. The same applies to a lesser degree to *Arabis hirsuta*.

These differences must be accounted for in the effectiveness of the roots. And that is where we came in several paragraphs ago. A lot could be learned simply by measuring water potentials in growing plants, yet this still seems beset by technical problems.

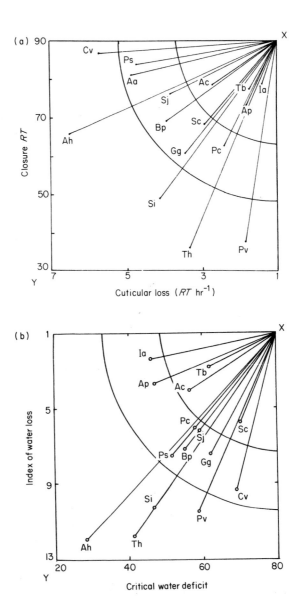

It can safely be said, then, that we still have a great deal to learn about chalk soil water relations. However, there is no doubt that on the whole chalk soils are far more hospitable to mesophytic plants than has been thought in the past (at least, by ecologists: farmers and foresters have known this for years—see Section 3.6.2.1 and Chapter 9). This interpretation renders less perplexing certain former paradoxes of plant geography, which can now be viewed in a new light. For example, some plants which show wide edaphic tolerance in the wetter north and west of Britain are restricted to the Chalk in the drier south-east of England, a pattern admirably explained by regarding chalkland water relations as particularly favourable. A good example is the orchid *Gymnadenia conopsea* (Fig. 5.17). A second group is typified by the valerian (*Valeriana officinalis*) with a distribution crisply summarised by Martin (1965) as "ditches, damp woods and chalk downs". The marsh thistle (*Cirsium palustre*) is another example, as are *Dipsacus* and *Scrophularia* mentioned in Chapter 4. There is, of course, always the possibility of ecotypic variation, and in fact in the valerian two distinct genetic populations have been recognised (Clapham *et al.*, 1962), though their ecology is by no means straightforward. Thus, plants of the Chalk and Oolite of the south of England are all tetraploid (2n = 28), while those of damp habitats in the south, and both damp and dry habitats in the north, are octaploid (2n = 56). Likewise, *Gymnadenia* in Britain is regarded by some botanists as consisting of three species, with *G. conopsea sensu*

Fig. 5.16. Indices of drought resistance for chalk grassland plants. From Fenner (1975). (a) Index of water loss computed for 16 species grown from seed. Closure *RT* = relative turgidity (%) at which stomatal closure occurred. Cuticular loss = loss in *RT* per hour after stomatal closure. X represents early stomatal closure with low cuticular loss, Y late stomatal closure with high cuticular loss. The distance, measured in arbitrary units, of any point on the grid from X is a measure of the shoot's resistance to water loss. This is the index of water loss. Ia, *Iberis amara* (2·3); Tb, *Teucrium botrys* (2·8); Ap, *Anacamptis pyramidalis* (3·7); Ac, *Ajuga chamaepitys* (4·0); Sc, *Scabiosa columbaria* (5·7); Pc, *Polygala calcarea* (6·1); Sj, *Senecio jacobaea* (6·2); Bp, *Blackstonia perfoliata* (7·2); Gg, *Gentianella germanica* (7·5); Ps, *Pimpinella saxifraga* (7·5); Aa, *Acinos arvensis* (7·8); Cv, *Clinopodium vulgare* (9·4); Si, *Senecio integrifolius* (10·3); Pv, *Polygala vulgaris* (10·5); Th, *Thesium humifusum* (11·8); Ah, *Arabis hirsuta* (12·0).

(b) Index of shoot drought resistance for 15 species, derived using the index of water loss explained above. The critical water deficit is the maximum water deficit (% water loss) from which 90% relative turgidity is regained. Grid and index are interpreted as in (a). Tb, *Teucrium botrys* (4·1); Sc, *Scabiosa columbaria* (5·2); Ac, *Ajuga chamaepitys* (5·7); Sj, *Senecio jacobaea* (6·7); Pc, *Polygala calcarea* (6·8); Ia, *Iberis amara* (6·9); Ac, *Anacamptis pyramidalis* (7·2); Gg, *Gentianella germanica* (7·5); Bp, *Blackstonia perfoliata* (8·0); Ps, *Pimpinella saxifraga* (8·7); Cv, *Clinopodium vulgare* (8·7); Pv, *Polygala vulgaris* (10·5); Si, *Senecio integrifolius* (11·5); Th, *Thesium humifusum* (13·3); Ah, *Arabis hirsuta* (15·1).

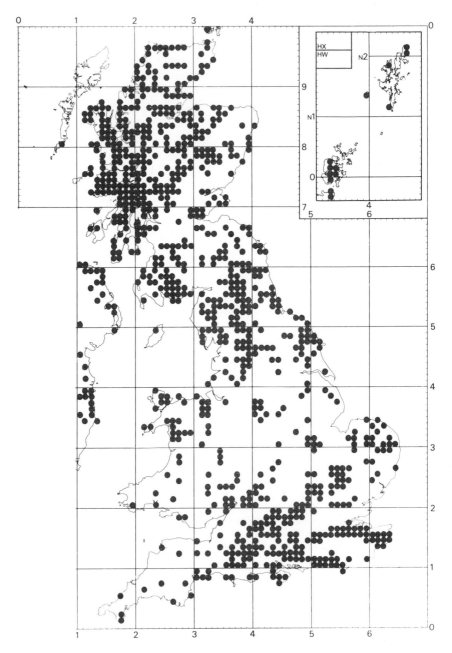

Fig. 5.17. Distribution of *Gymnadenia conopsea* in Britain. Interpretation and source as for Fig. 5.7a.

stricto and the fenland *G. densiflora* both occuring on chalk downland (F. Rose, personal communication).

5.4.2.3 Water–nutrient interactions

Most plants are caught in a vicious circle during times of drought. Their roots are embedded in either fairly nutrient-rich but dry topsoil, or moist but nutrient-poor subsoil, one reason perhaps why such a large number of species co-exist with none gaining dominance, at least as long as grazing is maintained and no nutrients deliberately added. Agronomic aspects of this phenomenon are discussed in Chapter 9. Plants which possess in addition to deep roots a horizontal surface root system are best fitted to take advantage of the most fleeting cycles of wetting and drying, when relatively large quantities of nutrients may be mineralised (Chapter 3). As already noted, this adaptation is particularly well seen in *Blackstonia perfoliata* (Figs 5.14 and 5.15b) and also *Thymus drucei* (Fig. 5.14).

Further advantages are to be seen in species capable of accumulating minerals, with or without the aid of mycorrhizal associates, and storing assimilates. Experiments by Rorison (1968) suggest that *Scabiosa columbaria* is able to thrive in calcareous soils very deficient in phosphate, not through possessing a particularly dense or extensive root system, nor through having either a high affinity or low requirement for phosphorus, but simply by being able to accumulate in its roots in times of excess a store of phosphorus which can be mobilised and drawn on subsequently "to ensure steady growth throughout the season". Regarding fungal symbionts, it was a remarkable find when Litav (1965) discovered mycorrhizal associations between fungi and shrub roots deep in the crevices of the solid chalk rock in which they were growing in Israel. Anderson noted fungal hyphae in or on the roots of most of the plants she excavated. She never had the opportunity to assess their significance, but it is now well known that mycorrhizal infection, especially of the phycomycetous (or vesicular–arbuscular) type, is common in many herbaceous plant species (Harley, 1969). It is tempting to contemplate whether mycorrhizal or nitrogen-fixing symbionts associated with more shallow-rooting plants might remain active at sufficiently low water potentials to allow some nutrient uptake during dry spells. Much the same can be said about the rhizosphere microflora.

5.4.2.4 Chemical agencies

We saw in Chapter 3 that there are two major features which govern the chemical properties of chalk soils. These are (i) an alkaline reaction resulting from an abundance of calcium carbonate, and (ii) virtual or complete saturation of the exchange complex with divalent calcium ions (Ca^{2+}). The strongly oxidising conditions which normally prevail are common to all freely drained soils.

Referring to calcium first, de Silva (1934) drew attention to the fact that certain calcicoles "flourish in soils in which there is no detectable amount of calcium carbonate". By growing selected plants in nutrient solution culture, he established the importance of exchangeable calcium in the ecology of species such as *Mercurialis perennis* (see Chapter 8), which are typical not only of chalk but also of clay soils of high base status in which calcium ions are predominant. More recent work has confirmed that some plants of calcareous soils grow poorly in concentrations of calcium which are adequate for species of acid soils (e.g. Jefferies and Willis (1964) for *Origanum vulgare*). In some species of wide edaphic tolerance, such as *Festuca ovina* and *Trifolium repens*, calcicolous ecotypes exist which have a much greater demand for calcium than their calcifuge counterparts (Snaydon and Bradshaw, 1961; Snaydon, 1962b). Acid soils of lower exchange capacity are normally deficient in calcium, though it should be noted that wherever calcium ions occur in relatively large quantities there is always the risk of a shortage of other cations, notably potassium, through antagonism (Chapter 3).

Regarding soil reaction, it has long been recognised that, because the cell sap of plants is so effectively buffered, soil pH in itself is usually not of first importance as a determinant of plant distribution. Thus, when cultured in a properly balanced aqueous nutrient solution, the calcicole *Scabiosa columbaria* and the calcifuge grass *Holcus mollis* both grew as well at pH 7·6 as at 4·8 (Rorison, 1960b). But the side effects of pH in the field are legion, and it is these which are the prime chemical factors in the restriction of plants to particular soils. For example, as soil pH falls below 5·5, and especially below 4·5, certain polyvalent cations begin to accumulate which, though tolerated by extreme calcifuges, are strongly toxic to other plants. Manganese and especially aluminium are now well known to be potent factors in the restriction of many species to neutral and calcareous soils. Rorison demonstrated very clearly the inhibitory effect of aluminium on root initiation in *Scabiosa* (Fig. 5.18). Seedlings are particularly vulnerable, and so the effect is greatly augmented by competition.

From the opposite angle, many plant species unable to thrive on chalk and limestone soils fail through nutrient imbalance. In particular, many calcifuges are unable to obtain or metabolise sufficient iron, and exhibit symptoms of *lime-induced chlorosis* when grown in calcareous soils (Fig. 8.3). Clymo (1962) suggests that the significance of this phenomenon may have been overrated, but there can be little doubt that chlorotic plants are unable to photosynthesise at anything like their potential rate, if at all. Moreover, chlorotic leaves have been shown to be more susceptible to desiccation than ordinary green leaves (Grime and Hutchinson, 1967), and

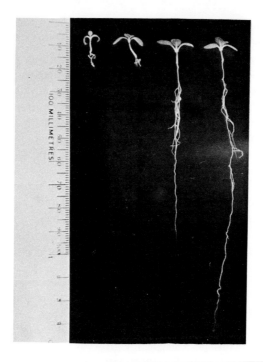

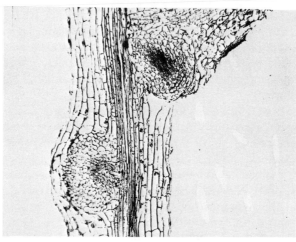

Fig. 5.18. The inhibitory effect of an acid medium on root growth in *Scabiosa columbaria*, shown to be due to aluminium. In (a, top) the seedlings on the left were grown in acid sand, those on the right in calcareous sand; the plants were all 28 days old. The bulging root initials illustrated in (b, bottom) developed in acid conditions, and failed to penetrate the cortex. From Rorison (1960a,b). (I. H. Rorison)

both leaves and roots of chlorotic plants are often markedly stunted (Grime and Hodgson, 1969).

Chlorosis appears to result in some species from the binding of iron in an unavailable form within the root, possibly by the same mechanism which serves to trap, and thus render harmless, aluminium in acid soils (Grime and Hodgson, 1969). In other species, chlorotic leaves contain as much, or even more, iron than green ones, and here it seems that something is interfering with iron metabolism. The bicarbonate ion, which inhibits root growth in calcifuges, appears to be significant in this respect (Woolhouse, 1966a,b; Lee and Woolhouse, 1969a, b). Some calcicoles appear either to possess a relatively small requirement for iron, or to utilise their supply very efficiently, or both. Thus, when cultured plants in which chlorosis had been deliberately induced received a foliar feed of iron, *Scabiosa columbaria* regained its normal green colour within four days, while *Holcus mollis*, one of the least susceptible of calcifuges to lime-induced chlorosis, took ten days to recover (Rorison, 1960b).

This ability to make do with an absolutely minimal supply of an essential nutrient is shown by some calcicoles for phosphate. The ability of *Scabiosa* to accumulate phosphate in its roots in times of relative plenty has already been mentioned, but the calcicolous grass *Brachypodium sylvaticum* is even more remarkable. This species typically grows in calcareous soils in which the concentration of soluble phosphate is barely measurable, yet Abeyakoon and Pigott (1975) could find no evidence whatever for any ability either to increase the solubility of its major source of mineral phosphate, apatite, or to obtain organic phosphate from soil humus through the phycomycetous mycorrhiza with which it is invariably infected. It must simply be highly efficient in its use of what small quantity of phosphate it is able to take up.

Nitrification of ammonium to nitrate-N normally proceeds more rapidly in neutral and calcareous soils than it does in acid soils, which tend as a consequence to contain most of their nitrogen in the form of the ammonium ion, NH_4^+ (Jackson, 1967), although well-drained acid soils are by no means as deficient in nitrate as is sometimes suggested (Davy and Taylor, 1974b; Havill *et al.*, 1974). Nevertheless the possibility has been raised on a number of occasions that this may be yet another factor distinguishing calcicoles from calcifuges, with plants of the former group making better growth on nitrate-N, the latter faring better on ammonium-N, and more widely distributed plants showing relative indifference to form of N.

The results of some experiments by Gigon and Rorison (1972) with plants in solution culture lend support to this hypothesis, although some of the responses were not as clear cut as might have been expected (Fig. 5.19). Plants from a typically strongly calcicolous Derbyshire population of

Scabiosa columbaria failed completely when provided with ammonium-N in a pH of 4·2 or 5·8; they made some growth in this form at pH 7·2, but the roots were tenuous and the shoots stunted. In nitrate-N, growth was consistently better whatever the pH but, even so, final dry weights were not significantly greater than those of the plants raised in ammonium-N at pH 7·2. Much more striking differences were shown, however, by *Scabiosa* plants grown from seed taken from a Swiss population which exhibited slightly less extreme calcicolous tendencies, reflecting the mildly acid reaction of the soil from which it originated. Contrasting responses were seen in the growth of *Deschampsia flexuosa* which, predictably, made better growth in ammonium-N than in nitrate-N, hardly growing at all in nitrate-N at a pH of 7·2. But a warning note was sounded by the result for the ubiquitous *Rumex acetosa* which, despite being raised from seed collected from plants growing in a sandy soil of pH 4·2, grew far better in ammonium-

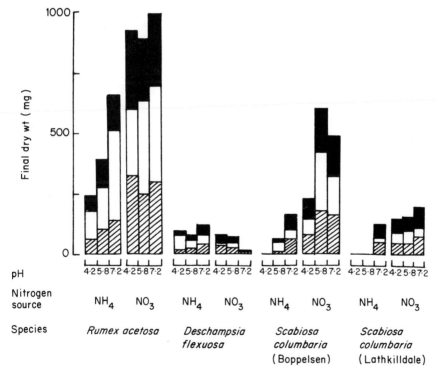

Fig. 5.19. Dry weight (mg) of plants of *Rumex acetosa*, *Deschampsia flexuosa*, and two populations of *Scabiosa columbaria* after growth in either ammonium or nitrate nitrogen in acid and alkaline media. See text for details. From Gigon and Rorison (1972).

N at pH 7·2 than in more acid media, and performed altogether more impressively in nitrate-N whatever the pH.

Gigon and Rorison suggest that the toxic effect of ammonium-N on calcicoles at low pH may be due to an ionic antagonism, in which the uptake of K^+ ions is inhibited. Moreover the uptake of NH_4^+ ions accentuates the acidity of the rooting medium (and so aggravates the situation), for an influx of NH_4^+ is countered, at least in some plants, by an efflux of H^+ ions. The uptake of NO_3^- ions, on the other hand, is balanced by an efflux of bicarbonate (HCO_3^-) ions, so that species utilising nitrate-N enrich their medium with bicarbonate which in turn raises the pH (Kirkby and Mengel, 1967).

An inability to utilise nitrate-N has been regarded by plant physiologists as evidence for the absence of the enzyme nitrate reductase, which reduces nitrate to nitrite in the first stage of nitrogen metabolism within the plant. Many ericoids and other strong calcifuges are known to lack this enzyme, and to rely on ammonium-N as their source of N. Yet, neither the work of Gigon and Rorison, nor a more extensive survey of plants of chalk and limestone grassland, acid heathland and moorland by Havill *et al.* (1974), reveal any such clear-cut differences in nitrogen metabolism; many calcifuges, even *Deschampsia flexuosa* and *Galium saxatile*, possess nitrate reductase and are perfectly capable of utilising nitrate.

In this work on nitrate reductase the activity of the enzyme is assayed by monitoring the production of nitrite within the tissue under test. In the normal course of events in the plant, this nitrite is reduced in turn by nitrite reductase for synthesis into amino-acids and proteins. The possibility that calcifuges are unable to cope with nitrite seems never to have been investigated with the degree of sophistication afforded to nitrate. Evidence that some calcicoles are relatively tolerant of nitrite (which is known to occur in calcareous soils—see Section 3.8.3.3) was obtained in sand culture experiments performed at Reading (C.J. Smith and A. H. Bunting, unpublished). It was impossible to culture plants exclusively on nitrite, for a small proportion of nitrate always occurred in the rooting medium through spontaneous oxidation, but there was no doubt of the toxicity of the nitrite ion (NO_2^-), which acted by greatly restricting root growth. It was thought at first that this might be due to direct competition for oxygen between respiring roots and nitrite ions, but D.S. Jenkinson (personal communication) ascribes the response to interference with nitrogen metabolism, thus:

$$R-CH_2-NH_2 + HNO_2 \rightarrow R-CHOH + N_2 + H_2O.$$

In fact, this is the principle of the van Slyke reaction which is used to measure nitrite in plant and soil analysis. The calcifuges *Deschampsia flexuosa* and *Agrostis tenuis* suffered much greater inhibition of roots than *Poterium sanguisorba*, *Helictotrichon pubescens* and a chalkland population of

Festuca rubra, while *Bromus erectus* was particularly tolerant of nitrite. The roots of the last species developed a thick black coating in the nitrite treatments, but whether this played any part in the response was not determined.

Experimental work in this fertile field of plant ecology continues, but for every piece of evidence gained to add support to the leading hypotheses, as much if not more information invariably arises either to contradict it, or to necessitate hanging the most laboured and involved conditions on its interpretation. A brave attempt has been made by Grubb *et al.* (1969) to classify calcicoles on the basis of distribution in the field and behaviour in culture, and their proposed scheme is shown in Fig. 5.20.

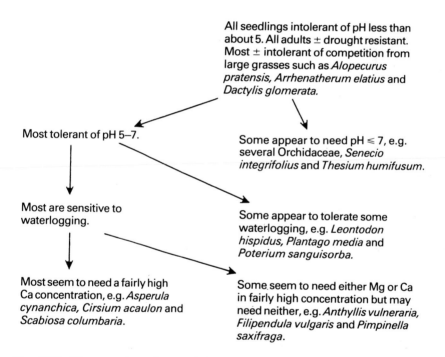

All seedlings intolerant of pH less than about 5. All adults ± drought resistant. Most ± intolerant of competition from large grasses such as *Alopecurus pratensis, Arrhenatherum elatius* and *Dactylis glomerata.*

Most tolerant of pH 5–7.

Some appear to need pH ≤ 7, e.g. several Orchidaceae, *Senecio integrifolius* and *Thesium humifusum.*

Most are sensitive to waterlogging.

Some appear to tolerate some waterlogging, e.g. *Leontodon hispidus, Plantago media* and *Poterium sanguisorba.*

Most seem to need a fairly high Ca concentration, e.g. *Asperula cynanchica, Cirsium acaulon* and *Scabiosa columbaria.*

Some seem to need either Mg or Ca in fairly high concentration but may need neither, e.g. *Anthyllis vulneraria, Filipendula vulgaris* and *Pimpinella saxifraga.*

Fig. 5.20. The tentative scheme of Grubb *et al.* (1969) for the classification of calcicoles of chalk grassland.

This section cannot be concluded without reference to an almost totally unexplored branch of plant autecology: the pathology of individual species. Ecologists are familiar with the preponderance of fungal pathogens (indeed, fungal activity in general) in soils of acid reaction. Gardeners know only too well the consequences of trying to grow the calcicolous crucifers, particularly brassicas and wallflowers, in acid soils: they often develop grossly distorted and ineffectual roots as a result of infection by the club-root slime-mould *Plasmodiophora brassicae*. Raising the pH of the soil by the application of lime is the only really effective means of control without recourse to fungicides. Comparable examples must exist in the natural situation, but documentary evidence is lacking. More than 50 years ago, Salisbury (1921) pointed out that calcicolous species of *Iris* are liable to severe attack by the pathogenic fungus *Heterosporium gracile* when grown on soils deficient in calcium: a fact which may help to explain the curious distribution of *Iris foetidissima* in Dorset and Kent referred to in Chapter 4.

5.4.2.5 Chalk heath

Chalk heath is immediately distinguishable from "ordinary" chalk grassland by its intimate admixture of calcicoles and calcifuges (Fig. 5.6). Gorse, heather and ling are usually present, though these become more obvious and take on greater significance in the absence of grazing or mowing (see Sections 6.5.3.1 and 7.2.1). In the herbaceous vegetation, an abundance of *Agrostis tenuis* is a good indication of chalk heath, and the diminutive annual *Aira praecox* a less obvious one. Heath grass (*Sieglingia decumbens*) and tormentil (*Potentilla erecta*) are diagnostic in the south-east except near the sea, but occur in "ordinary" chalk grassland both further west and in the Lincolnshire Wolds. The strongly calcifuge *Galium saxatile* was recorded very locally in chalk heath by Tansley and Adamson (1926). Chalk heath is distinguished from true heathland, of course, by the presence of the calcicoles, although some of these, such as *Koeleria cristata*, *Scabiosa columbaria*, *Pimpinella saxifraga* and particularly *Helictotrichon pubescens* and *Bromus erectus*, are usually much less abundant in chalk heath than chalk grassland (P.J. Grubb, personal communication). In recent surveys, variants of chalk heath have been recognised according to whether they occur on the edges of the downland plateaux, on dip slopes, or in coombe bottoms (Ratcliffe, 1977).

Chalk heath was first described by Tansley and Rankin (1911), and given further attention by Tansley and Adamson (1925, 1926) in their South Downs surveys. From these early floristic accounts came a classic piece of dogma which became thoroughly engrained in the ecological literature for half a century. It even slipped through into the posthumous second edition of Tansley's book "Britain's Green Mantle" (Tansley, 1968, p.173). The assertion was that the acid surface layer is attributable to locally higher rates of leaching by rainfall, and that species of contrasting edaphic tolerance co-exist through differential root stratification, the roots of the calcicoles penetrating into the chalk subsoil, those of the calcifuges ramifying within the acid layer at the surface.

This explanation of chalk heath is now known to be untrue. For a start, a thin surface deposit of silt or clay—typically loessal silt—is now regarded as the cause of surface acidity, rather than leaching (see Chapter 3). Secondly, the root systems of the plants are not separated into discrete layers. Following their intensive study of chalk heath at Lullington, Sussex, Grubb *et al.* (1969) point out that this second proposal makes no allowance for seed germination, for, except possibly when moles, ants or earthworms bring chalky soil to the surface, seeds of all species germinate in a common (acid) medium. In any case, Anderson (1927) had already established the inherently shallow root distribution of many of the calcicoles found in chalk heath stands, yet the significance of these data never penetrated the contemporary discussions of chalk heath ecology.

Table 5.9

Soil pH around the necks of calcicolous plants growing in chalk heath [a,b]

Species	Soil pH
Asperula cynanchica	5·2 (L)
Cirsium acaulon	6·0 (L)
Hippocrepis comosa	5·5 (F)
Poterium sanguisorba	5·4 (L)
	5·4 (L)
Scabiosa columbaria	5·1 (F)
	5·5 (F)

[a] From Grubb *et al.* (1969).
[b] L=Laboratory determination; F=field determination.

In fact, Grubb's team were able to show that several species traditionally regarded as strongly calcicolous (notably *Asperula cynanchica, Viola hirta* and *Filipendula vulgaris*) had established themselves in soil with a pH between 5 and 6 (Table 5.9 and Fig. 5.21), and it was this discovery which led to the recognition of the relatively wide tolerance, even in a competitive situation, of many calcicoles, and to the proposal of a category of strict calcicoles in the narrow sense for species not normally found in soils of pH less than 7, and so absent from chalk heath. The calcifuges, too, displayed relatively wide edaphic tolerance. Not only were they able to grow satisfactorily in the same soil as the calcicoles, but many apparently healthy roots grew right into the chalk at a pH in excess of 7. Many more points of considerable interest and fundamental importance emerged from this work, and these are returned to under the heading of biotic influences in the next chapter.

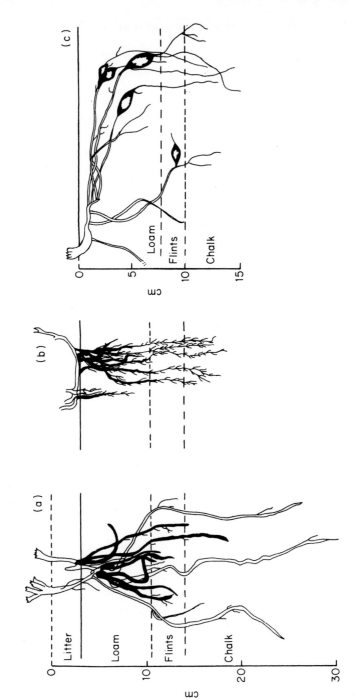

Fig. 5.21. Root systems of (a) *Poterium sanguisorba*, (b) *Scabiosa columbaria* and (c) *Filipendula vulgaris* based on drawings of fresh material. The longest roots of *Poterium* are left white in order to follow their course near their origin more easily. Note separate scale in (c). From Grubb *et al.* (1969).

5.4.3 Grasslands of formerly cultivated chalk downland

5.4.3.1 Changing downland husbandry

There remains one widespread and intriguing aspect of the physical environment which results in persistent and sometimes very conspicuous features of the botanical composition of chalk grassland. This is the effect of former disturbance of the downland sward by cultivation. Though clearly edaphic in its mode of operation, the cause is anthropogenic, and so this topic forms a convenient bridge between this chapter and the next, which covers biotic effects *per se*. (The flora of arable land is dealt with in Chapter 9.)

Numerous examples can be found in the historical records of farmers putting the plough into the turf on the high downs at times when it was economically expedient to raise corn. Sheep were sometimes folded onto roots which alternated with the grain crops, but yields were invariably poor, and crop failures frequent. Many areas were cultivated like this for a spell, and then abandoned and left to revert via a tumbledown fallow to grassland again (see Wells *et al.*, 1976). A more complete account of these major upheavals in chalkland farming is given in Chapter 9.

5.4.3.2 Open herb-rich swards

Many of these tracts of old arable land are easily recognised by the experienced eye of the countryman, and Cobbett, Jefferies and Hudson all devote a paragraph or two to them. Their most obvious features (Fig. 5.4) are the openness of the sward, the preponderance of herbs, especially those which form a basal rosette of leaves, and a corresponding paucity of the grasses which normally form the matrix of the undisturbed turf. Hudson wrote in 1900: "This kind of land, spoilt by the plough, is said by the shepherds to be 'sickly' ". Foresters report the failure of trees transplanted into these formerly arable soils (see Fig. 9.21). Sometimes land cleared of woodland can also give rise to these depauperate swards, and although these differ in some respects from the purely arable sites it is convenient to consider both together.

The first of Tansley and Adamson's lengthy papers on the grasslands of the South Downs (Tansley and Adamson, 1925) reported a survey of the Ditcham Park estate on the borders of Hampshire and Sussex. Here, the authors were more concerned with dynamic and successional aspects than they were in their more wide-ranging reconnaissance of 1926, in which they deliberately avoided all but apparently undisturbed downland. At Ditcham Park, two sites in particular supported the kind of sward under discussion here. One of these was Coulter's Dean wasteland, "clearly . . . a stage in succession of derelict arable", where interest centred on scrub encroachment (Chapter 7). The other was an area of Chalton Down, cleared of woodland some years previously, and obviously in the same category as the ex-arable grassland.

Tansley and Adamson coined the term "herbland" for this type of vegetation, but unfortunately they refer to it, almost as an afterthought, only in Table III of their paper, and it is not discussed at all in the text. The description of the vegetation in the rider to that table states that "the grasses are much less conspicuous [than in typical chalk grassland], and dicotyledonous herbs play the most prominent part". The grasses *Briza media, Koeleria cristata, Helictotrichon pratense* and *H. pubescens* were absent. On the other hand, species present here but found either not at all or only very rarely in the more typical grasslands included *Arenaria serpyllifolia, Moehringia trinervia, Cirsium palustre, Hypericum perforatum, Pastinaca sativa, Veronica chamaedrys* and the moss *Brachythecium glareosum. Clinopodium vulgare, Helianthemum chamaecistus* and *Prunella vulgaris*, though not exclusive to it, were found in much greater quantity in the herbland than in typical grassland. By 1936, *Helictotrichon pubescens* had appeared, and other grasses were less sparse, but the vegetation was still distinctly different from the downland sward proper (Hope-Simpson, 1941b).

In her survey of the chalk grasslands of the North Downs of Surrey and Kent, Cornish (1954) concentrated much of her attention on old fields formerly under arable cultivation, which were known to have been abandoned mostly between 40 and 70 years previously. Her comments tally exactly with those of earlier writers: "The most distinctive and striking feature is the unimportance of the usual grasses and complete mat of vegetation composed of the common chalk grassland herbs". Though always present (mainly as *Festuca ovina*), the fine-leaved fescues made a much smaller contribution to the sward than in undisturbed sites, while *Bromus erectus* was virtually absent from formerly disturbed ground, even when profuse stands of seed-bearing plants occurred in adjacent grassland. This inability of *Bromus* to invade tumbledown fallows was originally placed on record by Stapledon (1912), and has more recently been confirmed by Lloyd and Pigott (1967). Cornish noted that the soil in the former arable fields contained more "clay" and finely divided chalk particles than the typical undisturbed downland rendzina.

Some species had clearly re-established themselves in full force, and because of the smaller contribution from the grasses, gave an impression of even greater importance than in the undisturbed stands. *Poterium sanguisorba* was particularly notable in this respect. In addition, Cornish picked out 11 species as "abundant" (45 cm or less between individuals) in at least 25% more disturbed than undisturbed sites, and this list certainly conjures up the typical appearance of ex-arable grassland (Table 5.10). Others typical of the disturbed grassland included *Pastinaca sativa* and *Senecio erucifolius*, while the following species, "although not abundant, contribute much to the character of these old ex-arable sites": *Plantago lanceolata*, *Blackstonia perfoliata*, *Bellis perennis*, *Fragaria vesca*, *Medicago lupulina*, *Veronica arvensis*, and the grasses *Holcus lanatus* and *Dactylis glomerata*.

Table 5.10

Species noted as "abundant" (45 cm or less between individuals) in at least 25% more disturbed than undisturbed chalk grassland sites in the North Downs [a]

Species	Mean frequency in former arable sites
Linum catharticum	75
Lotus corniculatus	74
Viola hirta	69
Clinopodium vulgare	68
Origanum vulgare	68
Hypericum perforatum	65
Leontodon hispidus	57
Carex flacca	56
Thymus drucei [b]	55
Hieracium pilosella	51
Brachypodium sylvaticum	47

[a] From Cornish (1954).
[b] Recorded as *T. serpyllum*, agg., possibly including *T. pulegioides*.

All the species mentioned so far occurred in over 20% of the 56 sites surveyed by Cornish; in other words they scored a constancy of II or more. Among the rest (of constancy I, not "less than I" as Cornish puts it), only a few showed contrasting values of abundance between formerly cultivated and undisturbed grasslands: *Ranunculus repens* and *Silene vulgaris* were found more in the ex-arable sites, as were *Rumex acetosa* and the grasses *Agrostis stolonifera* and *A. tenuis*. These last three were particularly evident in the more recently abandoned arable sites (2–10 years old, average 7). Indeed, an interesting feature of these younger stands, not taken up by the author, is that *Agrostis tenuis* appears to take the place of *Festuca ovina*, and to behave as a pioneer. Species present in the mature grasslands but absent or virtually so from the ex-arable sites included *Koeleria cristata*, *Asperula cynanchica*, *Gymnadenia conopsea*, *Anthyllis vulneraria*, *Hippocrepis*

comosa and *Chrysanthemum leucanthemum*, though the last of these at least
is surely as typical of disturbed as of mature grasslands.

In connection with this last point, a word of caution must be interjected
about attaching too much significance to every difference in botanical composition between samples. The golden rules of any such exercise conducted
with a proper regard for objectivity are (i) not to rely on too few comparisons, where pure chance may account for some or all of the differences
observed, and (ii) to either avoid or allow for any bias which may enter into
the study. It seems that Cornish took care on both counts in her survey, yet
there is one remarkable feature about it: no less than 13 angiosperm species
found by Tansley and Adamson to be of high constancy in the South Downs
grasslands are not mentioned at all. These are the grasses *Helictotrichon
pubescens* (constancy IV), *H. pratense* (V), *Trisetum flavescens* (IV) and
Anthoxanthum odoratum (IV), and the herbs *Scabiosa columbaria* (V),
Trifolium pratense (V), *Achillea millefolium* (IV), *Campanula rotundifolia*
(IV), *Centaurea nigra* (IV), *Galium verum* (IV), *Plantago media* (IV),
Primula veris (IV) and *Ranunculus bulbosus* (IV). In fact, *Galium verum* is
almost absent from the mid-Kent Chalk where Cornish worked, and *Anthoxanthum* and *Achillea* are quite rare there, but the others are generally
common in old grassland (F. Rose, personal communication). Perhaps the
omission of these species reflects Cornish's preoccupation with the kind of
disturbed grasslands in which they played a minimal role, and which, with
one exception, Tansley and Adamson had deliberately avoided. Yet, 34 of
these North Downs samples came from mature and apparently undisturbed
grasslands.

In more recent investigations of plant communities of this kind, Lloyd and
Pigott (1967), working in the Chilterns, stress again the "great scarcity of
grasses and their failure to form a true turf", while on Salisbury Plain, Wells
et al. (1976) note that, though *Dactylis glomerata, Poa angustifolia, P.
pratensis* and *P. compressa* were not uncommon, "they never formed dense
swards". The openness of these communities and the preponderance of
dicotyledons is clearly shown in Lloyd and Pigott's data (Table 5.11) from an
old arable field at Pulpit Hill near Princes Risborough abandoned in 1941,
and now part of BBONT's Grangelands Nature Reserve (see also Chapters
5 and 10). These authors, like Cornish, express surprise at the absence of
Bromus erectus, despite an abundant source of viable seed adjacent to the
area in question. Preliminary trials in field and glasshouse suggested that
deficiencies of available nitrogen and phosphorus in the surface soil, accentuated by water stress, account for this lack of normal colonisation by
sward-forming plants. No one seems to have considered the possibility that
allelopathic agencies may be involved. This question, as well as other
aspects of plant interactions on these open sites, is discussed in the next

chapter. The association of dogwood (*Thelycrania sanguinea*) scrub with sites of this description is considered in Chapter 7.

Table 5.11

Herbaceous vegetation of a former arable field at Pulpit Hill, Buckinghamshire [a,b]

Species	Plot		
	1	2	3
Agrostis stolonifera	4	11	3
Catapodium rigidum	—	2	—
Dactylis glomerata	1	3	2
Festuca rubra	18	4	15
Acinos arvensis	1	1	+
Crepis capillaris	4	3	5
Daucus carota	1	2	2
Euphrasia nemorosa	+	—	—
Fragaria vesca	5	—	4
Gentianella amarella	—	+	—
Hieracium pilosella	20	17	22
Hypericum perforatum	3	6	+
Leontodon hispidus	6	4	8
Linum catharticum	6	2	3
Medicago lupulina	+	—	—
Plantago lanceolata	2	—	2
Polygala vulgaris	—	—	+
Potentilla reptans	1	3	+
Prunella vulgaris	13	18	8
Ranunculus bulbosus	1	—	2
Senecio jacobaea	1	3	2
Thymus pulegioides	14	20	13
Viola arvensis	—	—	+
V. hirta	—	+	1
Bare ground	29	26	30

[a] From Lloyd and Pigott (1967).
[b] Figures represent percentage cover measured by contacts with a needle 1·75 mm in diameter placed at random 100 times in triplicate 50 cm × 50 cm plots; plus sign indicates that the species was present in that plot but not hit by the needle.

The general absence of turf-forming grasses and the tendency of the components of the sward to die back annually to ground-level rosettes favours the growth of certain bryophytes, and even lichens (see below), by providing them with an almost completely open habitat during their period of maximum growth in winter. Cornish found that the percentage cover of the bryophytes was always large, although for some unknown reason there were never very many species. As with the angiosperms, some bryophytes which might have been expected to establish and grow on the open soil of the disturbed grassland were not found, and Cornish cites in particular *Neckera crispa*, *Entodon orthocarpus*, *Aloina aloides*, *Dicranum scoparium* and *Ditrichum flexicaule* as notable absentees. A general rule of succession was for acrocarpous mosses such as species of *Weissia* and *Barbula* to be overcome by pleurocarpous species within a few years. Among the latter *Campylium chrysophyllum* and *Eurhynchium swartzii* often formed distinct carpets.

5.4.3.3 Lichen-rich grasslands

Cornish mentions the presence of lichens in some of her ex-arable stands (and see Fig. 5.4); here and there these can take on quite remarkable abundance. A lichen-rich sward occurs, for example, on part of Watlington Hill on the Chiltern escarpment ploughed in the 1930s. But the most spectacular examples of this type of grassland, if it can be called grassland (Fig. 5.5), have been discovered recently on the Porton Ranges by Wells *et al.* (1976), who note its similarity in some respects to Watt's Grasslands A and B of Breckland (Section 5.3.2). They describe it thus (p.600 of their paper):

> Physiognomically, the vegetation was a mosaic of lichens, mosses and prostrate forbs, never more than 1–2 cm high, interspersed with tufts of *Festuca ovina*, *Trisetum flavescens* and other grasses and forbs which were usually less than 8 cm high. The community was open; cover varied from 60 to 80% in summer but was less in winter when some of the herbs died back. Much of the "bare ground" was covered with angular flints, mostly 3–10 cm in diameter, mixed with small water-worn pebbles, fragments of chalk rock and snail shells. Lichen cover varied considerably. It reached 80% in some patches 5 m across but in other areas it was less than 10%. *Cladonia rangiformis* accounted for *c.* 85% of the total lichen cover.

No less than 20 further lichens were identified, including 10 from flints and chalk rubble. Among the mosses, *Camptothecium lutescens*, *Campylium chrysophyllum* and *Hypnum cupressiforme* were "frequent and sometimes locally abundant".

The commonest angiosperms were *Hieracium pilosella*, *Leontodon hispidus*, *Prunella vulgaris* and *Thymus drucei*, and there was an "abundance of legumes, especially *Anthyllis vulneraria* and *Lotus corniculatus*, which in some areas formed roughly circular patches up to 6 m across". Small tufts of *Medicago lupulina*, *Trifolium campestre*, *T. dubium* and *T. repens* were also recorded. Among the other dicotyledonous species,

Cirsium acaulon, Galium verum, Scabiosa columbaria and *Viola hirta* were "conspicuous but never of high frequency". Characteristic grasses included *Trisetum flavescens* and *Phleum bertolonii. Helictotrichon pratense* and *H. pubescens* occurred only as isolated plants, their "low growing habit in this community contrasting with their vigorous form in neighbouring grasslands . . .". Annuals and biennials, clearly favoured by the openness of the stand, included *Linum catharticum, Arenaria serpyllifolia, Centaureum erythraea, Crepis capillaris, Erigeron acer, Erophila verna, Euphrasia nemorosa,* the forget-me-nots *Myosotis arvensis* and *M. ramosissima,* and the grasses *Aira praecox* and *Catapodium rigidum.*

The nature of these swards strongly suggests former exhaustive cropping, rabbit warrening, or a combination of the two (Section 9.3.2.2). A characteristic feature is the presence of numerous flints strewn about the surface. Intriguingly, these often give the appearance of having been emplaced quite recently, the turf beneath them exhibiting no signs of etiolation or compression, or baring of the soil surface. This can only mean that the flints are regularly moved about, perhaps by inquisitive rooks or grazing rabbits.

5.4.3.4 Ex-arable communities on deeper soils

The curious swards described in the foregoing paragraphs, whether dominated by herbs, bryophytes or lichens, appear to be typical only of the shallowest soils. Where the soil is deeper, moister and more fertile, and where grazing is slight or absent, a totally different type of vegetation develops, typically dominated by *Arrhenatherum elatius.* This type of community is also characteristic of some road-verges.

Characteristically lank and ungrazed (and so strictly referable to Section 6.5; see Fig. 6.16), *Arrhenatherum elatius* grassland is floristically poor. *Helictotrichon pubescens* is often co-dominant, but scattered patches of other tall grasses such as *Dactylis glomerata* may also occur, and closer scrutiny usually reveals considerable quantities of *Festuca rubra* as an "understorey" species, and the moss *Pseudoscleropodium purum* beneath this. Other mosses found to exceed 20% frequency in the *Arrhenatherum* grasslands of the Porton Ranges (Wells *et al.,* 1976, p.598) included *Acrocladium cuspidatum, Campylium chrysophyllum, Hypnum cupressiforme* and *Rhytidiadelphus triquetrus.* Herbs are few in number and are represented by robust species able to hold their own in the tall grassy herbage such as *Centaurea scabiosa, Knautia arvensis* and more locally *Succisa pratensis.* Large individuals of the grazed-turf flora (e.g. *Leontodon hispidus* and *Poterium sanguisorba*), as well as wayside species (e.g. *Agrimonia eupatoria, Cirsium arvense, C. vulgare* and *Linaria repens*) may also be found.

5.4.3.5 Successional relations

Using historical evidence from local residents, old records and maps, and aerial photographs, Wells and his colleagues were able to estimate the age of the Porton grasslands, and to relate this to differences in soil characteristics and botanical composition. They then formulated a scheme indicating possible successional relations according to past management and soil nutrient status, and this is shown in Fig. 5.22.

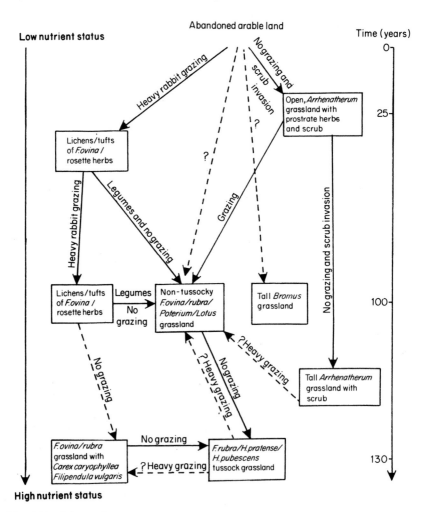

Fig. 5.22. Schematic summary of the possible relationship between grassland types, management, soil nutrients and the passage of time on former arable land on the Porton ranges (Wells *et al.*, 1976).

5.5 Small-scale Differences in Plant Distribution

5.5.1 Descriptive surveys

Many examples of local variations in the botanical composition of chalk grassland apparently attributable to differences in topoclimate and soil have been detected in the course of surveys which were entirely or essentially descriptive. Tansley and Adamson (1926) noted the distinct physiognomy of the vegetation of the north-facing slopes of the South Downs escarpment, which differed from typical chalk grassland in having a greater proportion of mosses and tall grasses of the tribe Aveneae, as well as containing the tussock-forming grasses *Deschampsia caespitosa* and "*Festuca pratensis*" (more commonly, according to J. F. Hope-Simpson, *F. arundinacea*) and herbs such as *Primula veris*, *Succisa pratensis* and *Rumex acetosa*. Hope-Simpson (1941a) and later Watson (1960) confirmed the influence of aspect on bryophyte distribution (Fig. 5.23).

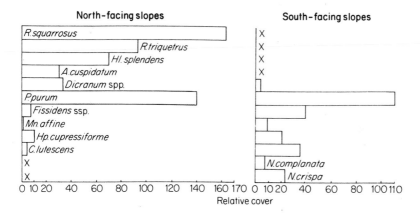

Fig. 5.23. The distribution of twelve mosses on four north-facing and five south-facing downland slopes in southern England, from an exploratory bryophyte survey by Watson (1960), taking in Lullington Heath, Kingley Vale, Old Winchester Hill and Aston Rowant NNRs (see Table 10.9, and reference to A. S. Thomas's work in the text). Cover values are derived from a combination of point-frame and quadrat data obtained from 30 m transects. The genera indicated by letter only are *Acrocladium* (A), *Camptothecium* (C), *Hylocomium* (Hl), *Hypnum* (Hp), *Mnium* (Mn), *Neckera* (N), *Pseudoscleropodium* (P) and *Rhytidiadelphus* (R). X indicates absent. The absence of *Neckera crispa* from north-facing slopes, of which it is generally characteristic, was an anomaly of the survey. Refer to the original paper for details of methodology and a critical discussion of the whole exercise.

More recently, Wells (in Ratcliffe, 1977) has proposed a number of distinct *facies* within some of the chalk grassland types listed in Table 5.3: for example, a *Poterium sanguisorba–Helianthemum chamaecistus* facies characteristic of south-facing slopes in Kent, East Anglia and the Chilterns, where the climate is relatively continental, and a *Leontodon hispidus* facies, "widespread, but best developed in Wiltshire and Dorset". Another facies well seen in the Wessex chalklands is distinguished by an abundance of devil's bit scabious (*Succisa pratensis*). Archibald (1949) recognised "*Linum–Briza–Koeleria*" and "*Helianthemum–Thymus*" types of fescue grassland in the Chilterns. There is clearly a case here for liaison with phytosociologists.

Thomas (1962) has indicated the possible significance of outcrops of hard bands in the Chalk to the more strongly calcicolous element of the chalk grassland flora, although in northern France Liger (1952) could find no notable floristic differences between swards of contrasting bands of Middle and Upper Chalk. Nevertheless, as we have already seen, topography can cause marked variations in soils and vegetation over a common geological base, and this is particularly well seen on a small scale in the pattern of ridges and depressions which run across the former arable land at Grangelands, below Pulpit Hill, Buckinghamshire. Here, thin, skeletal Icknield-type rendzinas have developed on the dry, gritty beds of Middle and Upper Chalk exposed on the ridges, and these support the open, herb-rich flora described earlier. Deeper soils with better water and nutrient relations, referable to the Coombe Series (Bradley, 1968), occupy the gulleys, and here quite a different flora occurs, dominated by *Arrhenatherum elatius* and containing rather a poor complement of species, including tall herbs such as *Centaurea scabiosa*. *Arrhenatherum* grassland can sometimes be seen as circular patches a few metres across within other grassland types, perhaps coinciding with pockets of drift deposits, or places where the ground has been disturbed at some time. Examples of patchiness attributable to biotic factors are dealt with in Chapter 6.

Of course, not all the heterogeneities of chalk grassland (or any other biome) are as obvious to even the trained eye as these examples. The finer the scale and the more subtle the degree of these variations, the more difficult it is to recognise them and to assess their ecological significance. Here, the investigator is obliged to turn to more specialised quantitative techniques such as association, pattern and multivariate analysis.

5.5.2 Association analysis

The extent to which plants encountered in a survey are positively or negatively associated with one another can be determined by drawing up a 2 × 2 contingency table of all the species recorded. Correlation between pairs of species is then sought by means of a X^2 (chi-squared) test, which pinpoints those pairs which occur together to either a greater or lesser extent than could be expected purely by chance. X^2 values for each species are then summed. The species with the largest sum of X^2 values is immediately apparent, and the quadrats are split into two groups: those with, and those without this key species. The process is repeated as many times as is deemed appropriate, but at most until no further significant correlations remain. Obviously, the scale of any such operation is important: too large a sampling unit will pool too much detail and overestimate positive association, while too fine an analysis has the opposite effect, and overestimates negative association. This method has been used both to substantiate descriptive accounts based on visual assessment, and to detect groupings less obvious to the human eye. Measurements of environmental factors can be included in the analysis. Useful résumés are provided by Gittins (1965c) and Kershaw (1973). Gilbert and Wells (1966) applied this method, among others, to the results of A. S. Thomas's surveys of changes in chalk (and other) grasslands following the outbreak of myxomatosis in the 1950s (Chapter 6). They were able to confirm several associations already familiar to field botanists, though the emergence of some spurious examples emphasised the limitations of transect recording in association analysis.

5.5.3 Pattern analysis

Detailed accounts of the basis and applications of pattern analysis can be found elsewhere, particularly in the writings of its chief protagonists, Greig-Smith (1964) and Kershaw (1973), but briefly it operates as follows. A quadrat of convenient size ("block size 1") is placed on the turf, and a list is drawn up of the species present, together with some quantitative attribute of each, such as the number of individual plants, vegetative or reproductive shoots or flowers in the quadrat, or an estimate or measure of the contribution to cover, or to the fresh or dry weight of the total herbage inside the quadrat. After sufficient samples have been taken, the data are tabulated, and means and variances calculated. This procedure is then repeated after doubling the area of the quadrat (block size 2), then again (block size 4), and so on, usually up to block size 128. A nest of contiguous quadrats provides a convenient way of doing this. Alternatively, transects of increasing multiples of an original length can be used.

The principle of the method is best illustrated by "patchy" species made up of distinct morphological units, such as *Trifolium repens* (see Kershaw, 1973, Fig. 8.4). The block size which corresponds with the individual leaf tends to score "all or nothing", and so the data exhibit a peak of variance at that size for that species. Likewise, individual shoots generate a peak of variance at the block size which coincides with their mean area, and so on. Patchiness within a turf caused by morphology and habit of growth is thus readily demonstrated by pattern analysis, although not all peaks are susceptible to such simple explanation. Exactly the same technique can be applied

Table 5.12

Attributes used in pattern analysis of Bromus erectus *grassland* [a]

Species	Attribute	Code	Measurement
Bromus erectus	Density	D	Number of tillers per sampling unit.
	Performance	Pf	Mean length (to next whole cm) × mean breadth (to nearest 0·5 mm) of 2nd and 3rd, or 3rd and 4th youngest leaf of a random tiller. One tiller per sampling unit.
	Bulk	Dp	D × Pf.
Carex flacca	Density Performance Bulk	D Pf Dp	As for *B. erectus*, except that any two mature leaves measured.
Poterium sanguisorba	Density	D	Number of rosettes per sampling unit.
	Performance	Pf	Mean leaflet area, measured as mean length × mean breadth (each to nearest 0·5 mm) of penultimate leaflet pairs from mature leaves. Two leaves (i.e. two pairs of leaflets) assessed per sampling unit.
	Bulk	Dp	D × Pf.
Cirsium acaulon *Hieracium pilosella*	Density	D	Number of rosettes per sampling unit.
Brachypodium sylvaticum			Number of tillers per sampling unit.
Briza media			Number of inflorescences per sampling unit.

[a] From Austin (1968a).

to environmental factors, and if the pattern of plant distribution and that of, say, a certain soil chemical factor are found to be correlated (for example by covariance analysis), strong evidence accrues for the significance of that factor in the distribution of the species, and guide lines for further experimental work are revealed.

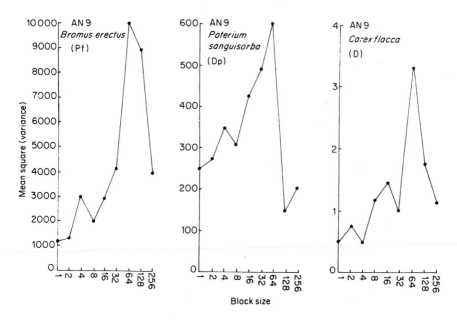

Fig. 5.24. Examples of mean square: block size graphs from chalk grassland. Block size 1 = 5 cm square. Four replicated strips of these units made up one analysis (AN 1, etc.–AN 9 shown here), and analyses were paired. Pf, Dp and D are defined in Table 5.12. From Austin (1968a).

Austin (1968a) analysed pattern in a number of chalk grassland stands on the North Downs near Sevenoaks, Kent, in which the dominant grass was *Bromus erectus*. He selected for study the more important and readily measured species in the sward, including *B. erectus* itself, *Brachypodium sylvaticum*, *Briza media* and *Carex flacca* among the graminoids, and *Poterium sanguisorba*, *Cirsium acaulon* and *Hieracium pilosella* from the herbs. Attributes measured for each species are summarised in Table 5.12 and examples of the influence of block size on the variance of these attributes is shown in Fig. 5.24. Not all these results were investigated in detail, but primary peaks attributable to the size of the morphological unit occurred at block sizes 1–2 for *Carex, Poterium* and *Cirsium*; at 4 for *Briza*; between 2 and 8, centring on 4, for *Bromus*; 8 for *Hieracium*; and 16 for *Brachypodium*. Secondary and tertiary peaks were attributed to competitive and environmental effects, and correlation and covariance analyses performed on these data suggested that the major cause of pattern was phytosociological dominance by *Bromus*, although it is by no means clear to the uninitiated exactly how this conclusion was reached. However, a meticulous dissection of carefully chosen patches of uniform turf, coupled with an ingenious method for recording the direction of growth of the stolons and tillers involving the construction of a polygon of vectors, showed a fairy-ring pattern which had obvious significance in the demography of the turf. Of this, more later.

Hall (1971) conducted a similar investigation further east along the North Downs escarpment near Maidstone, on chalk grassland which had been invaded by *Brachypodium pinnatum*. The interrelationships between *B. pinnatum* and the two grasses associated with it, *Bromus erectus* and *Festuca rubra*, were analysed, as well as concentrations in the soil of nitrogen, iron and boron, which earlier work (Hall, 1967) had suggested might repay further study. Pattern analysis (Fig. 5.25) revealed distinct peaks in variance in all plants and soil factors tested. A combination of covariance analysis (applied where there were common scales of pattern between species and soil factors) and garden experiments led to the rather surprising suggestion that *B. pinnatum* is a poor competitor, especially against *Bromus erectus*, but, interestingly, that it is able to take advantage of relatively high boron levels, which in turn are deleterious to the growth of *Festuca rubra*. Yet, an outstanding feature of the sward was the lack of any dominant species. Hall concluded that this could be explained by small-scale mosaics of relatively high and low N and B levels (and probably other factors as well), which ensure that no one species is normally able to dominate. No significant effects of iron were found.

5.5.4 Multivariate analysis

Ordination of data can, like pattern analysis, help to make sense out of matrices of species counts and environmental measurements. Perring's work (Figs 3.12 and 5.10) is, in fact, a simple example of a two-dimensional ordination of species and environmental factors along axes which coincide with real and immediately recognisable influences: slope and aspect. Grime's triangular ordination of species and stands (Figs 4.1 and 4.2) and Fenner's grids (Fig. 5.16) are other examples. A technique becoming more widely used is principal component analysis, in which field data are ordinated on a basis which is entirely abstract. Plant stands are arranged in two- or three-dimensional representations of "hyperspace", and those similar in species composition tend to cluster together on the ordination. Causes of similarity may be pinpointed by including environmental measurements in

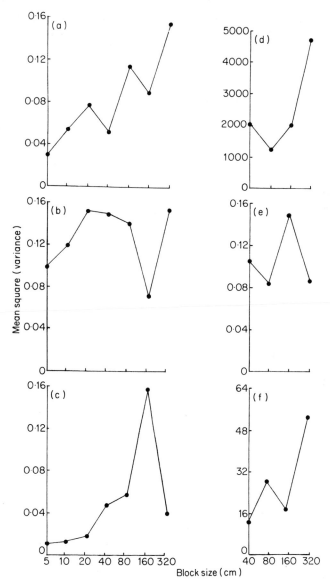

Fig. 5.25. Pattern analysis of plant and soil factors in chalk grassland invaded by *Brachypodium pinnatum*. Block size is expressed as 1 cm lengths of transect. Graphs show dry weights of (a) *Brachypodium*, (b) *Bromus erectus* and (c) *Festuca rubra*; (d) indicates nitrogen, (e) boron and (f) iron level in soil. From Hall (1971).

the ordination, provided they are measured in, or can be transformed to, units of a comparable order of magnitude. Bradley subjected the results of his survey of Grangelands (Section 5.5.1) to a stand ordination, which helped to consolidate his descriptive data and drew attention to features worthy of further investigation. It is also possible to ordinate species, and to extract and display species clusters with similar ecological requirements. For more details of this and comparable methods, the reader should refer to appropriate texts on multivariate analysis, for example those by Gittins (1969, and see below), Shimwell (1971a), Kershaw (1973), and Goldsmith and Harrison (1976).

Where some sort of pattern or trend can be seen in, or expected from, the original data, the chances are good that something meaningful will come out of the ordination, though its very simplicity may make it rather pointless, except as an educational exercise. On the other hand, the more equivocal the input of data, the more difficult it is to make sense of the results. Two contrasting examples serve to illustrate this point. The first, by Gittins (1965a,b), bears only indirectly on the theme of this book, for it was actually carried out on Carboniferous Limestone grassland in Anglesey, but it convincingly extracted the eutrophic grasses *Phleum bertolonii* and *Dactylis glomerata* (or stands dominated by these species), which were associated with more fertile pockets where sheep tended to lie-up at night. In turn, the stands composed of less-demanding species were separated into those typical of the deeper, neutral soils, and those of the shallower soils where the limestone was closer to the surface.

The second example, though conducted specifically on chalk grassland (Austin, 1968b), is less open to clear and unambiguous interpretation, and demonstrates the futility of expecting ordination necessarily to extract something sensible out of too homogeneous or extensive an input of raw data from the field. A mass of data was fed to the computer, including measurements of the performance of various species (see Table 5.12) and of these environmental attributes: aspect, angle of slope, estimated evapotranspiration, and the depth, pH, water content and exchangeable potassium, calcium and available phosphorus content of the soil. A small sample of the results is shown in Fig. 5.26. In fact, it was impossible for Austin to draw more than the most tentative conclusions from this exercise, concerning the effect of competition from the dominant *Bromus erectus*, and the influence of "variation in a soil property correlated with soil depth, available phosphorus and exchangeable calcium, and in a factor related to aspect and the ratio of calcium to potassium".

It is a fact of life that "good" field ecologists and "good" statisticians are not always the same animal. In the past decade, the person in the field has been expected to grasp more and more advanced mathematical concepts, and several books have appeared on the market with this objective. Among these, perhaps that of Kershaw (1973) can be singled out for the extent to which the author guides the reader through the subject to what should be, but often is not, its logical conclusion: the solution of ecological problems. Nevertheless, many published papers on quantitative ecology are simply incomprehensible to the non-mathematician, and have either to be accepted in blind faith or, regrettably, disregarded as of statistical interest only. Admittedly, this argues for the simple expedient of team work, but it is plain

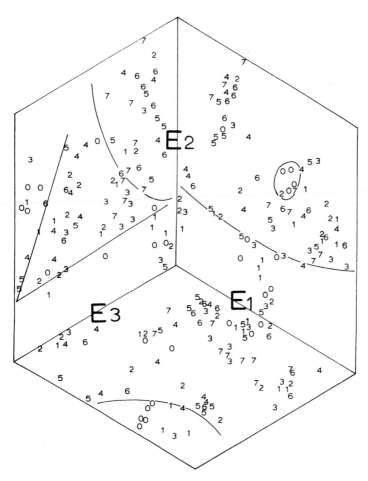

Fig. 5.26. Ordination of performance (Table 5.12) of *Bromus erectus*, expressed in octiles, on environmental principal component axes (E_1, E_2, E_3). From Austin (1968b).

enough that descriptive and quantitative ecology have yet to be profitably fused.

There is, too, the question of continuity. Most of the quantitative research referred to in the foregoing paragraphs was undertaken for higher-degree work, performed as much for familiarisation with techniques as for the ecological value of the results, and consequently in virtual isolation from other work. Participants have frequently departed for pastures new just at the stage when they might begin to take up issues of more fundamental ecological interest raised by their work, leaving the whole project in limbo. One of the best foundations for an objective approach to the study of chalk grassland was laid by Perring (Sections 3.7.3–3.7.5 and 5.3.2). Yet his many valuable pointers to further research have still to be taken up, even though they were made almost 20 years ago. It is encouraging to see, as in Bradley's survey for example (Sections 5.5.1 and 5.5.4), an increasing emphasis on the application of the results of work of this kind, particularly to problems of conservation.

6
Chalk Grassland: II.
Biotic Influences

6.1 Introduction

There are three main areas of interest under the heading of biotic influences upon chalk grassland: (i) the dynamics and interrelationships of the plants which constitute the sward, (ii) the effects of sheep and other grazing animals, as well as cutting and trampling, and (iii) the ecology of all the other animals not covered by (ii) with the main emphasis on the invertebrates. All three categories are tortuously interwoven, and in particular item (ii) grossly affects the other two. Nevertheless, it permits a logical expansion to deal with the relevant points in this order.

6.2 The Dynamics of the Grazed Chalk Grassland Sward

6.2.1 The time factor

The review so far has presented rather a static view of chalk grassland, but like any plant community it is a dynamic system, in a constant state of flux. Temporal changes in botanical composition can be recognised under two main headings, phenological and demographic, though there is some degree of overlap between the two. The first of these encompasses the seasonal waxing and waning of the component species, giving the familiar and striking contrasts in the appearance of the sward according to the time of year. Demography strictly covers fluctuations in population structure as new plants are recruited and others die off, but the term is stretched here to include competitive, allelopathic and other interactions both within and between species.

6.2.2 Phenology

Every year, the plants of each species begin or renew their vegetative growth either from seed or perennial meristems, in response to changing regimes of temperature and daylength. Buds open, shoots elongate, leaves expand, roots grow and flowers appear, although the exact order of these onto-genetic phases varies according to species. We saw in Chapter 4, for example, how the flowering shoots of coltsfoot precede the leaves by several weeks. The timing of each stage is likewise genetically controlled, each species appearing at its appointed time, though liable to some variation depending on its geographical location and the nature of the season (see Smith, 1968), as well as on management.

At the turn of the year, the sward owes its verdure—if indeed it exhibits any—to winter-green perennials such as the fine-leaved fescues, clovers, *Poterium sanguisorba*, *Ranunculus bulbosus* and *Hieracium pilosella*, as well as to certain mosses (notably the bulky *Pseudoscleropodium purum)* which become particularly conspicuous at this time, though declining as the summer advances. Subsequently shoots of other species such as *Bromus*

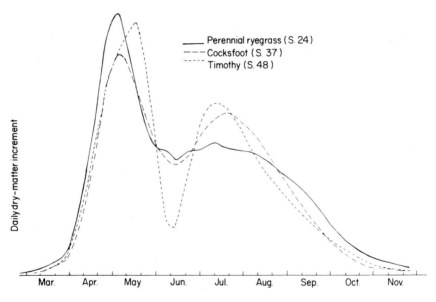

Fig. 6.1. Characteristic seasonal growth curves of three cultivated pasture grasses (perennial ryegrass, *Lolium perenne*; cocksfoot, *Dactylis glomerata*; timothy, *Phleum pratense*), measured at the Grassland Research Institute, Hurley, during 1964. The first peak coincides with ear emergence. From Anslow and Green (1967).

erectus, *Leontodon hispidus*, *Cirsium acaulon* and *Ononis repens* appear from overwintering buds, and the sward as a whole exhibits the characteristic flush of spring growth, well documented for the economic grasses and legumes (Fig. 6.1), but only scantily for wild species (Fig. 6.2; see also Wells, 1971; Williamson, 1976).

The earliest plants to flower in the spring are strictly hedgerow and

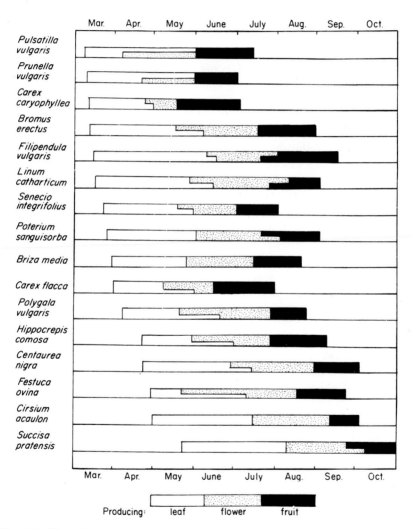

Fig. 6.2. Phenology of 16 chalk grassland species. Information from records kept during observations on heavily sheep-grazed chalk downland in Bedfordshire between 1964 and 1966. From T. C. E. Wells (1971).

wood-edge species, mentioned in other chapters, such as coltsfoot, sweet and wood violets (*Viola odorata* and *V. reichenbachiana*), primrose (*Primula vulgaris*) and dog's mercury, but strongly characteristic of chalk grassland is the slightly later flowering *Viola hirta*, sometimes represented by ssp. *calcarea*. *V. hirta* is noteworthy for the remarkable expansion of its leaves after the first flush of spring flowers, so that the plants grow into large, conspicuous clumps. According to Lousley (1950) the earliest flowers set very little seed, but the summer leaves conceal at their base a second flush of cleistogamous flowers "like little swollen fleshy buds" which set seed in abundance. The capsules bury themselves in the turf as their seeds mature.

Other downland plants which come into flower during April include the cowslip (*Primula veris*), ribwort (*Plantago lanceolata*), the woodrush *Luzula campestris* and the sedges *Carex caryophyllea* and *C. humilis* (the last strictly confined to the western chalklands and even there rarely setting seed). The most consistently abundant and conspicuous sedge in chalk turf, *C. flacca*, begins flowering during May, by which time many more species are following suit, such as *Plantago media*, *Poterium sanguisorba*, several more grasses and the common milkwort, *Polygala vulgaris*, in which individuals show a marked variation in flower colour from blue through mauve to pink and occasionally white. There is also a notable preponderance of plants with yellow flowers such as the bulbous buttercup (*Ranunculus bulbosus*), rock-rose (*Helianthemum chamaecistus*), the legumes *Lotus corniculatus* and *Hippocrepis comosa*, and the composites *Leontodon hispidus*, *Hieracium pilosella* (the latter with flowers of a distinct pale lemon colour), and the dandelions *Taraxacum officinale* and *T. laevigatum*. Of the last two, *T. laevigatum* (the lesser dandelion) is distinguished by a small appendage on each involucral bract, but small plants frequently occur without this modification and are referable to *T. officinale*.

Some of these vernal species have a relatively short flowering period and a few die down altogether above ground after flowering, notably *Ranunculus bulbosus*, of which there may be no sign from late July to early September except for "a small hole in the ground through which the corm is often visible" (Sarukhán and Harper, 1973). This phenomenon is known as aestivation and is seen in several woodland species (Chapter 8). Most constituents of the chalk turf continue to flower and vegetate, however, at least through the mid- to high-summer period of June and July, when many more species appear, ranging in size from the diminutive *Asperula cynanchica* to the robust *Chrysanthemum leucanthemum*. By mid-July, the sward is at its most diverse and floriferous, and cannot be done justice to by a mere recitation of names.

During August, flowers and foliage begin to take on a tired look as the cycle passes its peak for another year, though the umbellifers *Pimpinella saxifraga* and the much rarer *Seseli libanotis* typically flower at this time, and a further full month may pass before the biennial gentians (*Gentianella amarella* and *G. germanica*) and the diminutive autumn ladies tresses orchid (*Spiranthes spiralis*) appear. Several species display a particularly sustained flowering period, and if the season is sufficiently "open", these may flower well into the autumn. Common examples are *Blackstonia perfoliata*, *Campanula glomerata* and *C. rotundifolia*. Finally, a sufficiently damp season usually results in a brief flush of mushrooms and toadstools, such as species of *Agaricus*, *Psathyrella* and *Pareolus* in the turf itself, and *Strophoria aeruginosa* and *Coprinus* spp. on dung (T. C. E. Wells, personal communication).

An indication of the time and duration of the flowering periods, as well as vegetative and fruiting stages, of a number of chalk grassland species is shown in Fig. 6.2. It hardly needs emphasising that it is essential to allow for these seasonal changes in any major survey, and to collect records throughout the growing period. It is well known that errors can arise during subjective assessment through overestimating the relative contribution to the sward by species which happen to be the most showy. But even the most carefully executed quantitative analysis can miss important information if it is undertaken at the "wrong" time of the year. It is easy to see, for example, how *Ranunculus bulbosus* might be under-recorded or even missed altogether in a survey conducted in high summer, despite the fact that the species may have made a major contribution to the sward some six weeks previously.

Exactly which environmental factors these plants respond to ontogenetically is largely a matter for speculation, for very little experimental work of this nature has been done on wild plants, and most information which exists refers either to cultivated plants or their accompanying weeds. But a few generalisations can safely be made. As we noted before, vegetative growth normally proceeds once a certain threshold of temperature is passed. Provided the temperature does not rise too high, and that all other needs are satisfied, the rate of increase in dry matter per unit area per unit time will be directly related to the accumulated temperature, an integrated value expressed as "day-degrees", and determined by summing the mean daily excess of temperature above the minimum for growth for each species (where this is known). However, for seed germination and reproductive growth the situation may be more complex, and many species need either a period of low temperature (resulting in the process known as vernalisation), or a particular photoperiodic regime, before germination or flowering can take place. With regard to flowering, for example, *Chrysanthemum leucanthemum* is an obligate long-day plant (Howarth and Williams, 1968). Almost nothing is known of the phenology of root growth in wild plants. In cultivated grasses, root growth builds up to a peak in spring (its maximum rate slightly ahead of that of the foliage), declining or even ceasing as flowering begins, then increasing again to a smaller peak in autumn (Troughton, 1957). Observations by Elston (1963) showed a similar pattern in chalk turf dominated by *Festuca rubra* (Section 9.3.3.4).

Whatever their cause, the consequences of this dispersal in time of vegetative growth and flowering can be assumed to minimise competition between species for such factors as light, water and nutrients (and possibly pollinating insects), and to ensure maximum efficiency in the utilisation of these commodities by the community as a whole. The significance for grassland management is obvious, particularly where the aim is to conserve certain species, and we return to this point again in Chapter 10.

A point which so far has not been touched on is the variation of the contribution from each species from year to year: the well-known phenomenon of "good" and "bad" years for particular plants. The cause may be purely phenological, a "good" year resulting, for example, from ideal climatic conditions for vigorous vegetative growth, and optimal flowering, pollination and fruit-setting, operating on essentially the same plants that gave a poor show in the previous year, perhaps as a result of frost, drought or excessive wet. Alternatively, the cause may be demographic, the changes resulting from fluctuations in population structure. This aspect is considered separately in the next section, though clearly phenological and demographic effects overlap: for example, prolific seed production in one year may result in a large crop of seedlings in the next. It is equally clear that many year-by-year changes result from animal activity, but this aspect is deferred until later.

6.2.3 Demography

Intimately associated with phenology is the other major source of innate temporal variation within the community: the actual coming and going of individual plants. The study of population dynamics—demography—is a familiar concept in animal, and even woodland ecology, in which the aim is to build up a picture of the rate of turnover of individuals within a population from such information as birth rate (seed germination in plants), age structure, longevity, reproductive strategy and mortality. Information of this kind is known for some crop and weed species but relatively little is available on wild herbaceous plants.

There are four main reasons for this, three practical and one psychological. To take the last point first, few people seem to give a moment's thought to the age of the plants in a herbaceous community in the way that they might to a stand of trees. It would undoubtedly come as a surprise to many to learn that a patch of red fescue turf may contain plants which actually pre-date even the mature trees of the nearest beechwood by several hundred years. This has been suggested by Harberd (1961) who, on the basis of meticulous morphological and cytogenetic studies, concludes that the vegetative meristems of *Festuca rubra* remain indefinitely juvenile, and that clones of this species in some British localities could be as much as 1000 years old. This is impressive by any standards, but even on a less dramatic scale the thought of, say, a forty-year-old knapweed plant takes some getting used to.

Of the more practical explanations for the general dearth of information, the first stems from the difficulty of recognising individual plants in the intricate weave of shoots which constitutes the turf. Some plants, particularly those which form rosettes, present little difficulty, but for those which spread vegetatively, tedious and time-consuming techniques of turf dissection have to be resorted to, as practised by Harberd (1961) and Austin (Fig. 6.3). For these plants, it is usually necessary to adopt the vegetative ramet (the tiller of the Gramineae) as the functional unit. The second point is that, with the exception of certain species such as *Cirsium acaulon* (Fig. 6.4) which can be dated from morphological features, it is impossible to tell the age of individual plants unless they have been marked from their first appearance. This, too, can be surprisingly difficult, for grazing animals (and nosey people) pull off tags and knock over marker pegs, and the only solution is to map the plants carefully at regular intervals, ensuring that the permanent quadrats which contain them can themselves be relocated.

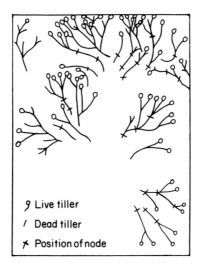

9 Live tiller

/ Dead tiller

✗ Position of node

Fig. 6.3. Microstructure of turf of *Bromus erectus* as revealed by dissection. From Austin (1968a).

The third point is amply demonstrated by the previous two but is worth stressing separately: this is that work of this kind is very demanding. Long-sustained records have to be kept, ideally for many decades, yet at the same time almost round-the-clock vigilance is required if all the relevant information, particularly on seedling establishment and mortality, is not to be missed. It is a daunting task, and for this reason alone it is not surprising that relatively few ecologists have been both able and willing to undertake such work, despite its obvious attraction and inevitable rewards.

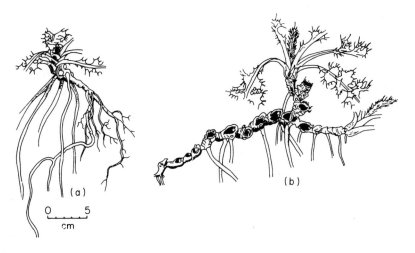

Fig. 6.4. Plants of *Cirsium acaulon*, (a) 1 year old and (b) over 13 years old. The circular scars result at the end of each growing season from the decay of the basal rosette. From Pigott (1968).

In Britain, this neglected branch of plant ecology has recently been taken up by J.L. Harper and his associates and given characteristically elegant treatment. We have already touched on one example in Chapter 4, in the reference to Ogden's work on coltsfoot. Following their work on buttercup (*Ranunculus*) populations in North Wales, Sarukhán and Harper (1973) emphasise

> the great importance of mapping and frequent records in any study of the dynamics of a plant population. The relative stability of plant numbers . . . conceals a very dynamic population flux which would have been completely obscured in a census of total population size made at yearly intervals.

Nothing encompassing quite such intensive recording or sophisticated analysis has yet been done on chalk grassland. Wells (1967) observed a population of autumn ladies tresses orchids (*Spiranthes spiralis*) in

Bedfordshire from 1962 to 1965 (see Chapter 10), while Watt (1940, 1962) noted fluctuations in both number and biomass of chalk grassland plants in Breckland (see below). Recording in both cases was done annually. But most information has to be gleaned from the autecological sections of the "Biological Flora of the British Isles", published serially in the *Journal of Ecology*, and from monographs such as that on the orchids by Summerhayes (1951).

6.3 Grazing Animals and Birds

6.3.1 Sheep

Chalk grassland—indeed, any mature pasture—grazed essentially continuously by sheep has a distinct appearance, neatly trimmed to within a few centimetres of the soil surface, and showing no rank patches or bare ground, other than the occasional lie-ups and the inevitable sheep-paths. Of course, this evenness is largely due to the judgement and skill of the shepherd or manager, who moves his animals about the down as the season allows or demands. Nevertheless, classic sheepwalk is remarkably resilient to all but the grossest changes in management, and Defoe's name for it, "carpet-ground", is delightfully apt. There are several good reasons for this, relating to the size and weight-distribution of the sheep, to its grazing habits and preferences, and to the way in which it disposes of its dung and urine. A typical downland ewe weights about 75 kg at maturity, and the relationship of this weight to the size of the hoof results in a pressure on the turf which is sufficient to keep it well consolidated without crushing it excessively or exposing or compacting the soil beneath (Table 6.1). Rabbits and cattle differ very markedly in this respect from sheep, and the effects of these other animals are considered in due course.

Table 6.1

Hoof stress on pasture by cattle and sheep [a]

	Cow	Sheep
Live weight (kg)	500	75
Total hoof area (cm^2)	320	80
Hoof stress, standing (kg cm^{-2})[b]	1·56	0·94

[a] From Frame (1975).
[b] Clearly the stress will be greater than this during walking or running, or on steep ground where a sliding component enters.

As in all livestock, individualists occur with their own special likes and dislikes: faecal analysis indicated that one wether in the grazing experiments at Aston Rowant NNR ate unusually large quantities of the moss *Pseudoscleropodium purum*, though this may have been because it had a deformed palate (Hawes, 1971). Sheep sometimes balk at long, lush grass, and only the seeds may be taken from coarse, stemmy herbage past maturity, but they graze ragwort (*Senecio jacobaea*), well known for its toxicity to cattle, with no apparent ill-effects (Harper and Wood, 1957). Sheep bite off the herbage by bringing the incisors of their lower jaw up against the hard, toothless pad of the upper jaw, and pulling. They can pull insufficiently anchored plants clean out of the ground. Compared with cattle, sheep distribute a relatively modest quantity of dung and urine evenly about the sward (Table 6.2), and the amount of scorching and fouling is therefore minimal (Fig. 6.5a).

Table 6.2

Approximate daily intake of water and production of dung and urine by sheep and cattle [a]

	Sheep	Cattle
Live weight (kg)	70	350
Water intake (l)	4–7	30–40 [b]
Urine produced (l)	1–5	10–25
Fresh faeces produced (kg)	1·8 [c]	34 [d]

[a] Data from R. V. Large (personal communication).
[b] More in lactating cows.
[c] Typically produced in numerous small pellets of about 2 g each.
[d] Produced in perhaps 12 or more pats.

As noted above, some sod-pulling may sometimes occur, but on the whole the only places where sheep actually damage the turf (apart from gateways and around water troughs) is along their well known sheep-paths and in the cavities they sometimes hollow out of slopes where they lie-up at night. Sheep-paths, which may be accentuated by soil creep, can lead to marked zonation in plant distribution: tall grasses, typically *Bromus erectus*, sometimes form a sort of miniature hedge along the lower edge of the path (see Tansley and Adamson, 1926, p.9; Thomas, 1959). Species typical of disturbed or trampled ground may colonise the paths themselves (see below). Dung and urine (and wool) may accumulate in the lie-ups and so enrich the soil there; in his work on limestone grasslands, Gittins (e.g. 1965a) found that the more nutrient-demanding (eutrophic) grasses *Phleum bertolonii* and *Dactylis glomerata* occurred mainly in these lie-up areas.

Over the past 40 years or so, virtually all of the chalk downland sheep flocks have disappeared, and those that remain are based on intensive leys, not native pastures. In many places, dairying and beef-rearing enterprises, arable agriculture or forestry have taken their place. In others, the downs have simply been left to nature (or the gamekeeper). These changes, which are described in more detail in Chapters 9 and 10, had been gathering momentum for the best part of two centuries, though the ecological effects of the dwindling sheep populations were largely offset until the mid-1950s by the rabbit.

6.3.2 Cattle and other domestic livestock

This is not the place to discuss practical aspects of downland beef fattening or dairying, or the various reasons for the substitution of sheep by cattle (see Chapter 9), but the effect of the change is plain enough, for reasons already stated: a longer (10–15 cm), coarser, floristically poorer sward results, partly from the way in which cattle graze, wrapping their tongues around the foliage and pulling, rather than biting it off, and partly from their far greater production of dung and urine (Table 6.2). Scorching and fouling (Fig. 6.5b) lead to death, or at least avoidance, of herbage, although the combined activities of dung-flies, dung-beetles and other invertebrates, as well as fungi and bacteria, ensure the release for recycling of the nutrients contained, and the ensuing enrichment of these patches of soil, particularly with N and K. Chinery (1973) notes the predatory yellow dung-fly *Scatophaga stercoraria* and the dor beetle *Geotrupes stercorarius* as common on cow-dung, and another dor beetle, *Typhaeus typhoeus*, on that of rabbits and sheep.

These nutrient-rich patches are rapidly taken over by eutrophic species of grass and herb. Some of these, such as *Dactylis glomerata* and *Taraxacum officinale* and others of comparable phenotypic plasticity, may result simply from the enormous increase in size and vigour of plants already present as small, depauperate tussocks, tillers or rosettes. Others, notably the stinging nettle (*Urtica dioica*) and the thistles *Cirsium arvense* and *Sonchus arvensis*, invade the sward anew either vegetatively or from seed. The finer species disappear, presumably mainly through competition from the coarser plants.

Poaching, especially in winter, is far more a feature of cattle-grazed pasture than it is with sheep because of the difference in the ratio of weight to total hoof area (Table 6.1). The turf may be seriously damaged if not totally destroyed, and the soil badly puddled and compacted. These areas are then open to colonisation by the familiar gateway and water-trough species such as *Plantago major*, *Matricaria matricarioides* and *Polygonum aviculare*. Among the named chalk soils, Cope (1976) mentions the relatively clayey

Fig. 6.5. The contrast in appearance and immediate effects of (a, top) sheep dung and (b, bottom) cattle dung on grazed turf. (C.J.S.)

Wallop series (Chapter 3) as particularly susceptible to poaching by out-wintered livestock.

Grazing horses create a distinct mosaic of vegetation, cropping some areas to within a few centimetres of the ground and favouring there a wealth of small herbs and grasses, but leaving other parts untouched (horse-sick), especially where they drop their dung, so that a mat of coarse grasses accumulates. Grazing pigs root about in the turf, and if confined can destroy it utterly within a few weeks with a combination of snouts and trotters, every bit as effectively as if it had been ploughed or rotavated. Further information on the effects of grazing by different kinds of livestock can be found in Norman (1957), Norman and Green (1958), Kydd (1964) and Duffey *et al.* (1974, Chapter 8).

6.3.3 The rabbit

The rabbit (*Oryctolagus cuniculus*) is not native to Britain. It was intro-duced for its meat and fur late in the eleventh century by the Normans, who kept the animals (then known as coneys) in warrens. Although there must inevitably have been escapes from time to time, the rabbit never really became numerous in the wild until the 19th century, when the old warrens were beginning to be broken up and turned into sheep grazings or arable land. Hedgerows, many of which were newly established, formed important centres for the survival and spread of the feral rabbits, while overwintering was made easier for them by the extension of the grazing season with improved grasses, the development of winter crops, and the control of natural predators, especially fox (*Vulpes vulpes*), stoat (*Mustela erminea*) and birds of prey, by gamekeepers.

In places, rabbits themselves continued to be deliberately managed for sport, even to the extent of installing game warrens. However, enlightened farmers naturally attempted to control the rabbits which their improved husbandry had inadvertently encouraged, and it was not until the golden age of British farming gave way to the periods of depression, particularly evident between the wars, that rabbits really increased. Sheail (1971), from whose informative and absorbing account much of this information comes, esti-mates the rabbit population of Britain as 50 million (four times the number of sheep) just before the 1939–1945 war, and 60–100 million by the begin-ning of the 1950s, when "one or two rabbits per acre were quite usual and in heavily infested areas, fifteen to twenty rabbits were quite common".

Then came one of the most spectacular mammalian population crashes ever witnessed, when myxomatosis was deliberately introduced from France as an early experiment in biological control (see Lockley, 1964). The myxoma virus which caused the disease was highly virulent and absolutely specific to the rabbit, and its spread, effected by the rabbit flea *Spilopsyllus cuniculi*, was greatly aided by the almost continuous distribution of rabbits from coast to coast. "The onset of the disease", writes Sheail, "was dramatic, its course was swift and deadly, and the results were sensational". In fact, by the winter of 1955–1956, 99 out of every 100 animals had died.

Myxomatosis has certainly not disappeared, but less virulent strains of the disease now predominate, and it has died down and stabilised at a relatively low level of occurrence comparable with endemic rabbit ailments such as coccidiosis. The recovery of the rabbit population has undoubtedly been favoured by factors other than just the diminishing potency of myxomatosis, however. A long series of mild winters, the combined effects of dressed corn and gamekeepers on natural predators, and the demise of the government-backed Rabbit Clearance Society scheme, may all have contributed to that steady increase in numbers. Locally, and not least in the chalklands, rabbits are now once again to be seen in concentrations reminiscent of the early 1950s.

Despite the inroads into farm and garden crops which earn the rabbit its status as a pest (Fig. 9.16), its total extermination never seems to have been taken really seriously in Britain. It has always provided sport for those with a hankering to carry a gun under their arm, and for the more sentimental of us, the rabbit, like the grey squirrel, is regarded with affection for its gentle, endearing appearance. Traditional bunny rhymes and tales have been augmented recently by Richard Adams' celebrated story of Watership Down. From the aesthetic point of view, too, the rabbit has long been regarded as a valuable aid to landscape conservation. Thus, the British Ecological Society (1944) opposed the wholesale extermination of the rabbit because it helped to maintain the appearance of such habitats as open grass downland. Certainly, when natural food abounds, the activities of the rabbit may in many respects resemble and supplement or even replace those of sheep, in maintaining an open, floriferous sward, and in preventing the invasion of scrub.

Unfortunately, it is a very small step from attractive, close-cropped, simulated sheepwalk to the absolute destruction of the turf which is the mark of intensively populated rabbit territory (Fig. 6.6). Much of our knowledge of the grazing habits and preferences of rabbits stems from the pioneer work of Farrow in Breckland (e.g. Farrow, 1917), continued for many years by A. S. Watt, and extended into the South Downs by Tansley (e.g. Tansley and Adamson, 1925, especially pp. 211–218) and more widely by Thomas (1960, 1963). Southern (1940) and Lockley (e.g. 1964) have contributed classic accounts of the social organisation and behaviour of the rabbit.

Regarding the herbaceous flora, Tansley and Adamson observe: "It does not appear that moderately heavy rabbit attack has much effect on the floristic composition of chalk pasture. . . . A sample of a rabbit infested area shows much the same list of species as one in which rabbits are scarce or absent"; though it is acknowledged that on "steep northern exposures and other places shaded from direct insolation" mosses, which in any case are "generally especially prominent in the herbage" tend to become dominant under the influence of rabbits.

Fig. 6.6. An intensively rabbit-grazed sward on the south-facing slope of Lodge Hill, Buckinghamshire. Cushions of *Sedum acre* and drying crusts of mosses including *Bryum caespiticium* predominate, interspersed with depauperate plants of *Crepis capillaris* and *Rumex acetosella*. All the inflorescences of the grasses have been eaten off. (C.J.S.)

Such generalisations are unwise, however, without reference to both the time factor, and to the possible presence of other animals. Thus, the longer a piece of chalk grassland is subjected to grazing by rabbits alone, so the ultra-short pile of the turf progressively lifts away from the surface of the mineral soil, the rabbits riding rather than consolidating it as they graze. It is this raised mat which gives the turf its cushioned resilience which is so highly praised by hikers. But it renders the sward highly vulnerable not only to drought and to "scorching" where the rabbits urinate, but to scuffing by feet and hooves, and particularly to damage by the numerous scrapes which the rabbits dig, apparently in their search for roots to eat. Obviously, where rabbits simply supplement sheep, this is less likely to occur—a point made long ago by the Sussex bailiff who told Hope-Simpson (1941b, p. 239): ". . . the treading down of the turf by the sheep has a very important effect in keeping it intact and preventing rabbit scraping and subsequent frost from destroying it".

Exactly what plants get eaten by rabbits depends on how much herbage is available and how large a population is being supported. Rabbits tend to favour young, succulent buds, shoots, leaves and inflorescences, but they are quite partial to dry and even dead grass, as well as to the roots which they dig for and the bark which they strip from shrubs and trees. It is widely accepted (see, for example, Farrow 1917; Tansley and Adamson, 1925) that certain species are normally avoided, even in the thick of dense rabbit colonies. Examples are nettles (*Urtica*), burdocks (*Arctium*), ragwort (*Senecio jacobaea*), common mullein (*Verbascum thapsus*), nightshades (*Atropa belladonna* and *Solanum dulcamara*), stonecrop (*Sedum acre*), thyme (*Thymus*), sandworts (*Arenaria*) and forget-me-nots (*Myosotis*). Yet, surprisingly little really objective work has been done on the diet of wild rabbit populations, and recent studies by Williams *et al.* (1974) and Bhadresa (1977), who compared the botanical composition of herbage on offer to rabbits with that of the epidermal fragments voided in the faeces, indicate the importance of time of year on dietary preferences, and that previously unsuspected species may prove to be unpalatable. Bhadresa discovered, for example, that the grass *Poa pratensis* and the moss *Dicranum scoparium* were undoubtedly disliked by rabbits.

Some of the herbs which become more obvious under the influence of rabbits may already have been present in the turf for some time. For example ragwort may persist for many years under sheep grazing purely as a vegetative rosette, throwing up its flowering shoots only when the sheep are removed. Similarly, initially quite modest quantities of the rock-rose (*Helianthemum chamaecistus*) can steadily extend their long, straggling shoots through the most closely grazed fescue turf, and a generous scatter of their attractive yellow flowers is a sure sign of rabbits in large numbers. The same applies to ladies bed-straw (*Galium verum*) and ground ivy (*Glechoma hederacea*).

Table 6.3

A classification of annuals and biennials of chalk grassland proposed by Grubb (1976), according to their characteristic microhabitat in the turf, their probable natural habitat before the first forest clearances, and their major adaptations [a,b]

Group	Microhabitat in chalk grassland	Natural habitat	Major adaptations
A	Small gaps in short turf; a few also able to establish in tall swards.	Basic grassland on edges of woodland (e.g. by cliffs), marsh and fen, and on sand-dunes.	Able persistently to reinvade small gaps in continuous turf. Small size of most species important (minimal demand for water and nutrients, plants "fit between" perennials); species of tall grassland large-seeded. Many features not understood (e.g. mycorrhiza, minute size of some seeds, persistence in soil of others).

Examples Annuals: *Blackstonia perfoliata, Centaurium erythraea, Erigeron acer* [c], *Euphrasia nemorosa, E. pseudokerneri, Medicago lupulina* [d], *Rhinanthus minor* agg. Biennials: *Arabis hirsuta* [d], *Carlina vulgaris, Daucus carota, Gentianella amarella, G. anglica, G. germanica* [e], *Linum catharticum, Picris hieracioides* [d], *Pastinaca sativa, Senecio jacobaea* [d], *Seseli libanotis* [d].

Group	Microhabitat in chalk grassland	Natural habitat	Major adaptations
B	Mostly tops of active ant-hills, rarely mole-hills or rabbit scratchings.	Cliffs and sand-dunes, rarely ant-hills.	Able to establish and persist on open, very light-textured soil. Very small size (really minimal demands for water and nutrients); all normally annual; grow in cool periods; have appreciable drought resistance but mature before most intense drought; need for after-ripening prevents germination when soil surface drought-prone in summer. Some species benefit from excessive drainage and warm soils.

Examples All annuals: *Aira caryophyllea, Aphanes arvensis, Arabidopsis thaliana* [c], *Arenaria serpyllifolia* [c,f], *Catapodium rigidum, Cardamine hirsuta, Cerastium semidecandrum, Erophila verna, Myosotis discolor, M. ramosissima, Saxifraga tridacty-lites, Trifolium dubium, Veronica arvensis, Vulpia bromoides.*

Table 6.3 continued

Group	Microhabitat in chalk grassland	Natural habitat	Major adaptations
C	Wayside sites, cleared scrub, some around rabbit burrows.	Woodland clearings; a few also on cliffs.	Able to reach "unpredictable", short-lived sites caused by disturbance. Large seed output; seeds with specialised dispersal mechanisms and/or long-dormant in soil; mostly biennial; compete effectively for light and nutrients (tall shoots, deep roots).

Examples Annuals: *Galium aparine, Myosotis arvensis* [d]. Biennials: *Carduus nutans, Cirsium eriophorum, Cynoglossum officinale, Hyoscyamus niger* [e], *Inula conyza, Onopordum acanthium, Reseda lutea* [d], *R. luteola, Verbascum lychnitis, V. nigrum* [d], *V. pulverulentum, V. thapsus.*

Group	Microhabitat in chalk grassland	Natural habitat	Major adaptations
D	Bared calcareous soils, mostly on rabbit scratchings.	Sand-dunes and/or cliffs.	Able to grow on slender nutrient supply on stony chalk soil. Life-cycles poorly known and adaptations little understood. All grow actively through the summer; large seeds may aid early development of deep root-systems in spring.

Examples Annuals: *Acinos arvensis, Ajuga chamaepitys* [d], *Iberis amara.* Biennials: *Echium vulgare, Teucrium botrys* [e].

[a] Reproduced by permission of the author and Applied Science Publishers Ltd.
[b] This scheme is plainly tentative but it does pinpoint numerous opportunities for research.
[c] Occasionally biennial.
[d] Occasionally perennial.
[e] Occasionally annual.
[f] This species also found in group A sites.

Most of the characteristic plant associates of rabbits arise anew from seed, however, although the exact process of invasion warrants closer investigation along the lines recently proposed by Grubb (1976) (see Table 6.3). Among the smaller plants, the sandworts and forget-me-nots are very characteristic. The sandworts (*Arenaria serpyllifolia* and *A. leptoclados*) are diminutive greyish-green plants with typically caryophyllaceous flowers resembling miniature stitchworts. They are not at all easy to tell apart, least of all in the depauperate state in which they frequently occur, but *A. leptoclados* (which is regarded by some botanists as merely a subspecies of *A. serpyllifolia*) is more straggly in habit and more slender about the leaves. It has slightly smaller flowers (3–5 mm, compared with 5–8 mm in *A. serpyllifolia*) and is more strictly annual. The fruits and seeds help in identification: in *A. serpyllifolia* the capsules are firm with rounded sides when ripe, and contain black seeds 0·5–0·7 mm in diameter, while in *A. leptoclados* the capsules are more straight-sided, thin-walled, and contain smaller (0·3–0·5 mm) dark red seeds. The forget-me-nots are likewise able to flower in the most remarkably depauperate state. The common *Myosotis arvensis* occurs widely, but more specific to dry, disturbed chalky soils are the smaller-flowered *M. discolor* and *M. ramosissima*, especially the latter. *M. discolor* is readily distinguishable by the yellow colour of its flowers as they open (soon turning blue), and *M. ramosissima* by the marked extension of its inflorescences at the fruiting stage.

Other characteristic species include the composites *Crepis capillaris* and *Erigeron acer* and the crucifers *Arabis hirsuta* and *Iberis amara*, all annuals, biennials or short-lived perennials, the longer-lived *Sedum acre*, and acrocarpous mosses such as *Bryum* spp. and *Tortula ruralis*, among which even *Rumex acetosella* may grow (Fig. 6.6). *Erigeron acer*, the blue fleabane, has slender blue ray florets encircling a yellow centre of disc florets which develop a pappus unusual for its red colour. *Arabis hirsuta*, the hairy rock-cress, is an inconspicuous little plant sending up usually one slender leafy shoot bearing a dense cluster of white flowers each 3–4 mm across. In fruit, the plant takes on an even narrower look, for the fruits (siliquas), which are all held erect parallel to the main stem on stalks up to 8 mm long, themselves elongate to as much as 50 mm in length. *Iberis amara* (the wild candytuft) is more conspicuous than *Arabis*, but much more local, and it is one of the few British plants virtually restricted to the Chalk. Its flowers, which may be pink or mauve as well as white, are larger (up to 8 mm) than those of *Arabis*, and altogether more showy, for they are arranged in a corymb (head) and in each flower the two outer petals are at least twice as long as the inner pair. *Iberis* will flower freely throughout a sufficiently mild winter. The distinct succulent-leaved shoots of the stonecrop form great carpets in heavily rabbit-infested areas, and at its peak flowering period during high summer this species makes a most spectacular display of dazzling yellow, visible from a great distance. With the exception of the annual *Catapodium rigidum*, it is noteworthy how small a part the Gramineae seem to play in the colonisation of these bared patches.

The larger herbaceous species of rabbit ground, typified by the biennial mulleins, burdocks, thistles (e.g. *Cirsium vulgare* and *Carduus nutans*) and the occasional henbane (*Hyoscyamus niger*), and perennials such as *Sonchus arvensis, Atropa belladonna* and *Solanum dulcamara* (see Chapter 4) can grow to impressive proportions on the mounds of soil thrown up around the burrows and enriched by the incorporation of urine, droppings and fur. Many of these showy plants are a great attraction in their prime both for their inherent beauty and interest, and for the insects which visit them. But as the season progresses, and they begin to seed, the whole scene takes on an appearance of desolation and neglect, accentuated on the steeper slopes by soil erosion. Describing heavily rabbit infested ground at Windmill Hill, Hampshire, Tansley and Adamson (1925) write:

> Even on the upper part of the north slope of this hill, where the declivity is not more than 12 or 15°, a great deal of soil is . . . exposed, and the rain washes the earthy constituents away and disintegrates the chalk, so that the surface becomes covered, largely or entirely, with small chalk fragments. The bare white patches thus formed show up on the hillside from a great distance away.

Where the rabbits burrow into the chalk itself, they usually avoid the harder strata. Thus, at Shirburn Hill on the Oxfordshire Chiltern scarp, Bourne (1931) observed that the extensive rabbit warren there (still to be seen) was almost entirely confined to the "rubbly brown chalk with calcareous clay" immediately underlying the Chalk Rock (see Section 7.3.4).

6.3.4 Other mammals

The brown hare (*Lepus europaeus*) is widespread on the chalklands and has similar tastes to the rabbit, but it is not nearly so numerous and is far less destructive, although it will graze off young tree and shrub saplings, typically about 20–30 cm above the ground (see Chapter 9). The hare's form, a shallow hollow in the ground, is in no way comparable to the complex and populous rabbit warren, and renders the young much more vulnerable to predators and agricultural machinery.

Commonly found in closely grazed grassland, the mole (*Talpa europaea*) excavates its runs in the mineral soil usually a few centimetres beneath the surface, throwing up the spoil at regular intervals to produce the familiar mole-hills which, temporarily at least, completely smother the surface of the turf. The mole's diet consists mainly of earthworms, but these are supplemented where necessary by insects, and this is especially true in chalk soils where earthworms may be quite scarce (Chapter 3). Indeed, in the thinnest rendzinas, moles may be forced to excavate elaborate networks of runs in order to catch sufficient food, and the large concentrations of mole-hills which result when this happens can give an exaggerated impression of population size (Mellanby, 1971).

The need for all but the longest-lived plants of herbaceous grassland vegetation to reproduce regularly from seed has already been noted. Though seedlings may frequently be seen establishing themselves in the closest-knit swards with no sign of any bare soil, this process must be greatly aided by the formation of mole-hills. Buried seeds are brought to the surface and these, together with any which arrive on the wind, are able to germinate and establish under conditions of minimal competition from the mature turf, though some of the more robust plants may push through from underneath, and others spread vegetatively from the surrounding turf. This disturbance of the soil must be especially important where annuals and biennials form a significant proportion of the flora, as in Breckland (Watt, 1974).

Information to substantiate these ideas is scarce, but during observations on chalk grassland in Oxfordshire and Wiltshire, King (1972) found these species colonising mole-hills: *Festuca rubra, Helictotrichon pratense, Carex flacca, Plantago lanceolata, P. media, Campanula rotundifolia, Poterium sanguisorba, Leontodon hispidus, Galium verum, Cirsium acaulon, Veronica chamaedrys* and *Sedum acre* (most of which had grown from root fragments, or spread vegetatively from the surrounding turf), and the annuals *Arenaria serpyllifolia, Linum catharticum* and *Fumaria* spp., which established from seed. A longer-term study on the colonisation of mole-hills in Jurassic limestone grassland in Cantal, France, is quoted in some detail by Elton (1966, p. 116). Note that mole-hills differ from ant-hills in several important ways. For example mole-hills consist of "proper" soil, complete with viable propagules, and they are thrown up only once and are not subject to repeated heaping-on of more soil (King, 1977a; see also Section 6.4.4.2).

6.3.5 Downland birds

It is less easy to talk of typical birds of the downs because of their mobility. Indeed, during times of migration almost anything might be seen on the wing, for it is now well established that the chalk scarps mark major flyways for migrating birds. Breeding species likely to be encountered will depend very greatly on the nature of the surrounding soils, vegetation and land-use. Scrubland, in particular, greatly increases the diversity and number of breeding birds (Chapter 7). On the other hand, where trees and hedges are quite absent, birds may be forced to nest on the ground which are not characteristically found there. Thus Sheail (1971) cites examples of jackdaws (*Corvus monedula*) nesting in old rabbit holes on the Yorkshire Wolds. Proximity to open arable land increases the likelihood of putting up great whirrings of partridges (*Perdix perdix*) or of hearing the curious but unmistakable call of the cornbunting, *Emberiza calandra* (see Fig. 7.12). The lapwing (*Vanellus vanellus*), skylark (*Alauda arvensis*) and meadow pipit (*Anthus pratensis*) are likewise birds of the champaign, and as readily discerned by the ear as by the binoculars. Of the birds of prey, the kestrel (*Falco tinnunculus*) is commonly seen hovering over the downs, while the buzzard (*Buteo buteo*) is becoming more common again over the more westerly chalklands.

But two birds, normally summer visitors, must be singled out as especially characteristic of chalk downland: the stone curlew (*Burhinus oedicnemus*) and the wheatear (*Oenanthe oenanthe*). Both are illustrated in Fig. 6.7. The stone curlew, which nests on open ground among turf or arable crops, provides a most remarkable example of camouflage, eggs, fledglings and adults all blending perfectly with their background, especially, as often occurs, if this includes surface-strewn flints. The adult birds actually lie down, head on ground, at rest, and when they do move they keep the head low. These features, coupled with their nocturnal feeding habits, make stone curlews elusive birds at the best of times, but the species has greatly decreased in the last 20 years, probably above all because of the incompatibility of its life cycle with the bustle of activity which accompanies present-day spring cereal cultivation on the downs.

The wheatear, though smaller, is much more conspicuous than the stone curlew, and is more widely distributed. On the downs, it shelters and nests in rabbit burrows, where, according to Coward (1950), the fledglings "group

(a)   (b)

Fig. 6.7. (a) Stone-curlew, *Burhinus oedicnemus*, and (b) wheatear, *Oenanthe oenanthe*, characteristic birds of closely-grazed chalk downland. (a) About one-fifth life-size, (b) about three-eighths.

themselves at the mouth of the hole when waiting for food, but scuttle back on absurdly long legs if danger threatens". The wheatear is mainly insectivorous, Coward noting cinnabar larvae and an adult small heath butterfly as examples of food taken. Hudson (1900) records a curious tradition of eighteenth and nineteenth century shepherds of the Sussex Downs of catching wheatears for the local gourmets as the birds assembled in late summer for their migration to continental Europe. More details of this can be found in Chapter 10.

6.4 The Invertebrate Fauna of Grazed Chalk Grassland

6.4.1 Sources of information

Even when it is closely grazed, chalk turf teems with invertebrate life— Hudson's "fairy fauna"—and it is a daunting task indeed to attempt to catalogue it in a way which is informative yet digestible. Information on molluscs is contained in a classic paper by Boycott (1934) and in a more recent publication by Evans (1972) already referred to in Chapter 1. Reviews by Edney (1954) and Sutton (1972) fulfill similar requirements for the isopods (woodlice). A comprehensive annotated list has been compiled by Sankey (1966) of all the invertebrates likely to be found on the chalk of the Surrey North Downs, while geographically broader reviews of the insects and arachnids of the English Chalk are to be found in Duffey and Morris (1966) and Ratcliffe (1977). Duffey's detailed work on the spiders of the limestone grassland of the Wytham Estate, Oxfordshire (Duffey 1956, 1962a, 1962b), is almost entirely applicable to chalk grassland. Valuable accounts of the invertebrate ecology of calcicolous grasslands are contained in Elton (1966) and Duffey et al. (1974). The reader is strongly recommended to follow up all these texts, particularly the last two which contain extensive bibliographies, for here it is possible only to be very selective.

6.4.2 Molluscs

The molluscs are well represented on chalk, particularly by the snails, whose long-persistent shells are often very abundant and, as we saw in Chapter 1, of great value to chalkland archaeologists. Their impact on even the casual observer is captured in these words of the Wiltshire naturalist John Aubrey, writing 300 years ago (Aubrey, 1685):

> When I had the honour to waite on King Charles [II] and the Duke of York to the top of Silbury Hill, his Royal Highnesse happened to cast his eye on some of these small snailes on the turfe of the hill. He was surprised with the novelty, and

Table 6.4

Chalk grassland snails and slugs, and their ecology [a]

Name	Synonym(s)	Comments
Pomatias elegans [b]	—	Markedly calcicolous and restricted to the south. Shaded and moist habitats (e.g. scrub and open woodland) are favoured, with broken ground and friable soil into which it can burrow. A good indicator of former disturbance of the soil surface. See also Kerney (1972). Illustrated in Fig. 6.12.
Carychium tridentatum	—	Favours ungrazed grassland, scrub (especially hawthorn) and the leaf litter of deciduous woodland. Very sensitive to drying out. Virtually ubiquitous.
Cochlicopa lubrica [b]	—	Ubiquitous.
C. lubricella [b]	—	Widespread, but particularly characteristic of chalk grassland. Occurs in drier and more exposed habitats than *C. lubrica*, but not restricted to them. Sometimes found in damp grassland and heavily shaded woodland.
Vertigo pygmaea [b]	—	Widespread, though never abundant, in dry habitats. Common in very closely grazed or mown chalk turf, but occurs also in longer grass.
Abida secale	*Pupa secale*	Very local, mainly restricted to dry, open grassland, scrub and light woodland on Chalk and Oolite. Typical of juniper sere (Fig. 7.8) in Chilterns. Characteristically associated with *Vallonia costata*.
Pupilla muscorum [b]	*Pupa muscorum*, *P. marginata*	Common in grassland, scrub and open woodland, but mainly in the south. Particularly favoured by bare ground, especially around rabbit burrows, but not arable land. Often very abundant, yet occasionally rare or absent.

Species	Synonyms	Notes
Vallonia costata [b]	—	This and the other *Vallonia* spp. noted below are generally restricted to the south and east. *V. costata* sometimes occurs in profusion on chalk downland, yet is frequently rare or absent. In the Chilterns it is (with *Abida secale*) more characteristic of juniper than of hawthorn scrub, especially in the driest sites, and it is occasionally to be found there in open beechwoods.
V. excentrica [b]	—	Characteristic of grazed and ungrazed chalk grassland and hawthorn scrub, frequently in association with *V. costata*, though these two sometimes exhibit an antipathetic distribution similar to that of the *Cepaea* spp. (Fig. 6.8).
Ena obscura	—	Found in chalk grassland where the local or topoclimate is relatively damp.
Punctum pygmaeum [b]	—	Widespread in open, dry habitats such as chalk grassland, where it typically occurs in association with *Cochlicopa*, *Vitrea*, *Nesovitrea* and *Vitrina*.
Discus rotundatus	—	Predominantly a woodland snail (Fig. 8.16), but occasionally found in open downland.
Vitrina pellucida [b]	—	Common in short chalk grassland (also marshes, etc.).
Vitrea crystallina	—	Ubiquitous, but mainly in damp places (hedges, woods and marshes) rather than dry turf.
V. contracta [b]	—	Widespread. Common on Chalk.
Aegopinella nitidula [b]	*Retinella nitidula*, *R. pura*	Usually woodland species, but also common in tall grassland and hawthorn scrub on Chalk.
Nesovitrea hammonis [b]	*Retinella radiatula*, *Hyalina radiatula*	Widespread, but rarely common.
Oxychilus cellarius [b]	*Hyalina cellaria*	Very common on open downland, and in tall grassland, scrub and woods.

Table 6.4 continued

Name	Synonym(s)	Comments
O. helveticus	H. helvetica	Like Ena obscura, found in relatively moist chalk grassland.
Deroceras reticulatum [b]	Agriolimax agrestis (pre-1940)	Slug.
Arion intermedius [b]	—	Slug.
Euconulus fulvus	—	Widespread, but more typical of woods and marshes than chalk grassland.
Candidula intersecta [b] C. gigaxii [b] Cernuella virgata [b]	Helicella caperata H. gigaxi, H. heripensis H. virgata	Common calcicolous xerophiles of calcareous and maritime grassland, as well as arable land; fairly recent introductions.
Helicella itala [b]	—	Generally more local than either Candidula or Cernuella, but very characteristic of short chalk turf. Very susceptible to increase in height of the grassland canopy, e.g. following the cessation of grazing (Section 6.5.3) and declining for this reason.
Trochoidea elegans	H. elegans	Rare, introduced xerophile, confined to extreme south-east England.
Monacha cartusiana	Helix cartusiana	A rare xerophile of extreme southern and eastern England.
M. cantiana	Helix cantiana	Common, though with a marked southerly and easterly distribution.
Trichia striolata	Hygromia striolata	Fairly widespread, but more typical of road-verges and arable land than chalk turf.

Species		Notes
T. hispida [b]	Hygromia hispida	Ubiquitous, usually represented on chalk downland by var. nana.
Cecilioides acicula [b]	—	Subterranean. Characteristic of recently cultivated grassland.
Arianta arbustorum	—	More typical of damp habitats, but found on chalk downs in thick grassland and on north-facing slopes.
Cepaea hortensis [b]	Helix hortensis	Widespread. Preferring somewhat drier sites than Arianta, but damper than C. nemoralis (see Fig. 6.8; see also Cameron, 1970a,b). Illustrated in Fig. 6.12.
C. nemoralis [b]	Helix nemoralis	Woods, hedges, marshes and downland, but on Chalk found in warmer and drier habitats than C. hortensis. Very variable in colour and banding patterns. Illustrated in Fig. 6.12.
Helix pomatia	—	Very local, occurring mainly in the Chilterns and North Downs (as well as on Oolite). This is the impressively large (up to 40 mm diameter), edible "Roman snail".
H. aspersa	—	Much commoner than H. pomatia, but still with a marked southerly distribution. The common garden snail; not particularly characteristic of chalk grassland. Also edible.

[a] Compiled mainly from Boycott (1934) and Evans (1972), with assistance from M. P. Kerney and J. Royston. Species are arranged in phylogenetic order. Nomenclature according to Waldén (1976). See Kerney (1976) for distribution maps.
[b] Species most characteristic of chalk grassland, rarities excluded.

commanded me to pick some up, which I did, about a dozen or more, immediately; for they are in great abundance. The next morning, as he was abed with his Dutches at Bath he told her of it, and sent Dr Charleton to me for them, to shew her as a rarity.

A list of the commoner snails (and slugs) of chalk grassland is shown in Table 6.4, where opportunity is taken to sort out the confusing interchangeability of many of the names.

Warm, humid, cloudy days after rain provide the best conditions in which to see these animals actually moving about on the turf, but how do they cope with the intervening dry spells? The slugs, of course, have absolutely no protection from desiccation other than to seek shelter under plants and stones and in the soil itself. Some snails do this, such as *Cepaea nemoralis* and *C. hortensis*, which "tuck themselves away under *Cnicus acaulis* [*Cirsium acaulon*] etc.; when they live on open downs" (Boycott, 1934). Indeed, some of the smaller species "take such good cover in dry, sunny weather that they may be difficult to find". Not surprisingly, the eggs of slugs and snails are extremely vulnerable to hot, dry weather, and a number of xerophilic species avoid this problem by breeding in autumn and winter (e.g. *Candidula intersecta* and *Cernuella virgata*—M. P. Kerney, personal communication).

Yet, many snails obstinately live out their lives fully exposed to sun and wind in the short turf, *Helicella itala* "perhaps to the most extreme degree . . . for it seems definitely to prefer the hottest and often the steepest part of a hillside". All of Boycott's apparently obligate xerophiles possess relatively thick shells, "which help them to tolerate the intense insolation to which they are often subjected and which they escape by climbing up the stalks of any available plants . . . so getting away from the overheated ground. . . . They make little attempt to burrow into the turf or get into any shelter". It is well known that adult snails conserve water by sealing off their operculum with a waterproof membrane of calcareous mucus known as the epiphragm. These hardy little downland snails must be particularly well adapted in this respect.

On a local scale, the distribution of molluscs may be distinctly patchy. In his survey of the Ivinghoe Hills, Stratton (1963) showed, for example, that *Cernuella virgata* was largely confined to the short turf alongside footpaths, while *Abida secale* was commonest in small patches of bare chalk overhung with thyme. Cain and Currey (1963a,b) have reported interesting patterns of distribution of the snails *Cepaea hortensis* and *C. nemoralis* in chalk grassland. On the Marlborough Downs, for example, *C. hortensis* predominates in the valleys and *C. nemoralis* on the upper slopes. The transition is often abrupt (Fig. 6.8). In his detailed and critical account of snail ecology, Evans (1972) has suggested that *C. nemoralis* might be more susceptible to the low night temperatures which prevail in the valley bottoms (see Chapter 2). This is certainly plausible, but the effect could be explained equally well by high maxima and large diurnal ranges in the valleys, or by the incidence of dew.

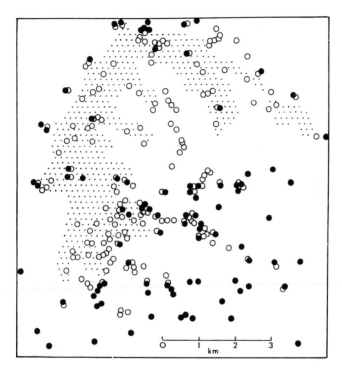

Fig. 6.8. Distribution of *Cepaea nemoralis* (open circles) and *C. hortensis* (black dots) on the Marlborough Downs. Land over 230 m (700 ft) is stippled. This diagram redrawn by Evans (1972) from Cain and Currey (1963a).

Some snails exhibit characteristic associations with other animals. The importance of snails in the diet of the song thrush (*Turdus philomelos*) is demonstrated by the well-known thrush's anvil—typically a flint in down-land habitats— against which the bird breaks open the shells with an unmistakable sound which can carry surprisingly far on a still day. *Cecilioides acicula* is commonly found on ant-hills (see Section 6.4.4.2), while an intriguing discovery at Ivinghoe was that some of the snails there were hosts of the parasitic fly *Ditaenia cineraella*, previously thought to occur only in marshy places (Stratton, 1963).

In close-knit turf (as distinct from cultivated arable or garden soil where food is relatively sparse), slugs and snails are essentially saprophagous, eating senescent and dying herbage and litter, even rabbit droppings, and the fungi which these materials contain. However, it seems to be generally accepted that seedlings are often taken, and several investigators, notably Grime *et al.* (1968), Grime and Blythe (1969) and Pallant (1969, 1972) among the more recent, have demonstrated certain preferences for living greenery which may have ecological significance (Table 6.5). Nevertheless, one is immediately struck by the large number of plant species which are apparently unpalatable in the fresh state, at least as mature plants (see also Section 7.7.2).

Table 6.5

The apparent palatability of some herbaceous plants to the snail Cepaea nemoralis [a]

Species	Palatability [b]
Anthoxanthum odoratum	−
Brachypodium pinnatum	−
Briza media	−
Bromus erectus	−
Carex flacca	−
Dactylis glomerata	−
Deschampsia caespitosa	−
Festuca ovina	−
Holcus lanatus	+
Cirsium arvense	+ +
Hieracium pilosella	+ +
Lathyrus pratensis	−
Leontodon hispidus	+ +
Lotus corniculatus	−
Mercurialis perennis	−
Ononis repens	+
Origanum vulgare	+ +
Plantago lanceolata	−
P. major	+
P. media	−
Potentilla erecta	−
Poterium sanguisorba	−
Ranunculus acris	−
Scabiosa columbaria	−
Senecio jacobaea	−
Thymus drucei	+
Urtica dioica	+ +

[a] From Grime *et al.* (1968), where full details of all 52 species tested may be found.
[b] + + = Palatable, + = moderately palatable, − = unpalatable.

6.4.3 Isopods

Much of what has been said about the molluscs applies equally to the isopods (woodlice). Here again is a group of invertebrates which, though abundant in chalk turf, is "comparatively ill-equipped for life on land" (Edney, 1954). "All the devices which permit such existence are in some degree makeshift." The major problem is excessive water loss from a relatively primitive respiratory system. Even the recent remarkable discovery that woodlice void excretory wastes as ammonia gas—surely the ultimate in water conservation—appears to bear more on the animal's energy budget than its water relations (see Sutton, 1972). Most woodlice escape the extremes of water stress by behavioural patterns which lead to their familiar aggregations in cool, moist places, and restrict their foraging to nocturnal sorties. The ubiquitous woodlouse *Philoscia muscorum* is a good example of such a species; its relatively rare appearance by day in the open turf belies its abundance, which is soon revealed by a more rigorous search in litter and soil.

Yet, it would be wrong to suggest that active woodlice can only be observed in chalk grassland by torchlight. Individuals of several species are readily found on the move in the open turf even in warm, sunny weather. None are more characteristic of these dry habitats than the pillbugs (the ones which roll up into a ball), of which the large and heavily armoured *Armadillidium vulgare* is the commonest. The even bigger *A. nasatum* is more restricted to southern England, and appears to prefer "hot, sunny slopes and screes" (Sutton, 1972). Several other species, including "ordinary" woodlice as well as pillbugs, show this more continental distribution too, and are absent from the northern chalk of the Yorkshire Wolds.

An interesting woodlouse is *Platyarthrus hoffmannseggi*, a diminutive, sightless, white species which lives in ants' nests, typically of *Lasius flavus* (see below). It is apparently attracted to this habitat by formic acid and possibly other ant secretions. The ants appear to tolerate, rather than encourage it, and it probably feeds on their faeces. Note that the millipede *Glomeris marginata* is a common associate of pill woodlice on chalk, and can easily be mistaken for one. The mature *Glomeris*, however, possesses 17 pairs of legs, while woodlice have 7 pairs. The two groups also differ in their arrangement of abdominal segments (cf. Figs 12 and 13 of Sankey, 1966).

6.4.4 Ants

6.4.4.1 The commoner chalk grassland ants, and their associates

The ants encountered in grazed chalk grassland are strictly soil-nesting animals, but the workers of several species may be seen on their hunting and foraging sorties through the sward. Where patches of bare ground occur, the common black ant (*Lasius niger*) is likely to be found, its nest betrayed by small piles of friable soil. The more elongated, dark reddish brown *Myrmica* spp. produce a more cryptic nest, from which the ants spread out rather thinly over their territory, so that individuals rather than groups catch the eye. More elusive because it is subterranean, yet easily the best known of the chalkland ants from the conspicuous mounds which it builds, is the yellow ant (*Lasius flavus*); we turn to this species in greater detail in the next section.

The remarkable symbiotic relationship between the large blue butterfly (*Maculinea arion*) and certain *Myrmica* species is well known, although this butterfly is now very rare in Britain, and certainly no longer occurs on the Chalk (Muggleton and Benham, 1975). However, other butterflies of this group (the Lycinidae, including the blues, coppers and hairstreaks) pass part of their larval stage in the care of *Lasius* spp., not inside the nest, but conveniently near it on their food plants. Ants have actually been observed to carry the larvae of chalk-hill and silver-studded blues (*Lysandra coridon* and

Table 6.6

Chalk grassland host species of subterranean aphids tended by Lasius flavus *and* L. niger [a]

Name of aphis	Host	Usual position (all underground)
Forda formicaria	*Carex* spp., *Brachypodium pinnatum*, other grasses	Internodes
Geoica utricularia (*G. eragrostidis*)	*Festuca* spp., *Agrostis* spp., other grasses	Internodes
Tetraneura ulmi	*Brachypodium pinnatum*	Stem bases; roots in early instars
Dysaphis bonomii	*Pastinaca sativa*	Root collar and basal petioles
Aphis chloris	*Hypericum perforatum*, *H. hirsutum*	Stem bases
A. poterii	*Poterium sanguisorba*	Stem bases
A. jacobaeae	*Senecio jacobaea*	Stem bases

[a] From Pontin (1963).

Plebejus argus) to the food plants, where they are "milked" for the sugar secretion produced by stimulating the so-called honey gland (Ford, 1957). Several aphid species display a similar association with ants, though here the eggs are taken into the nest for the duration of winter, after which they are returned to their food plants (usually on the roots) where, like the butterfly larvae, they are subsequently farmed. Pontin (1963) provides data on the relationships between *Lasius niger, L. flavus* and subterranean aphids, as well as a scale insect, which in turn feed on chalk grassland plants (Table 6.6). Some invertebrates inhabit ants' nests purely as scavengers, for example the woodlouse *Platyarthrus hoffmannseggi* (noted above), and many beetles (see Donisthorpe, 1927).

6.4.4.2 The ecology of ant-hills

It is *Lasius flavus* which is almost invariably responsible for the familiar, bulky ant-hills of chalk pastures, although these may sometimes be invaded by *L. niger*, and the latter occasionally constructs loose, rather fragile mounds, especially if the turf is allowed to grow away unchecked. Here, however, attention is directed specifically towards *L. flavus*. The mounds of *L. flavus* (Fig. 6.9) are built up year by year until they reach half a metre or

Fig. 6.9. A mound of the yellow ant (*Lasius flavus*) in chalk grassland at BBONT's Park Wood Nature Reserve, opened up by pheasants. (C.J.S.)

more in height. It is asserted (see Imms, 1957) that these nests are extended laterally always in an easterly direction, so that it is on the east side of the nest that the colony is most populous and active. However, though this may hold true in parts of the Alps, T. J. King (personal communication) is not convinced that it applies in Britain. Regarding age, King (in Wells *et al.*, 1976, and see below) estimates that the numerous large mounds on the Porton Ranges, Wiltshire, could be 80–100 years old; one has even been put at 150 years. Deserted hills may persist in woodlands long after the pasture in which they were formed has died out (Elton, 1966; Grubb *et al.*, 1969).

The maximum span of the mandibles of a *Lasius flavus* worker is 0·65 mm, so no soil particles larger than this in diameter are carried up to the top of the mound, hence the distinct appearance of ant-hill soils. Its fine, loose and friable nature (due, in part, to continued channelling within the mound by the ants) greatly decreases its thermal diffusivity, leading to excessive heating and drying of the surface layer, augmented on the sunward side by the high angle of incidence of the sun's rays (Section 3.6.3). Crusting follows heavy rain. Freshly heaped up ant-hill soil is richer in extractable K (Thomas, 1962b), but poorer in organic matter, and hence total N, than that of the surrounding turf (Table 6.7), although exact values depend on the age of the hill, the precise depth of sampling, and whether or not rabbits have been using the hills (see below). Like earthworms, ants ensure a constant return of chalky material to the surface, so discouraging the development of an acid layer; this is particularly well seen in chalk heath soils (Table 6.8).

The hills of *L. flavus* attract rabbits, which deposit urine and droppings there; certain grasshoppers lay their eggs in the surface soil of the mounds; and butterflies, particularly the wall (*Pararge megera*) with its well-known liking for hot spots, may settle to sun themselves. Destructive agents include pheasants (*Phasianus colchicus*) which delight in dust-bathing, and green woodpeckers (*Picus viridis*) and partridges (*Perdix perdix*) which feed on the ants. Some early observations on the ecology of ant-hills in chalk grassland were reported by Thomas (1962b), but more recent work by Grubb *et al.* (1969), Wells *et al.* (1976), and particularly the elegant and immensely detailed studies of King (1972, 1977a,b,c) have greatly expanded our knowledge in this field.

As the hill is built up, plants in the original turf are buried, and a characteristic flora develops. Some species are confined virtually or entirely to the ant-hills, while others are equally abundant in the surrounding sward. A third category is of plants which occur rarely if at all on the hills, even when they are present in the surrounding turf. There is often a marked contrast in botanical composition between northerly and southerly aspects of the hills (see Thomas, 1962b; King, 1972).

Using data from Oxfordshire, Wiltshire and Sussex, King (1977a) calculated an "affinity for ant-hills" index for 68 plant species. A small sample of the result is shown in Table 6.9, which also includes data from an additional survey at Aston Rowant NNR, Oxfordshire. Next, 206 ant-hills

Table 6.7

Analysis of soils from Lasius flavus *mounds and adjacent chalk grassland at Beacon Hill, Aston Rowant NNR, Oxfordshire* [a,b]

	Ant-hills	Pasture
Bulk density (g cm^{-3})	0·45 (16)	0·56 [d] (16)
Stones (%)	0·1 (25)	6·7 [e] (25)
pH	7·31 (25)	7·26 (25)
Organic matter (%)	14·2 (11)	16·85 [c] (11)
Calcium in calcium carbonate (mmol Ca (100 g)$^{-1}$)	131 (19)	142·5 (19)
Total nitrogen (%)	0·75 (11)	0·87 [c] (11)
Extractable phosphorus (g(g dry mass)$^{-1}$)	23 (6)	16 (6)

[a] From King (1977a), to which reference should be made for further details, particularly his Table 7.
[b] Numbers in parentheses indicate sample sizes.
[c] $0·05 > P > 0·01$.
[d] $0·01 > P > 0·001$.
[e] $P < 0·001$.

Table 6.8

The effects of ant-hills on soil pH in chalk grassland and chalk heath soils [a,b]

pH range	Chalk heath Active	Chalk heath Deserted	Grassland Active	Grassland Deserted
4·5–4·9	0	1	0	0
5·0–5·4	0	2	0	0
5·5–5·9	0	4	0	0
6·0–6·4	2	1	0	0
6·5–6·9	1	1	0	0
7·0–7·4	5	1	2	2
7·5–7·9	2	0	2	7
8·0–8·4	0	0	6	1
Total	10	10	10	10

[a] From Grubb *et al.* (1969).
[b] Figures indicate the number of samples in each pH category.

of a wide range of size and ant activity in a 0·6 ha plot at Aston Rowant were subjected to an analysis of their botanical composition and successional relations (King, 1977b). Finally, ten angiosperm species representing the main categories of ant-hill affinity were subjected to detailed autecological studies in order to account for their contrasting behaviour (King, 1977c). These were the winter annuals *Arenaria serpyllifolia* (possibly including some individuals of *A. leptoclados*—see Section 6.3.3), and *Veronica arvensis*, both with a percentage affinity for ant-hills of 100 (Table 6.9), the perennial chamaephytes *Cerastium holosteoides* (88), *Thymus drucei* (87) and *Helianthemum chamaecistus* (73), and the perennial hemicryptophytes *Plantago lanceolata* (23), *Leontodon hispidus* (14), *Carex flacca* (6), *Poterium sanguisorba* (4) and *Cirsium acaulon* (1).

Table 6.9

Examples of differences in the affinity of plant species for ant-hills [a]

| | | Comparative frequency [c] | |
Species	Affinity for ant-hills (%) [b]	Ant-hills	Surrounding pasture
Myosotis ramosissima	—	14·9	0·0
Veronica arvensis	100	13·6	0·4
Arenaria serpyllifolia	100	11·8	0·0
Cerastium holosteoides	88	—	—
Thymus drucei	87	—	—
Bryum spp. (mosses)	80	—	—
Helianthemum chamaecistus	73	—	—
Catapodium rigidum	—	3·5	0·0
Medicago lupulina	52	34·3	30·5
Linum catharticum	48	49·8	44·6
Euphrasia officinalis agg.	48	13·2	14·9
Carex flacca	6	—	—
Succisa pratensis	5	—	—
Poterium sanguisorba	4	—	—
Cirsium acaulon	1	—	—

[a] From King (1977a).

[b] Data obtained from 104 paired 0·25 m² samples from ant-hills and surrounding pasture at ten chalk grassland sites in Oxfordshire and Wiltshire, and three chalk heath sites in Sussex.

$$\text{Affinity for ant-hills} = \frac{100\,a_z}{a_z + p_z}$$

where a_z = percentage cover of species z on ant-hills, and p_z = percentage cover of species z in comparable samples of the surrounding pasture.

[c] Data obtained from 287 paired 0·25 m² samples from ant-hills and surrounding pastures at Aston Rowant NNR, Oxfordshire.

The light seeds of *Arenaria serpyllifolia* are produced in large numbers, and after a short period of dormancy, during which the consequences of any late-summer drought, and soil-heaping by ants and scraping by rabbits are avoided, germination (which is inhibited in the shade of the turf) occurs freely in the favourable light-climate of the ant-hill. Nevertheless, although very large numbers of seedlings emerge, mortality is high, mainly due to density-independent effects of the kind mentioned above, and to the washing away of seedlings from the loose ant-hill soil by heavy rain—*Arenaria* seedlings are very shallow rooted. Other agencies of mortality noted were shading by growing shoots of *Helianthemum* and mosses and by fallen leaves of beech and whitebeam (*Sorbus aria*), burial by rabbit droppings, grazing by slugs and snails, and in one case by a green woodpecker digging for ants. Delayed density-dependent mortality was also recorded, caused by self-crowding, with or without fungal infection. Yet, despite these setbacks, sufficient plants of *Arenaria* survive to flower in May–June of the next year and set seed, accounting for the obvious success of this species as a coloniser of active ant-hills. Behaviour was broadly similar in *Veronica arvensis* and *Cerastium holosteoides*, the latter possessing the additional advantage of being able to grow up through heaped soil by the elongation of its rootstock and the production of new rosettes at its tip.

The perennial chamaephytes *Thymus drucei* and *Helianthemum chamaecistus* are also very characteristic of ant-hills, but differ markedly from the previous species in relying almost entirely on vegetative growth rather than seed production. *Thymus* was found to flower and set seed abundantly on the ant-hills, but any seedlings quickly disappeared through unknown causes: one marked colony of 15 seedlings had all gone four days later, and another batch of 58 vanished within six days. *Helianthemum* flowers profusely, but its seeds exhibit dormancy and their dispersal is inefficient. Thus, although the seedlings which do appear are apparently more robust than *Thymus*, they rarely occur. The ability of these two species to invade active ant-hills and maintain themselves there depends on three important properties. Firstly, they are strongly drought resistant, due, in part at least, to efficient root growth, especially *Thymus* (Fig. 5.14). Secondly, as noted already, they are not grazed by rabbits. Thirdly, both have strong powers of recovery from sudden burial, and so are able to keep up with the mound as the ants continue to heap on more soil. In *Thymus* (which normally spreads along the surface of the soil), both axillary and apical shoots grow vertically when buried. *Helianthemum*, which invades new ground beneath the soil surface, spreads horizontally by apical growth when lightly buried with soil, but vertically by axillary growth when it is deeply buried. Either way, where initial colonisation has taken place at an early stage, the woody shoots of both these species typically permeate the whole of the ant-hill, and may well contribute to its structural stability.

Of the last five species tested (*Plantago lanceolata, Leontodon hispidus, Carex flacca, Poterium sanguisorba* and *Cirsium acaulon*) only *C. flacca* has the capacity for extensive vegetative spread, and its vigorous, far-reaching rhizomes might be expected to colonise ant-hills rapidly and efficiently. Indeed they often do just this, but the aerial shoots of this species are extremely susceptible to burial and die when even a small quantity of soil is heaped onto them. Moreover, the production of viable fruits is erratic, dispersal inefficient and germination poor. It is easy to see how it comes about that *C. flacca* is usually confined either to young ant-hills, or to the perimeter of larger ones.

The production of viable seed is rather poor in *Poterium*, variable in *Cirsium* (Section 5.3.2) and typically good in *Plantago* and *Leontodon*. Propagules of *Poterium* and *Plantago* fall vertically to the ground when ripe; those of *Leontodon* and *Cirsium* are wind dispersed, though more efficiently so from the tall heads of the former species than from the sheltered rosettes of the latter (in which the achenes frequently become detached from their pappi). Where fruits and seeds do reach the ant-hills, hazards to germination and seedling establishment of the kind already mentioned take their toll, however, and since at least one full season may elapse before flowering begins, if it occurs at all, few plants reach this stage, least of all *Poterium* and *Cirsium*.

In his thesis, King (1972) proposes an intriguing "time lapse" model in which all these happenings, somewhat stylised and greatly speeded up, are viewed from a position on the surface of the soil, with the ant-hills, continually being topped up with fresh soil, towering above the turf, and fruits and seeds dropping, catapulting and floating about the scene. Of course, as the hills mature or are deserted, their suitability for invasion by species of less marked ant-hill affinity increases (see Grubb *et al.*, 1969), and in the absence of grazing (see next section), much taller flowering stems develop which can overhang the ant-hills and greatly alter the spectrum of colonising species.

6.4.5 Other invertebrates

The three groups of invertebrates considered so far have been singled out simply because they are reasonably familiar to the non-specialist, and have obvious ecological interest and significance. Two rather different niches are represented. The snails and woodlice are essentially saprophagous (eating senescent or dead herbage) and so act mainly as comminuters and primary decomposers in concert with, or as a prelude to, the activities of the earthworms, springtails and mites. Ants parallel these activities to some extent, but also include a predatory element which is developed to a much greater degree in the spiders and certain beetles.

What of the rest? Grazed chalk turf harbours an impressive range of phytophagous insects. These are primary consumers which feed directly upon the living vegetation, though not all are exclusively phytophagous throughout their entire life cycle, nor are they necessarily confined at all times to the aerial shoots of the turf. Some may feed on roots (as noted in Table 6.6) and others may retreat at times into the soil itself, while species

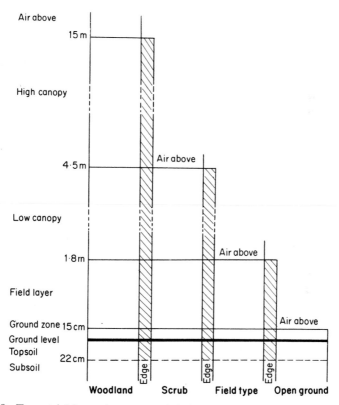

Fig. 6.10. Terrestrial formation types and their edges, showing vertical layering. See text for details. From Elton (1966).

which possess wings or other means of getting airborne are able to take to the air once they are mature (the "air above" category of Elton and Miller (1954)—see Fig. 6.10). Of these, the more mobile species may range quite widely and are bound to occur in a relatively large spectrum of habitats, especially in terrain which itself is physiognomically diverse. Thus the classic

Table 6.10

Butterflies singled out by Ford (1957) as characteristic of chalk downland, together with additional information on their life-cycles [a,b,c]

Species	Larval food-plant [d]	Overwinters as	Other comments
Chalk-hill blue (*Lysandra coridon*)	*Anthyllis vulneraria, Hippocrepis comosa, Lotus corniculatus*; also red and white clovers	Egg	Confined to southern Chalk and lime- stone. At least *L. coridon* requires very short turf for courtship behaviour.
Adonis blue (*L. bellargus*)	*Hippocrepis comosa*: also white clover	Larva of 2nd brood	
Silver-spotted skipper (*Hesperia comma*)	*Lotus corniculatus, Festuca ovina*	Egg	
Marbled white (*Melanargia galathea*)	*Dactylis glomerata, Phleum pratense, Festuca ovina* and other grasses	Young larva	Local, and virtually confined to Chalk and limestone, mainly in the south, but often abundant where it does occur. Favours tall grassland.
Grayling (*Hipparchia semele*)	*Deschampsia caespitosa, Festuca ovina*	Larva	Also found on heathland.
Dark-green fritillary (*Argynnis aglaia*)	*Viola canina*	Young larva	Widespread but rather local.
Meadow brown (*Maniola jurtina*)	*Poa pratensis* and other grasses	Larva	Widespread and very common.
Ringlet (*Aphantopus hyperantus*)	*Dactylis glomerata, Deschampsia caespitosa* and other grasses, including the woodland species *Milium effusum*	Larva	Favours long grass; perhaps more strictly a hedgerow/scrub species (see below).
Small heath (*Coenonympha pamphilus*)	*Cynosurus cristatus, Poa annua* and other grasses	Larva of 1st or 2nd brood	Widespread and abundant.

Species	Food-plants	Overwintering stage	Status and remarks
Common blue (*Polyommatus icarus*)	*Ononis* spp., *Lotus corniculatus*; also *Pimpinella saxifraga*, *Achillea millefolium*, white and red clovers	Larva of 2nd or 3rd brood	Widespread and abundant.
Small blue (*Cupido minimus*)	*Anthyllis vulneraria*	Fully-grown larva	Very local, and virtually confined to Chalk and limestone.
Brown argus (*Aricia agestis*)	*Helianthemum chamaecistus*	Larva of 2nd brood	—
Small copper (*Lycaena phlaeas*)	*Rumex acetosa*; also *Senecio jacobaea*	Larva of 3rd brood	Widespread. Adults favour *Chrysanthemum leucanthemum* and other tall composites for basking.
Dingy skipper (*Erynnis tages*)	*Lotus corniculatus*	Larva [e]	Rather local.
Grizzled skipper (*Pyrgus malvae*)	*Fragaria vesca*, *Potentilla reptans* and others	Pupa [e]	Rather local.
Small skipper (*Thymelicus sylvestris*)	*Holcus lanatus*, *Brachypodium sylvaticum* and other grasses	Larva [e]	Widespread and common.
Large skipper (*Ochlodes venata*)	*Dactylis glomerata*, *Brachypodium sylvaticum*, *Holcus lanatus* and other grasses.	Larva [e]	Widespread and common.
Essex skipper (*Thymelicus lineola*)	*Phleum* spp., *Brachypodium sylvaticum*	Egg	Local, but possibly overlooked by being taken for the small skipper.

[a] Life-cycle data gleaned mainly from South (1906), Stokoe (1944) and Goodden (1978).
[b] Coastal rarities are omitted. Of the species listed here, only the first three are strictly exclusive to chalk and limestone, but in many areas these formations underlie the only remaining habitats of suitable structure and extent. Much more work is needed on exactly what constitutes a "good" habitat.
[c] The proximity of scrub and hedges enhances the likelihood of encountering several additional species, such as brimstone, orange-tip, hedge-brown, Duke-of-Burgundy "fritillary", holly blue and green hairstreak (see Chapter 7). See also Sankey (1966, pp. 96–97) for data on 27 moths of chalk grassland.
[d] Where a species is widely distributed, only its chalkland food-plants are indicated.
[e] In hibernaculum of leaves drawn together by silken threads.

downland fritillary, the dark green (*Argynnis aglaia*), a fast and powerful flier, is bound to cross many more ecological boundaries than most of its loping or diminutive congeners. Conversely, the vanessid butterflies (peacock, small tortoiseshell, etc.) frequently stray over chalk downland, and through consuming nectar, transferring pollen, and even providing a meal for a bird or wasp, play a significant role in the downland ecosystem. Nevertheless, they do not belong in the same way as those insects which pass from egg to imago in the turf of the immediate vicinity, and it is these which are the primary concern here.

As with the molluscs and woodlice, it is possible to pick out examples of the phytophagous insects of chalk grassland which on the whole are reasonably well known to non-specialists, and which, though not necessarily the most important in terms of number or biomass, serve to illustrate ecological principles common to the whole group. The butterflies (Table 6.10) and some of the larger day-flying moths (all Lepidoptera) unquestionably fulfil this requirement, and even the most die-hard botanist can put at least a common name to those which are most characteristic of the downs: the blues, the skippers, the small heath, small copper and meadow brown butterflies, and the burnet moths (*Zygaena*). The cinnabar might be included here too, as much for its conspicuous larvae as for the moths to which these eventually give rise.

All these insects are leaf-eaters in the larval stage and nectar-feeders as adults, but they nevertheless illustrate a range of strategies. Thus at one extreme, the meadow brown and small heath butterflies are virtually ubiquitous throughout Britain. Indeed, the meadow brown is our commonest butterfly. Both species are known to feed on grasses which themselves are very widely distributed, the meadow brown particularly on *Poa pratensis*, and the small heath on *Poa annua*, *Cynosurus cristatus* and *Nardus stricta*. The meadow brown thus appears to be *stenophagous* (restricted to one food-plant) and the small heath *polyphagous*, but there is plenty of evidence to suggest that these butterflies can feed quite satisfactorily on other grasses. The large skipper, small copper and common blue are similarly widely distributed because of their polyphagous habits and the widespread occurrence of their food-plants—mainly Gramineae, docks (*Rumex* spp.) and Leguminosae, respectively. The brown argus has a more southerly distribution, but still occurs fairly generally: it feeds exclusively on *Helianthemum chamaecistus* on chalk and limestone, though it turns to common storksbill (*Erodium cicutarium*) on silicious soils.

On the other hand, the small, Adonis and chalk-hill blues, the silver-spotted skipper and the burnet moths (and others) are much more restricted to chalk and limestone grassland, some quite strikingly so. In the stenophagous species, this can be partly accounted for by the distribution of the food-plants, particularly the calcicolous legumes such as *Hippocrepis comosa* and *Anthyllis vulneraria*. Climatic and other factors unknown must play a part, however, for the full geographical spread of the food-plant is rarely fully exploited (Fig. 6.11), and some insects occur very locally despite feeding on plants which are widespread (see also Chapters 4 and 10).

Fig. 6.11. Comparative distributions of the chalk-hill blue butterfly, *Lysandra coridon* (each dot representing an area where the species is known to breed), and its food-plant the horseshoe vetch, *Hippocrepis comosa*. From Cox *et al*. (1976).

Other insect groups receive attention in the next section on ungrazed grassland, as well as in later chapters on scrub, woodland and conservation, but this is very much a field for the specialist, and on the whole there is still a great deal to learn (though it is to be hoped that amateur, as well as professional naturalists, continue to contribute to this knowledge). As an introduction to the following section, and as an aid to the digestion of the entomological papers which are available, representatives of the main groups of chalk grassland invertebrates are shown in Fig. 6.12.

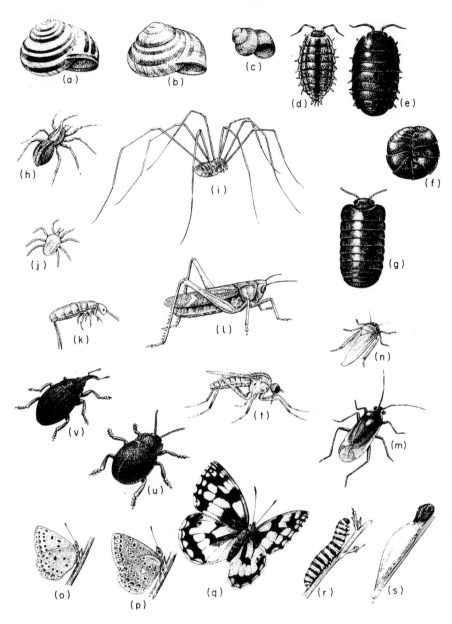

Fig. 6.12. Representatives of the invertebrate fauna of chalk grassland, not to scale. Molluscs: (a) black-lipped hedge snail *Cepaea nemoralis*, (b) white-lipped hedge snail *C. hortensis*, (c) round-mouthed snail *Pomatias elegans*. Isopoda: (d) woodlouse *Philoscia muscorum*, (e) and (f) pill woodlouse *Armadillidium cinereum*. Diplopoda: (g) pill millipede *Glomeris marginata*. Arachnida: (h) wolf-spider (Araneae,

6.5 Ungrazed Grassland

6.5.1 The cessation of grazing

The discussion thus far has concentrated on grazed downland more or less of the kind described in Section 5.1. What happens to this turf when grazing is withheld? The usual riposte to this question encompasses secondary succession, scrub encroachment and the re-establishment of climax woodland, but although there is a good chance that this will follow in due course, it may not always happen quite as the textbooks predict (see Chapter 7). In any case, there are many shorter-term changes to be seen in the herbaceous vegetation itself before any woody plants necessarily begin to assert themselves. Immediately following the removal of grazing animals, two stages can be recognised. The first is marked by a great flush of growth, and particularly of flowering, of the plants originally held in check by grazing. In the second stage, usually more protracted, there is a change in the botanical composition of the turf as those species capable of more robust growth begin to assume a dominant role, and the finer species are phased out. Many classical field experiments employing fenced enclosures, or more correctly exclosures, have illustrated these changes (e.g. Farrow, 1917; Tansley and Adamson, 1925; Watt, 1957, 1962), while the phenomenon was seen on a grand scale in the summers of 1954 and 1955 following the initial impact of myxomatosis (Thomas, 1960, 1963).

6.5.2 Short-term changes

Most of these studies have tended to focus attention on longer-term changes, and though they provide ample illustration of the displacement of floristically rich grazed swards by poorer stands dominated by coarse grasses (of which more anon), remarkably little quantitative information is avail-

---

Lycosidae) *Tarentula pulverulenta*, (i) harvestman (Opiliones) *Phalangium* sp., (j) harvest-mite (Acari) *Trombicula autumnalis*. Insecta: (k) spring-tail (Collembola) *Entomobrya autumnalis*, (l) striped-wing grasshopper (Orthoptera, Acrididae) *Stenobothrus lineatus*, (m) plant bug (Hemiptera, Heteroptera) *Capsodes gothicus*, (n) leaf-hopper (Hemiptera, Homoptera) *Limnotettix striola*, (o) common blue *Polyommatus icarus*, (p) chalk-hill blue *Lysandra coridon* and (q) marbled white *Melanargia galathea* butterflies, (r) larva of cinnabar moth *Callimorpha dominula* and (s) pupa of burnet moth *Zygaena* sp. (o–s Lepidoptera), (t) predatory fly (Diptera, Empididae) *Empis livida*, (u) lesser bloody-nosed beetle (Coleoptera, Chrysomelidae) *Timarcha goettingensis*, (v) weevil (Coleoptera, Curculionidae) *Miaris campanulae*.

able on the immediate consequences of discontinuing grazing. Still less has been published about the effects on the fauna. Morris's recent series of papers, which describes both botanical and entomological aspects of witholding grazing from chalk grassland, are thus particularly valuable, and substantiate earlier work by Elton and his associates on the Jurassic limestones at Wytham. Of course, this is a greatly simplified treatment, for it is more common for grazing to be relaxed gradually rather than to cease suddenly and completely. Lax grazing results in mosaics of grassland of contrasting structure and composition, some close-grazed, some rank. Moreover, it should be clear from what has been said so far that the term "ungrazed" means little without qualification of how long this state has been in force.

Working in the Barton Hills, Bedfordshire, Morris (1967a) monitored changes in flower and fruit production in certain herbaceous species of formerly heavily sheep-grazed chalk grassland, compared with adjacent plots to which sheep (and a few rabbits) still had access. The ungrazed areas had been fenced off in March 1965, and records were taken in the summers of 1965 and 1966. Thus, as Morris suggests, it would be more appropriate to describe the ungrazed areas as "recently grazed". Simultaneously, numbers of selected phytophagous invertebrates were recorded: the weevils *Apion loti* and *Miaris campanulae* (which feed in the fruits of *Lotus corniculatus* and *Campanula rotundifolia*, respectively), certain grass-feeding bugs and grasshoppers, and nectar- and pollen-foraging bumble-bees. Except for the grasshopper *Chorthippus brunneus*, which was more numerous in the short grass which it is known to favour (Elton, 1966), all these insect species showed predictable and in some cases substantial increases in numbers following the protection of the sward from grazing. Data for the weevils and their food-plants are shown in Table 6.11. An interesting feature is that a relatively much higher incidence of parasitisation of weevil larvae was found in the larger populations of the ungrazed plots. Later work (Morris, 1968, 1969) has shown that these differences were reflected in most invertebrate groups (Table 6.12). Other aspects of this and more recent work by Morris are returned to in Chapter 10.

There is more to newly rested grassland, however, than simply an increase in foliage, flowers and litter, and their associated animals which were there already. Many potentially tall-growing plants, able to persist for a long time under close grazing entirely as vegetative rosettes and similar structures, send up leafy shoots and flowering stems as soon as grazing is relaxed which form a much more substantial field layer (Fig. 6.10) than occurs in grazed grassland. Their flowers and fruits, and supporting structures, provide feeding sites for a whole host of specific phytophagous invertebrates (and their predators) which were not present at all in the grazed environment. An example for the lesser knapweed (*Centaurea nigra*) is shown in Fig. 6.13. Duffey *et al.* (1974) suggest that some 50 invertebrate species are probably linked with the reproductive phase of this plant alone.

Table 6.11

The effect of cessation of grazing on numbers of flowers and fruits of Lotus corniculatus *and* Campanula rotundifolia, *and of their associated weevils* Apion loti *and* Miarus campanulae, *in chalk grassland* [a,b]

Association	Attribute	Year	Grazed	Ungrazed
(a) *Lotus corniculatus/*	Number of *Lotus* flowers	1965	25·5	68·2
Apion loti [c]		1966	12·8	77·5
	Number of *Lotus* pods	1965	8·2	420·8
		1966	1·8	292·2
	Number of *Apion* larvae,	1965	0·2	42·0
	pupae and adults	1966	1·2	196·5
(b) *Campanula rotundifolia/*	Number of *Campanula*	1965	4·8	830·3
Miaris campanulae [d]	flowers	1966	32·0	1670·6
	Dry weight (total) of *Campa-*	1965	0·02	4·84
	nula seed capsules (g)	1966	0·18	9·95
	Number of *Miarus* larvae,	1965	0·8	95·7
	pupae and adults	1966	1·0	149·2

[a] From Morris (1967a).
[b] Data refer to numbers (or biomass) per 50 m² sample.
[c] Mean of four samplings in July and August.
[d] Mean of six samplings in August, September and October 1965, and five in August and September, 1966.

For others, the field layer provides, through shelter from sun and wind, a more equable microclimate than is found in open, grazed turf. Thus the snail *Candidula gigaxii* and the woodlouse *Porcellio scaber* are usually restricted to patches of taller grassland if they occur on downland. This type of sward appears, too, to be essential for the marbled white butterfly (*Melanargia galathea*). The female of this species, far from cementing her eggs individually to the underside of a leaf of a carefully chosen food-plant, simply drops them in mid-flight into the grass, and the vulnerable young larvae must gain considerable protection from a matted canopy as they seek out the grass plants on which they finally come to feed.

Another lepidopteran of these coarser swards is the day-flying six-spot burnet moth (*Zygaena filipendulae*), though here the key requirement seems to be for tall grass stems on which the pupae are (very conspicuously) attached. Any structures such as these—and they serve just as well dead as alive—are essential to many snails, butterflies and others, which roost there

Table 6.12

Differences in invertebrate populations between grazed and ungrazed chalk swards [a,b]

Invertebrate group	Grazed	Ungrazed
Gastropoda	83	171
Isopoda	22	594
Chilopoda		
Geophilomorpha	345	432
Lithobiomorpha	601	2664
Diplopoda		
Polydesmidae	178	910
Julidae	—	2
Opiliones	1	1
Araneae	177	2023
Pseudoscorpiones	1	4
Acari	Not recorded	
Collembola	Not recorded	
Orthoptera		
Acrididae	—	13
Dermaptera	1	—
Heteroptera	184	1236
Homoptera		
Auchenorhyncha	80	764
Aphidoidea	17	73
Coccoidea	3	6
Thysanoptera	Not fully recorded	
Lepidoptera		
Adults	—	2
Larvae	82	76
Diptera	Not recorded	
Hymenoptera		
Parasitica	65	117
Formicidae	33	149
Symphyta larvae	—	5
Coleoptera		
Adults	654	1949
Larvae	612	566
Total	3139	11757

[a] From Morris (1968).

[b] Figures refer to total numbers extracted from quadruplicate turf samples 1/14th m^2 in area, taken on 14 occasions between November 1966 and October 1967. The ungrazed plots had last been grazed in March 1965.

overnight or in wet weather, or climb up from the heat of the ground layer in strong sunshine (Fig. 6.14). Birds climb up tall stems of the more substantial

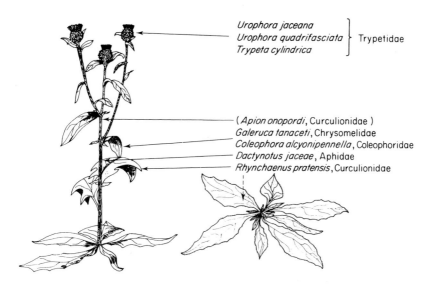

Fig. 6.13. Closely grazed herbaceous plants bear only roots, leaves and vegetative buds, while plants protected from grazing develop in addition to these structures reproductive buds, shoots, flowers and fruits. In this example, plants of *Centaurea nigra* in the ungrazed and grazed condition are shown, together with phytophagous invertebrates recorded on each. In this case, the record of *Rhynchaenus* on the grazed plant was considered doubtful, and that of *Apion* on the ungrazed plant was unconfirmed. From Morris (1971b).

plants such as the umbellifers to reach grass-seeds which would otherwise be inaccessible to them (Beven, 1964), while spiders use them as anchoring sites for webs and traps and for launching themselves into the air. In a classic series of papers, Duffey (1956, 1962a,b) provides numerous examples of the importance of the architecture of the canopy to its spider fauna, and his findings bear strongly on the whole subject of chalk grassland invertebrates.

6.5.3 Longer-term changes

6.5.3.1 Plant responses
In time, of course, the initial bonanza of floristic abundance and diversity subsides as the more robust components begin to exert dominance. Although this may occur through competition for water and nutrients, and

even allelopathy, there can be little doubt that shading is the major factor in the decline of the more diminutive plants of open turf such as *Thymus drucei* and *Asperula cynanchica* (Pigott, 1955; Watt, 1962), and *Euphrasia nemorosa* (Wells, 1969). Unfortunately, no one seems yet to have repeated in semi-natural swards what Stern and Donald (1962) accomplished in their work on competition in ryegrass–clover leys, but some figures of Duffey's

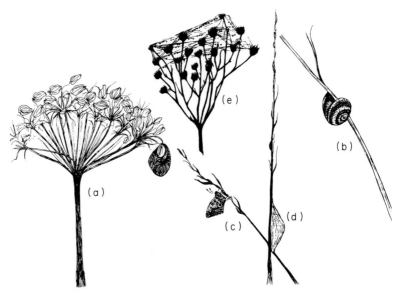

Fig. 6.14. Examples of invertebrates making use of tall stems, all life-size except for (c) which is about 75%. (a) and (b) Snails on hogweed and *Helictotrichon*, (c) chalk-hill blue butterfly resting on *Bromus erectus*, (d) burnet moth (*Zygaena*) pupa on *Arrhenatherum* and (e) spider's web on ragwort (most clearly seen in humid, foggy weather).

(Table 6.13) indicate plainly enough the vertical spread of biomass within the field layer, from which the shading power of the taller plants can readily be appreciated.

The overshadowed plants sometimes persist for a surprisingly long time. Indeed, some mosses, such as the delicate *Mnium longirostrum*, are clearly favoured by an overhanging canopy of grass leaves (and see Watson, 1960). *Cirsium acaulon* and the plantains (*Plantago* spp.) incline their leaves at a more vertical angle in response to shading and the former develops stalks to its inflorescences. Other species which a cursory survey suggests have died out may on closer examination be found, etiolated and depauperate, unflowering and inconspicuous, yet able to resume their former condition if the turf is grazed or cut once again. Nevertheless, persistent as they may be, much of the ecological significance of these plants is soon lost. Thus, Morris (1967a) found that the bumble-bee *Bombus lapidarius* switched its attention in ungrazed turf from *Cirsium acaulon* to the much more conspicuous *Centaurea nigra*, while in later work (Morris, 1969) a sharp decline in numbers of the bug *Agramma laeta* was attributed to the decreasing palatability of its food plants, particularly *Carex flacca*, which became brown and attenuated in the shade of the grass canopy.

Table 6.13

The vertical distribution of biomass in three types of limestone grassland [a,b]

| Height above ground | Grassland type[c] | | |
(cm)	A	B	C
0–10	90·5	86·0	82·50
10–20	19·0	18·5	3·75
20–30	12·5	6·0	2·25
30–40	5·0	2·0	1·25
40–50	1·0	1·0	1·00
50–60	0·5	1·0	0·75
60–70	Trace	0·5	0·25
70–80	Trace	0·5	0·25
80–90	Trace	0·5	0·25
90–100	Trace	0·5	0·50
100–110	—	—	0·75
Total	128·5	116·5	93·50

[a] From Duffey (1962a).
[b] Figures are the dry weight of herbage (g) collected from successive 10 cm layers of a column of vegetation 25 cm square.
[c] Grassland types: A, dense *Brachypodium pinnatum*; B, dense *Festuca rubra* with scattered tall plants of *Pastinaca sativa*; C, sparse vegetation over bare limestone, mostly annuals and tall biennials and perennials such as *Pastinaca sativa*, *Inula conyza*, *Hypericum perforatum* and *Dipsacus fullonum* ssp. *fullonum*.

It is, of course, the taller grasses which typically come to dominance within that first decade of neglect. Of these, the fine-leaved fescues (especially *Festuca rubra*), *Bromus erectus* and *Brachypodium pinnatum* are the most common, though exactly what factors cause one or other of these to occur in a particular place are not known. We saw this in Chapter 5, from which it will also be recalled that other grasses, notably *Dactylis glomerata* and the Aveneae, may also assume local dominance, according to aspect and

Fig. 6.15. Plants of contrasting size from adjacent closely grazed (and trampled) and infrequently grazed turf at Pink Hill, Buckinghamshire. (a) *Dactylis glomerata*, (b) *Poterium sanguisorba*, (c) *Plantago media*—the larger plant was growing on an abandoned ant-hill—and (d) *Leontodon hispidus*. The differences here were assumed to be phenotypic, but where differential grazing or trampling regimes are sustained ecotypic variations can arise.

soil conditions, when grazing is relaxed. Commonly associated with these are the taller herbs, such as *Centaurea nigra, C. scabiosa, Succisa pratensis, Knauita arvensis, Galium mollugo, Poterium sanguisorba, Pastinaca sativa* and others. Some of these grow up from previously strongly repressed plants, and may show remarkable phenotypic plasticity. Examples of *Dactylis* and *Poterium* from contrasting grazing regimes are shown in Fig. 6.15. Others invade only after grazing ceases: *Arrhenatherum elatius* (Fig. 6.16) is well known in this respect (Hope-Simpson (1940b) though see Section 4.2.2.4).

Fig. 6.16. Tall, species-poor Arrhenatheretum on a Coombe series soil near Saunderton, Buckinghamshire. The field was cultivated until 1939 (descendants of the last crop, lucerne, still appear sporadically, as can be seen in the foreground), but it is now managed as an open space by the National Trust, normally being mown once or twice a year. Despite its lack of botanical interest, this tract provides an ideal habitat for the marbled white butterfly (*Melanargia galathea*) and the burnet moths (*Zygaena* spp.). (C.J.S.)

A common feature of these older grass-dominated swards is the development of tussocks and hummocks. These are especially noticeable in winter,
and best seen in *Bromus erectus* and the fine-leaved fescues. In the absence
of grazing, each flush of leaves is blown over in the direction of the prevailing
wind. Ford (1937) recognised three zones in the hummocks of *Bromus
erectus* with distinct microclimates (Fig. 2.18) and faunal distributions to
match. This structuring is essentially permanent, "except in the face of

Fig. 6.17. An almost pure stand of *Festuca rubra* at Swyncombe Down, Oxfordshire
protected from grazing for over a decade, and exhibiting the curious hummocky
structure described in the text. (C.J.S.)

strong contrary winds, when it breaks down, with disastrous effects on the
fauna". Similar hummock formations have been observed in long-ungrazed
Festuca rubra grassland at Swyncombe Down, Oxfordshire (Smith *et al.*,
1971) (Fig. 6.17) and at the Porton Ranges, Wiltshire (Wells *et al.*, 1976).
Substantial quantities of dead litter accumulate in the hummocks, and the
links between the few living tillers and the mineral soil become very tenuous.
Indeed, moribund hummocks can be lifted clean off the soil surface.

Where the downwind "tatty ends" of the hummock decompose, the
mineral soil becomes exposed, and other species are able to establish—the
only sites in this competitively highly exclusive turf where invasion is
effective. Curiously, colonisation at Swyncombe was invariably by dicoty-

ledonous species, such as *Crepis capillaris, Sonchus arvensis, S. oleraceus, Pastinaca sativa* and *Cirsium vulgare*, which were consequently restricted to the hollows between the hummocks. Parts of the old turf were later invaded by stinging nettles (*Urtica dioica*), and these had a dramatic effect on the hummocks: presumably because of the marked alteration of the C:N ratio, the matted litter decomposed rapidly and completely (within a few weeks during a mild autumn spell). The whole plot is now quite bare of its original vegetation, and is being colonised anew by an assortment of ruderal species, including *Solanum dulcamara*.

It was in his Grassland A (Section 5.3.2) that Watt observed the interactions between populations of *Festuca ovina* and *Hieracium pilosella* and their hummock-and-hollow effects upon micro-relief, which form a part of his classic paper on pattern and process (Watt, 1947). As each plant of *Festuca* increases in size, it accumulates loose soil particles blown around its roots and tillers, building up a hummock rather like a small ant-hill, which is then invaded by *Hieracium*. In due course the *Festuca*, and then the *Hieracium*, senesce and die, whereupon the hummock begins to disintegrate, wind and rain eventually eroding the soil away again back to base level (Fig. 6.18).

It was impossible to tell whether this was a purely passive cycle of events, or whether *Hieracium* was instrumental in the death and degeneration of the *Festuca* tussock. Watt considered *Hieracium* simply to be the more effective competitor for water, but in a detailed study of dry chalk grasslands of similar habitats in northern France, Guyot (1957) found that *Hieracium* consistently suppressed other species, and though there was no confirmation of the production of a toxin, an allelopathic effect was thought to be distinctly possible. In the Breckland enclosure, *Bromus erectus*, according to Watt more drought-tolerant than either *Festuca* or *Hieracium*, first appeared in 1948, and increased until, in 1960, it was regarded as "a new and more permanent challenge to the dominance of *Festuca*". This is, in fact, what would be expected from Guyot's studies, which showed an open mosaic of micro-associations of tussocks, clumps and cushions of *Festuca ovina, Carex flacca, Thymus serpyllum, Hieracium pilosella* and many others, forming an intermediate stage between the initial colonisation of bare soil and the eventual establishment of a complete cover of *Bromus erectus*. In Breckland, it seemed possible at one stage that invading seedlings of *Pinus sylvestris* from neighbouring plantations might transform Grassland A into pinewood, but this has not happened. In 1978 the vegetation was dominated by *Thymus* spp., with *Bromus* "frequent to abundant".

Regarding *Bromus erectus* itself, Austin's (1968a) painstaking pattern analysis and turf dissection (Section 5.5.3) led to the discovery of a comparable sequence of events in closed *Bromus* grassland on the North Downs

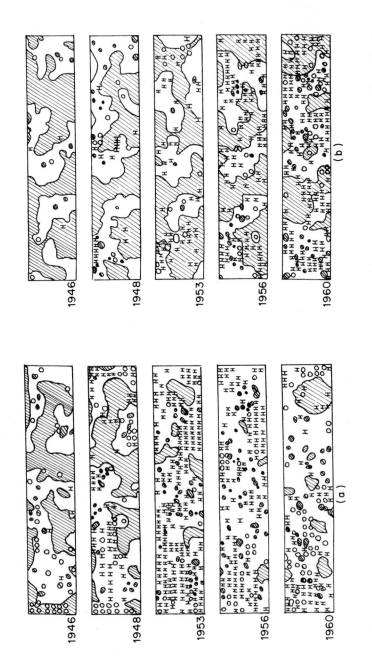

Figs 6.18a and b. Charts of part of (a) enclosed and ungrazed, and (b) slightly rabbit-grazed plots of Breckland Grassland A, showing the positional relationships between *Festuca ovina* (shaded) and *Hieracium pilosella* (H) and their changes in time. Other species are indicated with open circles. From Watt (1962).

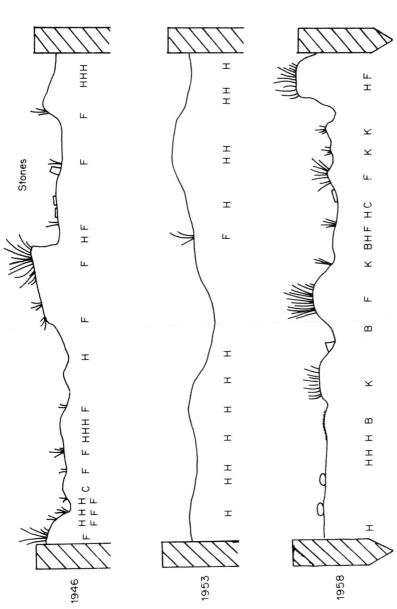

Fig. 6.18c. Profile of microrelief of surface soil at or near the same place as in Figs 6.18a and b in 1946, 1953 and 1958, showing base-levelling under *Hieracium pilosella* (H) and the initiation of the hummock-and-hollow profile under *Festuca ovina* (F). Other species shown are *Koeleria cristata* (K), the lichen *Cladonia alcicornis* (C) and bryophytes (B). Length of transect 50 cm. From Watt (1962).

of Kent. Circular patches develop which spread outwards in a fairy-ring pattern, leaving as they mature a hinterland, initially devoid of plants and filled with persistent leaf litter. As the ring of growth extends outwards, the central area no longer receives fresh litter and the original material begins to decompose. The formerly sterile degenerate phase now begins to be colonised by mosses, including *Fissidens* spp. and *Pseudoscleropodium purum*, in the first stage of a micro-succession. These are followed in turn by *Poterium sanguisorba*, *Lotus corniculatus*, *Origanum vulgare* and *Carex flacca*. Diagrams of these patches are shown in Fig. 6.19.

The growth of a tall canopy of plants in grassland in which *Lasius flavus* ant-hills occur results in the desertion of the hills (see below), and the replacement of the more characteristic plants by a much less specialised

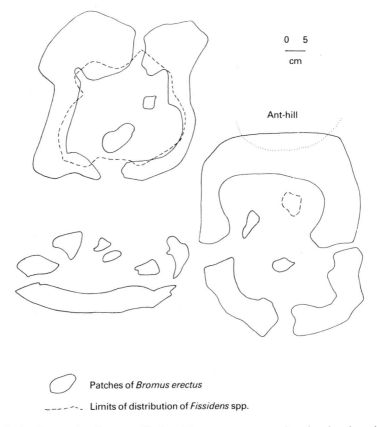

⬭ Patches of *Bromus erectus*

╌╌╌╌╌ Limits of distribution of *Fissidens* spp.

Fig. 6.19. Composite diagram of isolated *Bromus erectus* patches showing the colonisation of hinterland areas by mosses, in this case *Fissidens* spp. From Austin (1968a).

flora. This is especially marked where flowering culms of tall grasses such as *Bromus erectus* and *Arrhenatherum elatius* overtop and drop their seeds on to the hills. Another striking effect of protection from grazing is seen in the vegetation of chalk heath, for the calcifuge shrubs are able to grow up and their litter then acidifies the soil to a much greater degree than when the herbage was grazed (Grubb *et al.*, 1969; Grubb and Suter, 1971). Figure 6.20 summarises diagrammatically the successional relations of this "tall chalk heath" vegetation.

6.5.3.2 *The animal life of rank pastures*
It is commonly observed that as a pasture becomes increasingly rank and floristically poor, its entomological interest declines, though it is perhaps more correct to refer simply to change rather than impoverishment of the invertebrate fauna. Evidence is lacking. Nevertheless, some differences are obvious. As noted above, *Lasius flavus* ant-hills become deserted as soon as they are overshadowed by tall vegetation, apparently because the tempera-ture falls below the optimum for brood production (Pontin, 1955). Mole activity declines. Watt (1962) reported the persistence and even an increase in moles in his enclosed plots in Breckland, but this proved to be temporary.

On the other hand, a tangle of grassy litter soon attracts the field vole (short-tailed bank-vole,*Microtus agrestis*), as well as the occasional frog or slow-worm. The vole makes a labyrinth of runs and tunnels through the matted herbage, feeding not only on the leaves and stems of the grasses, but on the young shoots and bark of any woody plants endeavouring to grow up. The vole is thus very effective at maintaining this type of rough grassland and, though inconspicuous, it is no less significant than sheep and rabbits in checking the growth of the coarser grasses and delaying the appearance of shrubs and trees which would normally be expected in the course of second-ary succession (Chapter 7). Not surprisingly, the vole is a serious pest of tree plantations when downland is turned over to forestry (Chapter 9).

In a very informative study of the influence of grassland type on breeding birds which is considered in more detail in Chapter 7, Venables (1939) found that the thickening up of chalk grassland following the relaxation of grazing favoured skylark and meadow pipit, but discouraged the true birds of the open turf, the lapwing, wheatear and stone curlew (Section 6.3.5).

6.5.4 Succession, eutrophication and stability

This last section, on the consequences of removing the grazing animal from chalk downland, has been concerned essentially with one phenomenon: seral eutrophication (see Green, 1972). This process is implicit in the clas-sical ecological theories of succession and pedogenesis: left alone, soil and

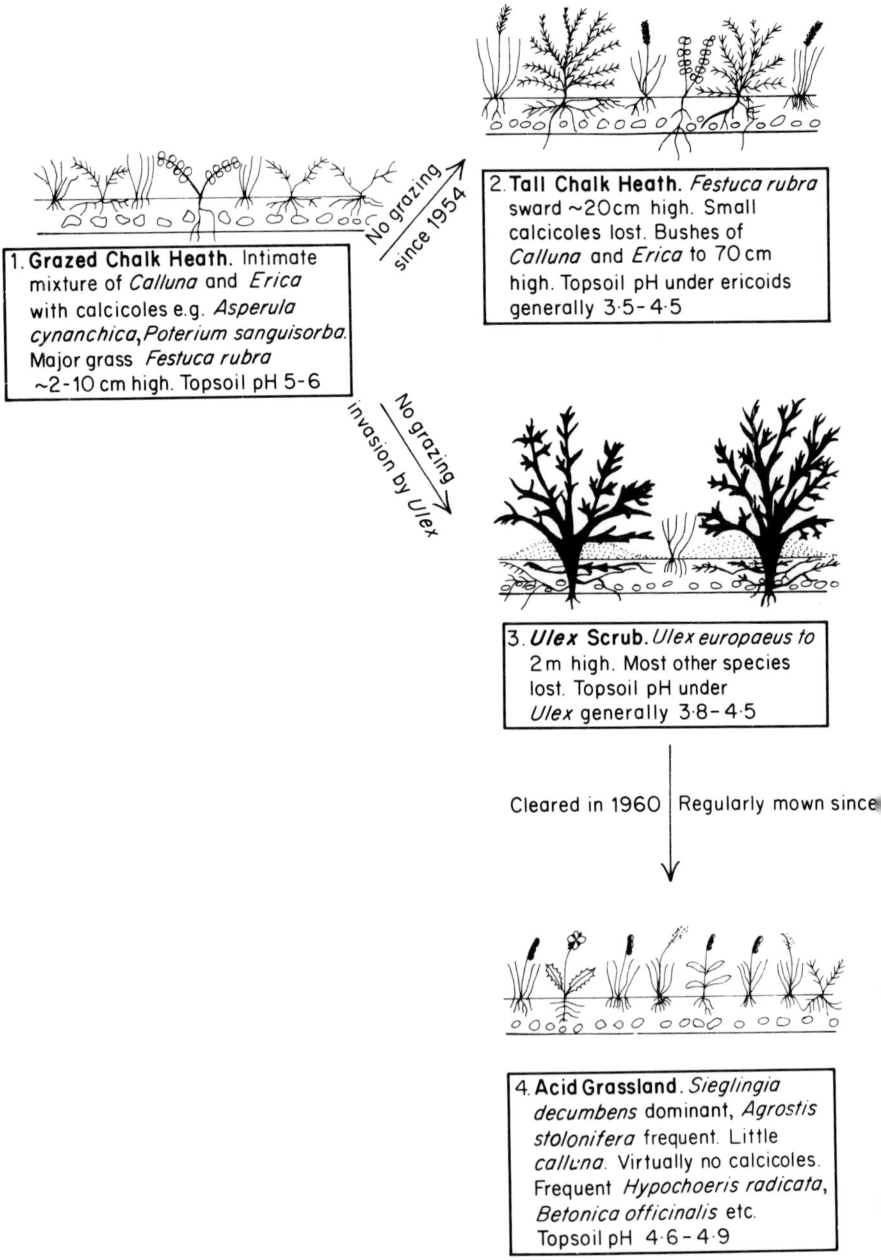

1. **Grazed Chalk Heath.** Intimate mixture of *Calluna* and *Erica* with calcicoles e.g. *Asperula cynanchica, Poterium sanguisorba*. Major grass *Festuca rubra* ~2-10 cm high. Topsoil pH 5-6

No grazing since 1954 →

2. **Tall Chalk Heath.** *Festuca rubra* sward ~20cm high. Small calcicoles lost. Bushes of *Calluna* and *Erica* to 70 cm high. Topsoil pH under ericoids generally 3·5- 4·5

No grazing invasion by *Ulex* →

3. **Ulex Scrub.** *Ulex europaeus* to 2 m high. Most other species lost. Topsoil pH under *Ulex* generally 3·8- 4·5

Cleared in 1960 | Regularly mown since

4. **Acid Grassland.** *Sieglingia decumbens* dominant, *Agrostis stolonifera* frequent. Little *calluna*. Virtually no calcicoles. Frequent *Hypochoeris radicata, Betonica officinalis* etc. Topsoil pH 4·6 - 4·9

Fig. 6.20. The development of "tall chalk heath", chalk-heath scrub and acid grassland from chalk heath, according to management. From Grubb and Suter (1971).

vegetation accumulate energy, organic matter and mineral nutrients until an equilibrium is reached between synthesis and degradation, the exact expression of which varies with a range of environmental variables, but in which climate is of overriding importance. The idea was established in Chapter 3, and is followed through its more advanced stages in Chapters 7 and 8. Another fundamental element of ecology which has been introduced is the question of community stability and the importance of diversity. This line of argument is taken up in Chapter 10.

6.6 Chalk Meadowland and the Grass-verge Habitat

Chalk grassland has been represented throughout the last two chapters as either more or less continuously grazed on the one hand, or recently or long deserted by the grazing animal on the other. Although mention was made of the effects of grazing at different times of the year (about which more in Chapter 10), no account has been taken of meadowland, in which the herbage is allowed to grow up until mid-summer (traditionally Lammas), when it is cut for hay. Livestock are only turned on to graze the aftermath. Not surprisingly, a stand quite distinct in composition and structure from that of the grazed sward develops under this regime when it is maintained for many years. Nowadays, however, virtually all grassland thus managed consists of sown leys of productive cultivars rather than the unimproved sward (Chapter 9).

The hayfield or meadowland biome has, however, been perpetuated in the grass-verge (Fig. 2.9), typically mown regularly but infrequently so that a tall-grassland community results. The flora varies greatly, of course, depending on substrate and topography, previous disturbance, whether any fill has been used, whether it has been seeded, how frequently it is mown, whether the mowings are left or removed, and whether or not growth-retarding chemicals are used. Many stands will resemble grassland types described elsewhere in this book, such as the chalk-pit and primitive-grassland communities (Chapter 4), arable weeds (Chapter 9) and tall-grass stands of which the *Arrhenatherum elatius* type (Section 5.4.3.4) is very common. Tall herbs feature conspicuously, including the familiar seasonal march of white-flowered umbellifers (*Anthriscus sylvestris* in early spring, then *Chaerophyllum temulentum* followed by *Heracleum sphondylium*) and the beautiful mallows (*Malva* spp.) and chicory (*Cichorium intybus*) of high summer. Hooper (1970b) has suggested that some of these species may thrive only in the shelter of a hedge, but there is a great lack of even the most fundamental descriptive surveys of these road-verge communities. Many are now recognised for their conservation value, and under this heading are briefly referred to again in Chapter 10.

7
Chalk Scrub

7.1 What is Scrub?

"Scrub" is a term which is widely used but rarely defined. It can be readily pictured in the mind's eye, and described loosely as a sort of temperate savanna or "bush grassland" (see Pratt *et al.*, 1966), or simply as "bushland" (Duffey *et al.*, 1974). Tansley (1939) defines scrub as a community of plants "dominated by shrubs or bushes", these two terms being regarded as synonymous, but he goes on to show how difficult it is to be more specific. For one thing, it is not easy to define a shrub, and for another, the overlap between scrub and both preceding (grassland) and subsequent (woodland) seral stages makes it impossible to say with any certainty exactly when the scrub phase has begun or ended.

Thus, Tansley describes the dominant species as "broadly of stature intermediate between trees and herbs", and he continues:

> Very many shrubs have, or may have, multiple stems, a tendency much enhanced by coppicing; but others if left uncut form a single trunk, and there is in fact only a more or less arbitrary distinction between shrubs and small trees. Shrubs of less than 2 m but more than 25 cm in height fall into Raunkiaer's class of *nanophanerophytes*, but many of our shrubs grow taller than this and have to be included in the next class, the *microphanerophytes* (2–8 m) which also includes small trees.

Yapp (1955) and Elton and Miller (1954) have used 8 m as an upper limit for scrub, though Elton (1966) later came to regard the scrub formation type as being comprised of "all woody structures rising to between about 6 and 15 ft" (roughly 2–4·5 m). Elton also lays emphasis on the ecological significance of the low, dense, stiff, branching habit of living shrubs, and their long persistence when dead, but of this more later. An added complication is the fact that chalk scrub usually contains a proportion of large, or potentially large, trees, of which the whitebeam (*Sorbus aria*) is a good example.

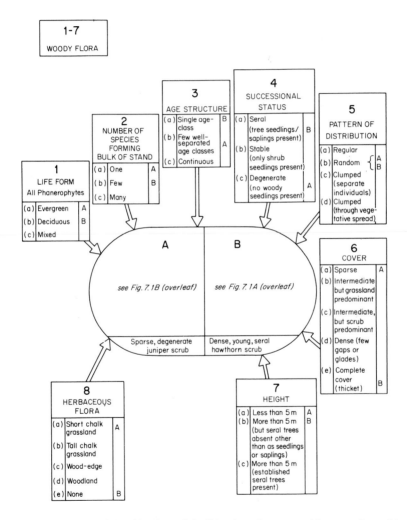

1-7

WOODY FLORA

1
LIFE FORM
All Phanerophytes

(a)	Evergreen	A
(b)	Deciduous	B
(c)	Mixed	

2
NUMBER OF
SPECIES
FORMING
BULK OF STAND

(a)	One	A
(b)	Few	B
(c)	Many	

3
AGE STRUCTURE

(a)	Single age-class	B
(b)	Few well-separated age classes	A
(c)	Continuous	

4
SUCCESSIONAL
STATUS

(a)	Seral (tree seedlings/ saplings present)	B
(b)	Stable (only shrub seedlings present)	
(c)	Degenerate (no woody seedlings present)	A

5
PATTERN OF
DISTRIBUTION

(a)	Regular	
(b)	Random	A / B
(c)	Clumped (separate individuals)	
(d)	Clumped (through vegetative spread)	

A

see Fig. 7.1B (overleaf)

Sparse, degenerate juniper scrub

B

see Fig. 7.1A (overleaf)

Dense, young, seral hawthorn scrub

6
COVER

(a)	Sparse	A
(b)	Intermediate but grassland predominant	
(c)	Intermediate, but scrub predominant	
(d)	Dense (few gaps or glades)	
(e)	Complete cover (thicket)	B

8
HERBACEOUS
FLORA

(a)	Short chalk grassland	A
(b)	Tall chalk grassland	
(c)	Wood-edge	
(d)	Woodland	
(e)	None	B

7
HEIGHT

(a)	Less than 5 m	A
(b)	More than 5 m (but seral trees absent other than as seedlings or saplings)	B
(c)	More than 5 m (established seral trees present)	

Fig. 7.1. A suggested classification of chalkland scrub communities, together with an indication of how this might be applied to two types of scrub. Developed from an early scheme proposed by Venables (1939).

Fig. 7.1A

Fig. 7.1B

Regarding the overlap with other seral stages, Duffey *et al.* define scrub vegetation as "extending from the stage at which the area of woody plants exceeds that covered by grassland, to that when woody plants reach 7 m in height and are composed mainly of tree species." But this still leaves a substantial proportion of "grassland with bushes" in limbo, and in fact several additional features merit inclusion. These are summarised in Fig. 7.1.

7.2 Secondary Succession

7.2.1 The woody vegetation

It has been known ever since the first farmers cleared the primeval forests of Britain that the grasslands and arable fields of these islands can only be maintained by grazing animals and tillage, and that invasion by scrub and woodland inevitably follows cessation of these activities. Richard Jefferies wrote of his native Marlborough Downs in 1879:

> If the sheep and cattle were removed, and the plough stood still for a century, ash and beech and oak and hawthorn would reassert themselves, and these wide, open downs became again a vast forest, as doubtless they were when the beaver and the marten, the wild boar and the wolf roamed over the country.

This is all firmly entrenched in the popular idea of "nature claiming back its own", and many simple experiments with fenced enclosures (already referred to in earlier chapters) have helped to confirm and quantify this concept of secondary, or old-field succession, and the transient nature of seral scrub.

The classic demonstration of secondary succession on arable land comes from the clay loams of Rothamsted, where in 1882 a corner of Broadbalk Field was fenced off and its last wheat crop left unharvested. One half of the plot was then left to itself, the rest receiving various combinations of cutting and sheep grazing, and records were kept of the changes in botanical composition (e.g. Brenchley and Adam, 1915). The result, over 90 years later, is a tiny copse (Broadbalk Wilderness) and a plot of permanent pasture. Succession on chalk was a major interest of Tansley, and as early as the winter of 1908–1909 he had established rabbit-proof exclosures on the chalk downland of Ditcham Park on the Hampshire–Sussex border, in which the subsequent encroachment of briar, bramble and thorn in the absence of grazing animals was effectively demonstrated, and recorded in a series of charts such as those in Fig. 7.2 (Tansley, 1922). Secondary succession was seen on an unprecedented scale in the late 1950s after the rabbit population crashed from myxomatosis (Fig. 7.3; see also Chapter 6).

The woody species typically found in chalk scrub are listed in Table 7.1.

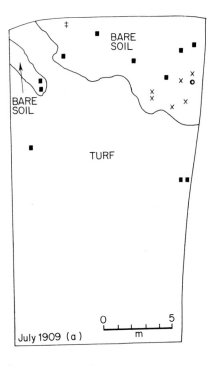

July 1909 (a)

0 ___ 5
m

■ *Crataegus monogyna*

x *Prunus spinosa*

‡ ⚹ *Rubus fruticosus* agg. (including *R.vestitus* and *R.ulmifolius*

⊢ *Rosa* spp. (including *R.canina* and *R.micrantha*

△ *Rhamnus catharticus*

♦ *Thelycrania sanguinea*

◊ *Clematis vitalba*

v *Viburnum lantana*

S *Solanum dulcamara*

▫ *Corylus avellana*

✱ *Calluna vulgaris*

○ *Fraxinus excelsior*

● *Fagus sylvatica*

т *Taxus baccata*

Q *Quercus robur*

Fig. 7.2a–c. Painstaking chart records of scrub encroachment in chalk grassland at Ditcham Park over a period of 11 years. The presence of *Calluna* and *Quercus* suggests neutral to slightly acid soil conditions in places. Redrawn with slight simplification from Tansley (1922).

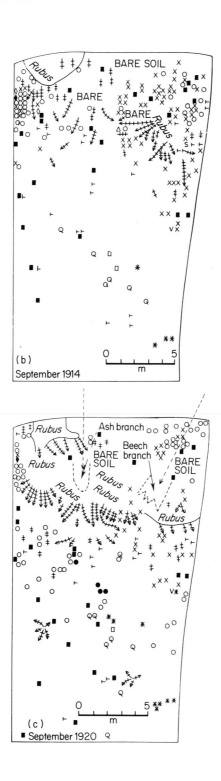

(b)
September 1914

(c)
September 1920

Fig. 7.3. Secondary succession on chalk downland, Whiteleaf Cross, Buckinghamshire. (a, top) In 1910 (Buckinghamshire County Museum) and (b, bottom) in 1979 (C.J.S.).

Table 7.1

Woody species characteristic of chalk scrub [a]

Botanical name	Common name	Comments
Buddleja davidii	Buddleia	Garden escape (usually on disturbed ground).
Buxus sempervirens	Box	Local. Potentially a climax species on steep, dry banks. Sometimes planted.
Clematis vitalba	Old man's beard, traveller's joy	Usually in advanced scrub unless a steep bank or cliff is available.
Corylus avellana	Hazel	Deeper soils
Cotoneaster spp.	—	Common bird-sown garden escapes.
Crataegus monogyna	Hawthorn, whitethorn, may	Extremely common.
C. oxyacanthoides	Midland hawthorn	Local; also hybrids between these two *Crataegus* spp.
Daphne laureola	Spurge laurel	Advanced scrub.
Euonymus europaeus	Spindle	Rather local. Very susceptible to barking by rabbits.
Juniperus communis	Juniper	Local and declining. Mainly on shallow soils.
Laburnum anagyroides	Laburnum	Occasional garden escape.
Ligustrum vulgare	Privet	—
Malus sylvestris	Crab-apple	Probably often an orchard or garden (or picnic) escape rather than genuine crab.
Prunus spinosa	Blackthorn, sloe	Deeper soils. Common on sea cliffs.
Rhamnus catharticus	Purging buckthorn	—
Ribes spp.	Currants	Including garden escapes.
Rosa spp.	Wild roses, briars	Including *R. agrestis, R. arvensis, R. canina, R. micrantha, R. pimpinellifolia, R. rubiginosa.* Ecology little known. Mainly on deeper soils.
Rubus caesius	Dewberry	Sometimes on very shallow soils.
R. fruticosus agg.	Bramble, blackberry	Most work devoted to taxonomy rather than ecology. *R. ulmifolius* and *R. vestitus* (recorded by Tansley as *R. discolor* and *R. leucostachys* see Watson (1958) and Fig. 7.2) are among the more calcicolous, though all seem to favour deeper soils.
R. idaeus	Wild raspberry	Including garden escapes; deeper soils.

Table 7.1 continued

Botanical name	Common name	Comments
Sambucus nigra	Elder, elderberry	Commonly associated with rabbits. Often with a marked epiphytic flora.
Sorbus aria	Whitebeam	Can grow into specimen trees.
Taxus baccata	Yew	Locally a potential climax species.
Thelycrania sanguinea	Dogwood	Spreads by suckering as well as from seed. Characteristic of former arable or cleared woodland on shallow soils.
Ulex spp.	Gorse, furze	*U. europaeus* occasionally found on chalk, but usually indicates the presence of superficial deposits. This and *U. minor* are characteristic of chalk-heath scrub.
Viburnum lantana	Wayfaring tree	—
V. opulus	Guelder rose	Rather local.

a Ericoids of chalk-heath scrub are not included here.

Hawthorn is very common, and not at all restricted to the Chalk. It often forms pure stands (Fig. 7.4), but also occurs in mixtures with any or all of the others, as in the "southern mixed shrub" communities, so-called because of their geographical distribution (Figs 7.5a,b). Juniper and dogwood (Figs 7.6 and 7.7), as well as elder and hazel, also occur in pure stands, as indeed do yew and, more locally, box, though these last two are considered separately in Chapter 8. In southern England, juniper scrub is found almost exclusively on the Chalk, but even here it is now much less widespread than it used to be and appears to be dying out. Dogwood, on the other hand, is an invasive species, particularly when attempts are made to control it by cutting. Elder is well known for its immunity to rabbits because of its repellent bark. As a consequence, elder bushes frequently grow up among the very burrows themselves, and pure stands of elder scrub are reliable indicators of past or present rabbit activity. The elder shrubs themselves often support a rich epiphytic moss flora. Hazel forms an important scrub type on chalk in places where it was formerly coppiced. It was frequently planted for this purpose (see Section 9.4.1). The unusual situation which arises during succession on chalk heath, in which litter from the calcifuge shrubs *Ulex europaeus*, *Calluna vulgaris* and *Erica cinerea* intensifies the acidity of the surface soil, has already been described. Ecological features of hawthorn, juniper and dogwood scrub are discussed in more detail later in this chapter. Much more work is needed on the detailed biology and ecology of the other types.

Various schemes have been devised to relate these different types of scrub to soil and climatic factors, and to describe the routes (seres) from grassland or arable land to woodland through which they are thought to develop in the course of succession. Watt's seral classification of chalk scrub is still widely

Fig. 7.4. Dense hawthorn (*Crataegus monogyna*) scrub, with some *Rosa*, at Bacombe Hill, Buckinghamshire. (C.J.S.)

quoted, and is shown, together with a more recent and comprehensive scheme, in Fig. 7.8, though we still know surprisingly little about the time-scale of these processes (Burges, 1960). Phytosociologically, deciduous scrub formations of chalk and other calcareous soils are allocated to the alliance Berberidion Br.-Bl. 1950 of the order Prunetalia Tx. 1952, which in turn is placed in the class Querco-Fagetea Br.-Bl. et Vlieg. 1937 (Shimwell, 1971a).

7.2.2 The herbaceous vegetation

As the canopy of the invading bushes develops, so the herbaceous flora changes in response to the lowered light intensity, increased shelter and humidity, and changes in soil water and nutrient regimes (Chapter 2).

Species characteristic of grassland give way to those more typical of wood-edge and woodland proper. The exact course of events will depend on a whole host of factors, however. For example, as is manifestly obvious from the foregoing chapters, the botanical composition of the turf at the point when the scrub begins to exert a dominating influence can vary according to soil type, topography, previous grazing regime and so on, so that there is no single convenient starting point.

As far as the actual replacement of grassland with woodland plants is concerned, we are still, as in other branches of field ecology, outrageously short of even the simplest records of observations. A rare example is to be found in Duffey *et al.* (1974, pp. 140–143). For the moment, we must make do with the arbitrary list drawn up in Table 7.2, but even from this an important point emerges which bears on conservation (Chapter 10): the herbaceous flora of the transient scrub habitat contains some of our rarest flowering plant and bryophyte species. Further reference is made to some of these in the following paragraphs. Cates and Orians (1975) have recently suggested that the herbaceous species of these later stages of succession are significantly less palatable, at least to certain slugs, than those of earlier stages. This is an interesting idea, worthy of further study (see also Sections 6.4.2 and 7.7.2).

7.3 The Establishment of Scrub

7.3.1 Mechanisms of invasion

Most of the factors which influence the encroachment of scrub and wood-land into chalk grassland and abandoned arable land are common to invasions of any kind, and have already received mention in earlier chapters. They include the number and viability of fruits and seeds and their agents of dispersal, the availability of germination sites, and the provision of suitable conditions for seed germination and seedling establishment. Much of what is said about these agencies in the following paragraphs rests on no more than intelligent observation: it has to, for quantitative information is surprisingly hard to come by. An important point often overlooked is that many woody plants may, in fact, already be established in the turf, closely and repeatedly coppiced by the grazing animal, but able to grow away rapidly and vigor-ously immediately grazing ceases.

7.3.2 The source and dispersal of fruits and seeds

The importance of the nature, proximity, size and viability of parent shrubs and trees is too obvious to necessitate elaboration. Equally obvious is the

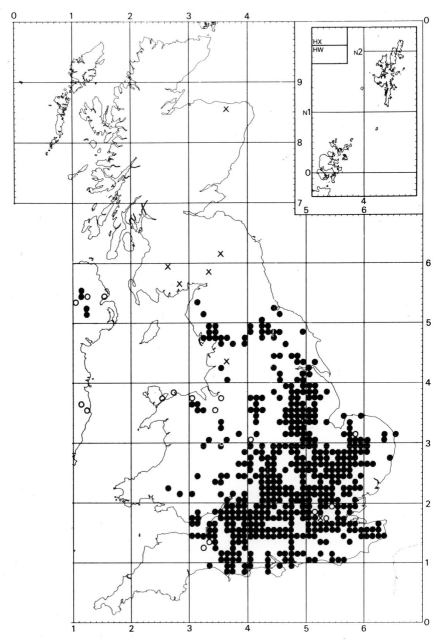

Fig. 7.5a. The distribution in England and Wales of *Rhamnus catharticus*, a characteristic species of the "southern mixed shrubs" chalk scrub formation. Interpretation and source as for Figs 5.7a,b. Crosses represent known introductions.

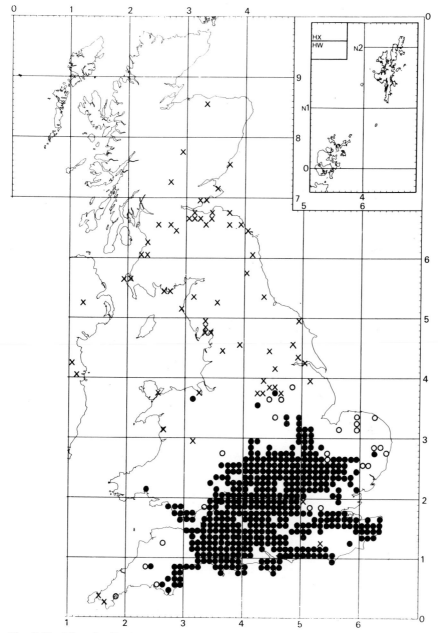

Fig. 7.5b. The distribution in England and Wales of *Viburnum lantana*, another "southern mixed shrubs" species. Interpretation and source as for Figs 5.7a,b. Crosses represent known introductions.

Fig. 7.6. The distinct (though increasingly rare) sight of a thriving juniper community, seen here near Aston Upthorpe in the Berkshire Downs. There is an obvious difference between the number and size of the bushes on the left (east-facing) bank and on the opposite side. This is probably the direct result of contrasting topoclimates (see Chapter 2), for, as noted in the text, cold conditions are known to favour juniper, but it could be an effect of differential grazing regimes. A typical downland plantation can be seen on the skyline. (C.J.S.)

Fig. 7.7. A dense stand of attractive but invasive dogwood (*Thelycrania sanguinea*) on former arable land at Lodge Hill, Buckinghamshire. The red-barked shoots spring up from underground suckers, and regenerate vigorously when cut back, though sometimes bushes die back, as can be seen at the extreme left of the picture. The dominant herbaceous species in the foreground is wild basil (*Clinopodium vulgare*). (C.J.S.)

idea, already stated in earlier chapters, that the more diverse the surroundings in geology, soils and land-use, the greater will be the variety of invading species, while a particularly vigorous and fecund population, or even individual tree or shrub, may result in a local preponderance of a single species. An overgrown hedge of hawthorn, for example, or plantation, hedgerow or parkland trees of sycamore, ash, birch or pine can lead to dense stands of these species, especially following a year of abundant fruit-setting, and where trees predominate, the shrub stage may be by-passed altogether.

It is a fact of observation that seeds dispersed by wind form a "seed shadow" of which the position and spread depend on the direction and speed of the prevailing wind during the peak period of dispersal. Heavier fruits and seeds not equipped for aerial dispersal tend to accumulate immediately under the canopy of the parent, and their subsequent spread is dependent on rodents, birds and sometimes other animals. Birds are most important in spreading the succulent-berried shrubs of chalk scrubland (Section 7.7.3.2), and their range is such that seeds may be spread several kilometres from source. However, enormous numbers of fruits and seeds must be damaged or destroyed, and those which are not tend in any case to be dropped or voided within the area of the bushes of origin where the birds feed, rest, roost and nest, rather than in the more open areas where, so to speak, they

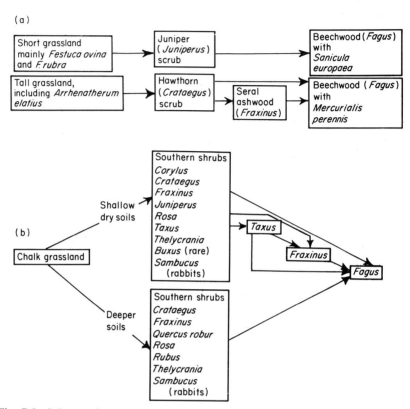

Fig. 7.8. Schemes for seral development from chalk grassland, through scrub, to woodland. (a) Watt's hawthorn and juniper seres proposed in 1934. For conditions thought to be associated with these two seres, see Table 7.4. (b) The more cautious scheme proposed by Duffey *et al.* (1974).

are needed. As a result, scrub normally advances only slowly from the nuclei in which it is already concentrated, a fact which is evident from Fig. 7.2. The more exposed, bleak areas of chalk downland may be as slow to scrub-up as they are, simply because they are out of range of a suitable wind-dispersed species, and unattractive to those birds most likely to distribute hawthorn, rose or wayfaring-tree seeds because of gustiness, turbulence or simply a lack of anything to perch upon. It is well known that a simple wire fence may be all that is needed to establish a bird-sown hedge.

Table 7.2

Herbaceous flowering plants typical of the woodland-edge habitat on chalk soils [a]

Grasses

	Dactylis glomerata
Arrhenatherum elatius	Festuca rubra
Brachypodium sylvaticum [b]	Poa trivialis
Bromus ramosus	Trisetum flavescens

Others

Achillea millefolium	Listera ovata
Ajuga reptans	Malva moschata
Agrimonia eupatoria	Ophrys insectifera [b,c]
Astragalus glycyphyllos [b,c]	Orchis mascula
Bryonia dioica (scrambler)	O. militaris [b,c]
Campanula trachelium [b]	O. purpurea [b,c]
Centaurea nigra ssp. nemoralis	O. simia [b,c]
C. scabiosa [b]	Origanum vulgare [b]
Chamaenerion angustifolium	Pastinaca sativa [b]
Clinopodium vulgare [b]	Platanthera bifolia [b,c]
Epilobium montanum	P. chlorantha [b,c]
Epipactis helleborine	Primula veris [b]
Eupatorium cannabinum	P. vulgaris
Galium mollugo [b]	Scrophularia aquatica [d]
Glechoma hederacea	S. nodosa [d]
Heracleum sphondylium	Sonchus arvensis
Hieracium exotericum [c]	Stachys sylvatica
Hypericum hirsutum	Tamus communis (scrambler)
H. montanum [b,c]	Tragopogon pratensis ssp. minor
H. perforatum [b]	Valeriana officinalis [c]
Inula conyza [b]	Verbena officinalis [c]
Knautia arvensis [b]	Veronica chamaedrys
Lathyrus nissolia [c]	Vicia sepium
L. pratensis	V. cracca
Leontodon hispidus [b]	Viola hirta [c]
Linaria repens [b,c]	

[a] Some species clearly overlap with earlier (grassland) and later (woodland) seral stages, but species more strictly typical of these other communities are not included.

[b] Calcicoles.

[c] Rare or local species.

[d] Especially on north-facing slopes.

7.3.3 Germination sites

Fruits and seeds must have somewhere to settle and establish. On open, recently disturbed soil, this is no problem, but turf is extremely resistant to invasion. The aerodynamic properties of some disseminules, such as the fruits of sycamore, cause them to fall in such a way as to lodge between the grass plants, whereupon they quickly stab a root down to the soil surface—a familiar sight on winter lawns. The feet of grazing sheep and cattle may help to press fruits and seeds through the turf into the soil. But normally, exposure of the mineral soil is essential for germination and seedling establishment. Watt calls these sites the "Achilles' heels" of the grassland sward.

The creation of germination sites in turf by earthworms, moles and similar agencies has been dealt with in Chapter 6, and by and large the same principles apply to the establishment of woody species. Likewise, rooting and scraping by rabbits, badgers, foxes and deer, as well as trampling—even poaching—by sheep and cattle are all important positive factors in woody succession even if, under normal circumstances, these or other animals do proceed to eat off or trample down most or all of the saplings they have helped to plant. In fact the most spectacular growth of scrub is associated with a cessation of grazing rather than simply no grazing: there is an important difference between the two situations. The consequences of Tansley's exclosure and the myxomatosis pandemic were both examples of what happens when a previously intensive grazing regime is suddenly withheld. As was noted previously, not only are new seedlings able to grow up, but so are older plants which may have been repeatedly eaten back to ground level under the former regime. On the other hand, scrub establishment in thick, matted, long-neglected and ungrazed turf may be a very much more protracted process.

Indeed, where the turf is very thick, seeds (especially of yew and hawthorn) may sometimes be seen lying on the surface, where they must be subjected to the greatest extremes of temperature and desiccation. Whether or not they can survive this environment is of little consequence, moreover, as long as they are thus prevented from reaching the mineral soil. In circumstances such as this, where all the usual agencies of soil disturbance may be presumed to be either absent or ineffective, it is interesting to contemplate the possible significance of the dead flowering stalks of the tall biennials, which rock in the wind and act like a gardener's dibber: as the stalks disintegrate, they leave a hole right down to the mineral soil in which seeds readily collect and, presumably, germinate.

7.3.4 Seedling establishment

It is not enough for a seed simply to germinate, and a thick mat of grassy litter can have three depressing effects upon seedling growth, two direct and the third indirect. Firstly the mat may temporarily become sufficiently moist to promote germination of any seeds caught up in it, but then dry up, leaving the seedlings suspended, literally high and dry, to quickly wither and die. Secondly, seedlings may establish in the comparative security of the mineral soil, but because of intense shading cease growth as soon as their food reserves are exhausted. The drawn, depauperate seedlings may thus slowly expire. Plants with the smallest seeds or slowest rates of growth must be most adversely affected in this way. Finally, a thick turf, as we saw in Chapter 6, can harbour large vole populations, and scrub seedlings may fail to establish in matted grassland simply because of insidious grazing by these active and voracious rodents.

In intermediate grazing regimes, particularly where the sward has been intermittently undergrazed, a number of shrubs may escape grazing and grow away, even though the surrounding turf is once again grazed heavily and any further woody encroachment prevented. Scrub of this kind, in which the woody plants are often similar in age, is especially typical of hawthorn and juniper. Moreover, the more effectively armed of these plants may recruit seedlings of other woody species around them as a result of protection from grazing, so that a pattern of increasingly large clumps or thickets develops, but still interspersed by close-grazed turf. Individual plants of briar and bramble may, through their ability to extend vegetatively, produce a similarly clumped type of scrubland. Bourne (1931) described a banded pattern of scrub development at Shirburn Hill on the Chiltern scarp which reflected lithological differences in the underlying chalk, but indirectly: rabbits selected the beds just below the Chalk Rock for burrowing because of their steepness of slope and ease of excavation, but, in order to graze, the animals would habitually move either up or down the hillside to gentler grades.

7.4 Juniper Scrub

7.4.1 Ecology of juniper

Although of wide edaphic tolerance throughout its full range of distribution in Europe (including Scotland), juniper is strikingly restricted to the Chalk in southern England (Table 7.3), where it forms a physiognomically very

Table 7.3

The geological influence on the distribution of juniper in southern England [a,b]

		Present	Extinct	Place names
Sedimentary rocks				
Eocene	Bracklesham/Bagshot Beds	1	6	1
	London Clay	0	7	1
	Reading/Thanet Beds	1	0	0
Cretaceous	Chalk	276	182	32
	Greensands	3	11	3
	Weald Clay	0	1	0
	Hastings Beds	1	5	0
Jurassic	Purbeck	0	1	0
	Portland	0	2	0
	Oxford Clay	0	1	0
	Cornbrash	0	3	0
	Great Oolite/Forest Marl	3	15	0
	Inferior Oolite	9	10	2
	Upper Lias	1	4	0
	Lower Lias	0	5	0
Triassic	Keuper Marl	0	1	0
Carboniferous	Pennant series/Upper Coal Measures	0	2	0
	Carboniferous Limestone	0	1	0
Devonian	Devonian Limestone	0	1	0
	Devonian Middle and Lower	3	4	1
Silurian	Ludlow and Wenlock Limestone	0	3	0
Igneous rocks				
	Serpentine	1	1	0
Total on chalk		276	182	32
Total on oolites		12	25	2
All other formations		11	59	6
Chalk with drift deposits		53	74	7
Other formations with drift		1	15	3

[a] From Ward (1973).

[b] Data are expressed per 1 km square.

distinct scrub community (Fig. 7.6). These southern plants belong to the so-called lowland type, *Juniperus communis* ssp. *communis*, distinguished by the longer leaves set almost at right angles to the stem, each tapering to a long sharp point. The habit of growth is usually somewhat spreading, though sometimes an erect, columnar form occurs, and there are intermediate types. As the community ages, long, prostrate branches of the more spreading bushes may be buried by the turf, giving the terminal branches the appearance of separate, younger plants. Others merge into clumps and thickets, though some stands retain a relatively tidy, even-spaced look. Juniper is, of course, a gymnosperm, and invariably dioecious. Male plants produce copious pollen from their solitary cones during early spring, and the berry of the female plant is formed by the coalescence of the fleshy scales of the female cone. Traditionally used in the flavouring of gin, the berries are green at first, maturing during their second or even third season into their attractive waxy-bloomed blue.

Referring to the distribution of juniper scrub in the Chilterns, Watt (1934) concluded that juniper occupies steeper (up to 34° 50′) and more "exposed" slopes than hawthorn, which typically occurs on less steep (up to 29°) and more sheltered slopes, though these effects of topography were considered to be indirect and to operate via soil depth and water relations. Thus the more exposed crests of the ridges (juniper) have thinner soils more prone to summer drought, while the more sheltered slopes (hawthorn) have deeper soils, more retentive of water and with less tendency to dry out (Chapter 3). Juniper soils were poorer in clay and humus and richer in calcium carbonate than hawthorn soils, while nitrification was inferred to be more active in the latter (Table 7.4).

All 68 soil profiles examined by Watt to substantiate these observations appear to have been taken from mature woodlands of the type to which each kind of scrub was assumed to have given rise, rather than from the scrubland soils themselves, though with juniper the dead bushes could plainly be seen among the trees. However, the results of surveys in the Chilterns by Littleton (1968) and Fitter (1968), and throughout southern England by Ward (1973) confirm Watt's association of juniper with thin soils and steep slopes. Both Fitter and Ward found more junipers on north- to west-facing aspects than any other and they were restricted in the main to altitudes between 120–150 m (Ward) and 180–210 m (Fitter), though both authors warn that these limits may simply reflect the nature of the last remaining sites in a sea of arable land and urban development in which juniper holds its own.

Although young juniper shoots are grazed by sheep and rabbits, this species is one of the first to be left once grazing is relaxed, and will then grow up unmolested, or at most receive only the occasional trim. Watt (1926,

Table 7.4

Possible differences between topoclimates and soils in hawthorn and juniper scrub [a]

Habitat characteristic	Juniper sere	Hawthorn sere
1. Exposure to prevailing westerly winds	Generally more exposed	Generally more sheltered
2. Angle of inclination of slope	Generally on steeper slopes, up to a maximum of 34° 50′	Generally on gentler slopes, up to a maximum of 29°
3. Chalk horizon	On both Middle and Upper, but typically developed on Upper Chalk or on harder strata	Definite preference for Middle Chalk or on softer strata
4. Mean soil depth (cm), with extremes in parentheses	31 (18–58)	51 (28–76)
5. [b] Percentage CaCO₃ in surface 15 cm of soil in mature beechwood	59·55	33·32
6. [b] Percentage loss on ignition in upper 15 cm of soil in mature beechwood	12·94	23·33
7. [b] Leaf litter in mature beechwood	Sparse	Forming a layer 2·5–5 cm thick
8. [b] pH values of upper 15 cm of soil in mature beechwood	7·9	7·6
9. [c] Soil water relations	Smaller water-holding capacity; less retentive and eustatic	Greater water-holding capacity; more retentive and eustatic
10. [c] Nitrification	Less active	Active

[a] From Watt (1934).
[b] Note that factors 5–8 were determined on mature woodland, not scrubland soils.
[c] By inference.

1934) has drawn attention to the strong positive association between juniper and yew. By growing up with juniper, yew seedlings which might otherwise be grazed receive some measure of protection. Tansley (1939) adds *Thelycrania, Viburnum, Euonymus* and *Rhamnus* (see Table 7.2), as well as ash and beech as having been thus nursed by juniper.

It was the general negative association of juniper and hawthorn which led to Watt's proposal of the separate juniper and hawthorn seres. No other clear-cut contrasts in woody vegetation were detected, but yew, holly, *Solanum dulcamara, Tamus communis, Sorbus* and *Rosa micrantha* were all more frequent in juniper scrub than in hawthorn. On the other hand, Ward (1973) found hawthorn to be the commonest woody companion of juniper, followed by yew, *Pinus sylvestris* (spreading from plantations), brambles, privet, elder (with which associated juniper is usually old), dogwood and beech. The situation needs clarifying.

Watt (1934) could find no qualitative differences between the herbaceous floras of ten samples each of juniper and hawthorn scrub in the Chilterns. Of the 48 species recorded from juniper scrub and 45 from hawthorn, 42 were common to both, and the remainder dismissed as "of no significance, accidental, or alien" (though see Section 7.7.2).

7.4.2 The decline of juniper scrub

7.4.2.1 The evidence

In recent years, concern has arisen over the apparent decline of the juniper on the Chalk of southern England. During extensive surveys by Ward (1973), many previously unrecorded locations of juniper scrub were discovered, particularly in Wiltshire, Hampshire and Sussex, a fact which tended to mask the loss of other juniper sites. But in all the counties surveyed, a distinct decline in juniper colonies was noted (Fig. 7.9). The loss of sites was found to be proportionately less marked in chalkland localities than elsewhere, but considering the relatively very small numbers on strata other than chalk (Table 7.3), this is perhaps not surprising. Specific examples are known from the Chalk (and elsewhere) where juniper scrub noted by Victorian naturalists, for example on the North (de Crespigny, 1877) and South Downs (Hudson, 1900), is now virtually or completely gone. A similar exercise was conducted by Fitter (1968) who surveyed former juniper sites in the Chilterns described by Druce (1926, 1927).

Why is this so? In part, of course, the loss of juniper sites can be accounted for by ploughing, afforestation, urban spread and, occasionally, road building (see Chapter 10). The extension of the M40 motorway from Stokenchurch to Waterstock in 1973, with its impressive cutting through the Chiltern escarpment, sliced with unparalleled irony clean through one of the

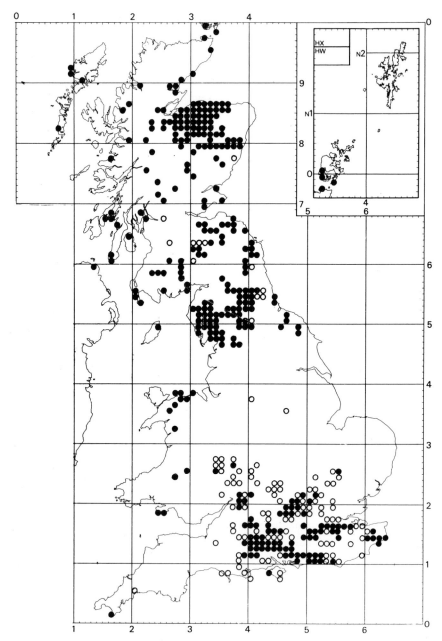

Fig. 7.9. The distribution of *Juniperus communis* ssp. *communis* in southern England (according to Ward, 1973) and elsewhere in Britain (Perring, 1968). Each spot represents at least one record in a 10 km × 10 km National Grid square since 1965 in southern England, and since 1930 elsewhere. Open circles indicate locations where the species occurred before these dates, but is no longer to be found.

fine juniper stands of the Aston Rowant NNR (Fig. 1.39) though here the situation has been made good by the propagation and replanting of cuttings from the original bushes (Section 10.3.3.5). But what about losses of individual junipers? Both the death of older bushes and the failure of young plants to regenerate from seed appear to be involved. What might the causes be?

7.4.2.2 The death of older bushes

Juniper is intolerant of shading, and quickly succumbs once a canopy of taller plants closes over it either during the course of natural succession (Fig. 7.10) or following the planting of downland with trees in afforestation

Fig. 7.10. Long-dead junipers, evidence of an earlier seral stage, persist among the beeches that overshadowed them at Windsor Hill, Buckinghamshire. (C.J.S.)

schemes. It is a different matter, however, when juniper bushes are seen to be dying off in the open. These moribund junipers (Fig. 2.8) have tempted some observers to suggest subtle changes in climate (Grose, 1957) or even industrial pollution (Bowen, 1965; Fitter, 1968) as the cause, but a glance through the older literature shows that there is nothing new in the phenomenon. Tansley (1939) refers to "the moribund appearance of much of the scattered juniper scrub of the chalk downs", an observation echoed by

Massingham (1940), while as long ago as 1900 Hudson drew attention to the fact that junipers were "not flourishing" on the South Downs.

The sudden death of whole branches has been attributed to the effects of frost, trampling and barking by livestock, and to parasitic fungi such as *Gymnosporangium clavariaeformae* and *Lophodermium juniperinum* (Ward, 1973). Many introduced coniferous species, as well as *Pinus sylvestris*, are known to die suddenly after as long as 40 years of healthy (but slow) growth on chalk (see Section 9.4.2.4). Perhaps juniper is similarly affected. On the other hand, it appears not to be a particularly long-lived species and many deaths may simply be due to natural senescence. The long persistence of juniper wood *in situ* without decomposing can give an exaggerated impression of death on a large scale.

7.4.2.3 Regeneration

It seems quite likely that the tendency for juniper to die off has been masked in the past by regular, if not prolific, regeneration of plants from seed, and that this regeneration is now much less widespread than it was. Indeed, it is possible to regard the junipers of the southern Chalk as the last stragglers of the extensive stands which dominated the vegetation of Postglacial Britain (Chapter 1). The association of juniper not only with the bleakest of downland localities, but with sparse herbaceous cover or even bare ground, has been noted by many observers. It is extraordinarily difficult to coax juniper seeds to germinate in the laboratory, but in the field they germinate mainly in early spring (Clapham *et al.*, 1962), and Pack (1921) showed that the seeds require a period of about 100 days' stratification, fully imbibed, at 5°C before germination is even initiated, after which a further period at this temperature must elapse for adequate growth to establish the seedlings, otherwise secondary dormancy ensues. Where discontinuation of grazing results in a thick, matted turf, juniper establishment is unquestionably inhibited, but it is interesting to contemplate whether this is entirely due to competition, or whether it might result from insufficiently low minima in the surface soil temperature compared with more open ground (Chapter 2).

All this assumes, of course, that there is a source of seed. Juniper is dioecious; berries are not produced every year; moribund bushes are less likely to fruit; and the seeds may be destroyed before they germinate. On this last point, the phytophagous mite *Trisetacus quadrisetus*, common on and specific to juniper, has recently been found by H. Bristow (personal communication) to entirely consume the seeds within the fruits, infecting 97% of the fruits tested in 1976, for example. There can be little doubt that this is a prime cause of the lack of regeneration of juniper, at least in the area of the South Downs around Brighton which Mrs Bristow has surveyed.

7.5 Hawthorn Scrub

Hawthorn (*Crataegus monogyna*) is very widely and abundantly distributed throughout the British Isles, especially as a hedge plant, and in the lowlands is the most abundant woody scrub species, colonising a wide range of soils which includes not only chalk but the less alkaline and heavy clays. Like juniper, hawthorn often forms almost or completely pure stands in which few or no other woody species grow (Fig. 7.4). In the south, the common species is sometimes accompanied by the rarer midland hawthorn (*C. oxyacanthoides*) distinguishable by its less deeply lobed leaves and twin stigmas which result in double-seeded berries (strictly drupes), though these are variable characteristics and hybrids between the two species occur. Although hawthorn seeds require one or two years to after-ripen before they will germinate (Davis and Rose, 1912) they are produced in such abundance, and are distributed so freely by birds that where grazing and cultivation are minimal constant regeneration of the species is assured, and bushes of all ages can normally be found in any locality.

In Watt's (1934) survey of the Chilterns, hawthorn occurred in sheltered places on deeper, more moisture-retentive soils than those supporting juniper (Table 7.4). Blackthorn (*Prunus spinosa*), ash, spindle (*Euonymus europaeus*) and dewberry (*Rubus caesius*) were found growing more often with hawthorn than juniper although none of these was exclusive to hawthorn scrub. Tansley (1922) found certain wild rose species, notably *Rosa micrantha* and *R. canina*, to be common associates of hawthorn, particularly in the early stages of grassland colonisation (see Fig. 7.2).

As noted above, the process of succession appears to be more gradual and to involve more seral stages than in juniper scrub. Often, a distinct stage dominated by ash intercedes (more so on the South Downs than the Chilterns) before climax beech finally establishes dominance. In places, the hawthorns form a closed canopy rather in the same way as yew (Section 8.7.2), in which the establishment of forest trees seems to be inhibited. "Thicket scrub" of this type has so far only been documented for abandoned arable fields on clay loams in Hertfordshire (Salisbury, 1918b), chalky-boulder-clay soils in Cambridgeshire (Ross, 1936) and acid (pH 4·4) superficial deposits on the Porton Ranges (Wells *et al.*, 1976), but a good example, complete with a typically woodland herbaceous flora, occurs on chalk soil (pH 8·0) at Newland Park (Smith, 1980).

7.6 Dogwood Scrub

Dogwood (*Thelycrania sanguinea*) is most readily recognised in early autumn when its foliage and bark have taken on a deep red colour, and its erect clusters of matt black, foetid-smelling berries are ripe. Although often accompanying privet (*Ligustrum vulgare*), *Rosa micrantha* and other southern shrubs, dogwood frequently forms pure stands on account of its ability to spread vegetatively by suckering. Indeed, in places this species has become a rampant invader of chalk scrub and is proving particularly troublesome in the management of nature reserves, especially where it has been repeatedly coppiced. In recent years, a tendency for older plants to die suddenly has been noted (Fig. 7.7). No explanation for this is yet forthcoming: it may simply have a short life span. But in contrast to the situation surrounding juniper, the condition may prove to be a blessing to the conservationist, rather than a cause for concern (see Section 10.3.4.5).

It has long been recognised that dogwood is strongly associated with succession on open, disturbed ground such as abandoned arable land or felled or coppiced woodland, especially where the soil is shallow or stony (Adamson, 1921; Tansley, 1922), but only recently has any experimental work been attempted to ascertain why this might be so (Lloyd and Pigott, 1967). The open, dicot-dominated swards so characteristic of certain formerly-cultivated chalk soils (Section 5.4.3.2) clearly must offer less impedance than a thick, grassy turf to the germination and establishment of invading shrubs, but why should dogwood be particularly favoured? According to Lloyd and Pigott, the answer may be in the rapid and abundant germination of dogwood seeds, especially compared with hawthorn. For a start, the dogwood berry contains two seeds, against the common hawthorn's one; but in addition to this, dogwood appears to lack the strong after-ripening requirement of hawthorn and so is more fitted to an opportunist strategy. After sowing 600 fruits of each species in experimental plots at Pulpit Hill, Buckinghamshire, in October 1962, Lloyd and Pigott recouped 354 seedlings of dogwood by the summer of 1964, but only 72 of hawthorn of which a mere 5 had appeared during the first summer (1963). This difference resulted, moreover, despite the treatment of the hawthorn stones with sulphuric acid, so that in normal field conditions the contrast between the two species should be even greater. Nevertheless, this work was very much of an exploratory nature, and there remains a nagging possibility that dogwood possesses some additional property, as yet unmasked, which enables it to establish so successfully on these hot, dry, impoverished soils of which it is so characteristic.

7.7 The Animal Life of Chalk Scrub

7.7.1 Changes in the fauna

The invasion of grassland by scrub greatly alters conditions for animals, and the nature and composition of the fauna changes accordingly. Among the invertebrates and birds in particular, open-ground and short-turf species give way to those of wood-edge and woodland habitats (though with the qualifications noted in Section 7.2.2). As with the plants themselves, these intermediate stages of scrub invasion provide a rich assortment of animals, a fact which apparently conflicts with established ideas of ecosystem stability (Section 6.5.4). Certainly, in the field of practical conservation, one of the greatest challenges is to maintain indefinitely what is inherently a transient and unstable combination of plant and animal communities (see Chapter 10).

The rabbit, of course, remains a potent factor in scrubland ecology, eating off young plants and barking the older ones (see Elton 1966, Plate 25), but enough has been said about this animal already. Woodland mammals are dealt with in Chapter 8. Here, two main groups of animals are considered: the invertebrates and the birds, though note that chalk scrub provides temporary shelter for many species which are not specific to, nor even characteristic of, chalk vegetation of any kind. A good example is the invertebrate fauna of juniper, especially in winter (Ward, 1977). Other examples are given in the next section on hedges. Many invertebrates enter the scrub fauna as a result of changes in the herbaceous plants rather than through any direct association with the bushes. These are briefly considered first.

7.7.2 Invertebrates of the herbaceous flora

The herbaceous plants which make their appearance as the scrub advances—the wood-edge species typified by *Origanum vulgare, Clinopodium vulgare, Glechoma hederacea, Viola hirta* and *Brachypodium sylvaticum* (see Table 7.2 for others)—can be expected to be colonised by animals, again mainly invertebrates, which depend on them. In fact, the more abundant wood-edge plants appear to support a rather poor phytophagous fauna (see Section 7.2.2), and relatively few insects have been recorded in the larval stage on the leaves of these plants. Sankey (1966) refers to the diminutive moths *Alucita hexadactyla, Pyrausta aurata* and *P. nigrata* as feeding on *Origanum* (which Grime *et al.* (1968) found palatable to the snail *Cepaea nemoralis*—see Table 6.6), while according to Stokoe (1944) the very local chequered skipper (*Carterocephalus palaemon*) feeds on *Brachypodium* and *Glechoma*. Both authors refer to the treble bar moth (*Aplocera plagiata*) on *Hypericum perforatum*. In the flowering season, of course, a wide range of pollen- and nectar-consuming insects (as well as their predators) are attracted to the heads of *Origanum*, and the richer the herbaceous flora, the more numerous and diverse these insects are likely to be, as reference to Table 7.2 and Chapter 6 will confirm. It is interesting to note (Table 6.4) that certain snails appear to favour juniper, and others hawthorn scrub, though this is probably due to contrasts in microclimates rather than any differences in herbaceous floras (Section 7.4.1).

7.7.3 Animals of the woody flora

7.7.3.1 Invertebrates

The woody plants themselves bring in their own invertebrate fauna. Some shrubs, probably because they are so abundant, support a very large number of species, while on others far fewer have been recorded. Duffey *et al.* (1974) provide a valuable summary of the numbers of phytophagous insect groups characteristic of most of the woody scrub species of lowland Britain (Table 7.5). Thus, *Crataegus* heads the list with 230 insect species recorded to date, with *Rubus* and *Rosa* scoring 107 apiece. Shrubs more restricted to the Chalk harbour fewer insect species: numbers range from 36 in *Sorbus* and 35 in *Ligustrum* down to a mere 6 in *Taxus* and 4 in *Buxus*. An extremely detailed account of the invertebrate fauna of juniper has recently been published by Ward (1977), and the possible significance of one of the mites in the decline of this species has already been noted.

For the most part, these insects are inconspicuous and unfamiliar, and known only to the specialist; often very little is known about their ecology. Some, however, are conspicuous and widely known, at least in the imago stage, such as the familiar brimstone butterfly (*Gonepterix rhamni*) whose larvae feed on buckthorn, and the beautiful privet-hawk moth (*Sphinx ligustri*) found in its no less impressive larval stage on *Ligustrum*. The attractive green hairstreak (*Callophrys rubi*) and holly blue (*Celastrina argiolus*) butterflies are very characteristic of open scrub, flying among the bushes and resting frequently on the foliage. On chalk, the larval food-plant of the former is typically rock-rose; the latter feeds on holly. But for both, the flower buds and young berries of dogwood provide an alternative (South, 1906). Many chalkland ecologists will have noticed the skeletonised leaves of *Virburnum lantana* caused by larvae of the beetle *Galerucella viburni*. Indeed, anyone who has reared these larvae in close confinement must have noticed the powerful smell of the foliage of the food-plant, one

Table 7.5

The approximate number of species of the principal phytophagous insect groups associated with genera of woody calcicolous scrub plants [a]

Plant genus	Macrolepidoptera	Microlepidoptera	Heteroptera	Homoptera Aphididae	Psyllidae	Coccidae	Aleyrodidae	Hymenoptera Symphyta	Cynipidae	Coleoptera	Thysanoptera	Diptera Cecidomyiidae	Agromyzidae	Total
Acer	7	12	3	7	0	2	0	3	0	2	1	4	0	41
Buxus	0	0	2	0	1	0	0	0	0	0	0	1	0	4
Clematis	13	0	0	0	0	0	0	0	0	2	0	0	3	18
Corylus	28	28	16	4	0	1	2	10	0	9	3	6	0	107
Crataegus	88	53	14	12	3	5	1	12	0	33	2	6	1	230
Euonymus	2	9	0	2	1	5	0	0	0	0	0	0	0	19
Ilex	3	2	0	2	0	2	0	0	0	3	0	0	1	13
Juniperus	5	8	6	1	0	2	0	1	0	2	1	1	0	27
Ligustrum	17	10	0	1	0	1	0	2	0	1	1	2	0	35
Lonicera	20	12	0	2	0	0	1	5	0	1	1	2	4	48
Malus	34	43	15	11	3	4	0	2	0	16	2	2	1	133
Prunus	57	43	5	6	2	4	0	13	0	17	4	6	0	157
Rhamnus	6	11	2	2	3	1	0	0	0	0	0	2	0	27
Rosa	27	25	2	10	0	4	0	22	10	3	1	3	0	107
Rubus	45	18	3	4	0	2	3	16	1	6	3	5	1	107
Sambucus	4	3	1	4	0	0	0	2	0	1	1	2	1	19
Sorbus	2	4	1	1	0	3	0	13	0	11	0	1	0	36
Taxus	2	2	0	0	0	1	0	0	0	0	0	1	0	6
Thelycrania	6	7	0	2	0	1	0	1	0	0	0	1	0	18
Ulex	9	12	6	2	2	2	0	0	0	12	7	0	0	52
Viburnum	3	3	0	4	0	1	1	3	0	1	0	1	0	17

[a] From Duffey et al. (1974).

possible reason why relatively few species are associated with it. Insects such as these which feed on the foliage of fairly abundant food-plants are in their turn reasonably abundant, at least in a good season. But where the food-plant is more local, and flowers or fruits rather than leaves or shoots form the source of food, an insect is inevitably both more restricted in its occurrence and more precarious in its station. A good example is again provided by Duffey *et al.*, who cite the juniper shield bug (*Cyphostethus tristriatus*) and the seed chalcid *Megastigmus bipunctatus*, both of which feed exclusively on juniper berries. An interesting point is that these, and others like them, may be useful indicators of the viability of scrub communities which are being deliberately managed for conservation. Their absence does not just make the reserve that much less interesting, but suggests that it might be too small, or too isolated, to be viable (see Ward and Lakhani, 1977; also Section 10.3.3.3).

Of course, all these phytophagous insects are susceptible in turn to predation by birds, spiders, wasps, robber-flies (Asilidae), lacewings (Plani-pennia), ladybirds (Coccinellidae) and other beetles, or to parasitism by Ichneumonidae and other wasps, which in their turn may be hyper-parasitised. It is a fascinating and instructive exercise to try to build up food chains and webs, from which ideas of energy flow and nutrient cycling can be gained (Fig. 4.7). Although in no way quantitative, a good example is that of Side (1955) for *Viburnum lantana* (not a particularly "rich" species, it will be recalled). His elaborate flow chart, redrawn by Ward, can be found in Duffey *et al.* (1974, Fig. 6.6).

7.7.3.2 Scrubland birds

Scrub is typically rich in bird life, many species nesting and breeding speci-fically in this type of habitat, or almost so. Examples are the yellowhammer (*Emberiza citrinella*), linnet (*Acanthis cannabina*), whitethroat (*Sylvia communis*) and the more local red-backed shrike (*Lanius collurio*). Others find food and shelter in scrub, a feature most impressively seen in winter when migrant redwings (*Turdus iliacus*) and fieldfares (*T. pilaris*) congregate in large flocks to roost in the bushes and feed off the berries. These, together with resident blackbirds (*T. merula*), mistle thrushes (*T. viscivorus*) and others are important agents in the dispersal of the berries, and in the spread of the scrub. Although based on acidophilous scrub, the survey by the London Natural History Society of the feeding ecology of the birds of Bookham Common, Surrey (Beven, 1964) contains a wealth of information of relevance to any scrubland ecosystem.

Reference was made briefly in Chapter 6 to Venables' (1939) bird surveys of the South Downs and elsewhere. Although the basis on which his counts of breeding birds were obtained was later criticised (Colquhoun and Morley, 1941), it would be unjust to dismiss this careful and detailed piece of work, which has a great deal in it of botanical as well as ornithological interest (see Fig. 7.1). It may well be that certain birds were scored twice or overlooked, but many interesting facts emerged which are unaffected anyway by absolute numbers, and which lent support to Lack's then recently formu-lated ideas on the psychological factor in bird ecology (Lack, 1937): the factor which determines whether or not a potential nesting site "looks right" to the bird.

Thus, Venables found that "it was rare for a bush-nester or -singer to be found breeding where there was not a great number of bushes (i.e. scrub) and for a ground-nester and air-singer to nest where there was not a big expanse of grass and much open air space immediately above it". Nevertheless, for some birds bushes provide no more than a songpost, and one isolated bush on which to perch may be all that is needed to encourage a tree pipit to breed in an area which in other respects still essentially resembles open downland. Another interesting aspect of bird behaviour to which Venables drew attention is that the willow warbler colonises scrub sooner than the closely related (and visually almost identical) chiff-chaff, because the latter demands a higher songpost for which a tree is normally required. Of course, all this relates to the natural order of events, and the situation may be altered by artefacts such as telegraph and electricity poles and wires, fences and so on and even race-horse jumps where these adorn the downs!

In thicker, more advanced scrub, especially where trees are beginning to invade, it is a fact of observation that the breeding bird population comes to resemble that of mature woodland. Even so, however, another important point established by Venables was that large trunks and high branches are not essential for woodland birds. Canopy-nesting species such as carrion crow, magpie and kestrel, as well as collared dove, green woodpecker and various tits, were all found in scrub vegetation.

The tricky problem of making accurate and reliable counts of individual birds and their breeding territories has now been largely surmounted by the introduction of standardised codes of conduct by the British Trust for Ornithology and International Bird Census Committee. These methods have been applied in censuses of birds in chalkland sites in the Chilterns (Williamson, 1967, 1975), at Kingley Vale NNR in Sussex (Williamson and Williamson, 1973) and on Salisbury Plain, Wiltshire (Morgan, 1975). Exactly how many individuals of each species descend upon a given patch of scrub in any one year will depend very much on the season. For those birds which overwinter in Britain this rests mainly on the severity of the previous winter, while migrants are, of course, affected by the conditions they have experienced in their winter quarters, and *en route*. Indeed, no element of the chalkland fauna can boast dependence on a wider range of circumstances than the birds which, twice a year, make those astonishing journeys in the face of every adversity from storm and drought to gun and trap. Moreover in any one locality, numbers, distributions and specificity to habitat will be greatly influenced by the nature of the surrounding vegetation and by inter- and intraspecific competition for nesting sites.

On the whole the recent censuses bear out predictions from earlier work and from general observation. Chalk scrub can certainly form a sanctuary for a large and varied breeding bird population, particularly if the scrub itself contains a mixture of species, and is heterogeneous in age structure and physiognomy (with here an open grassy patch, there a clump of trees, among the bushes). This effect is enhanced still further if the surrounding terrain supports a mixture of soil types and land-use. Such is the case at Ivinghoe and Steps Hills on the Chiltern escarpment near the Buckinghamshire-Hertfordshire border (Fig. 7.11) where in 1966 Williamson (1967) recorded a population of 659 pairs of breeding birds per square kilometre. Among these, willow warblers were characteristically numerous and meadow pipits unusually so. Their territories were mapped, and these are shown, together with some others for comparison, in Fig. 7.12.

Fig. 7.11. Steps Hill and Ivinghoe Beacon, Buckinghamshire: the location for studies in bird and snail ecology mentioned in the text. (C.J.S.)

Between 1968 and 1974 (excluding 1972), an annual average of 204 pairs of birds consisting of 36 species bred on the 24 ha observation plot at Steps Hill alone (Williamson, 1975), corresponding to a density of no less than 840 pairs per square kilometre. Throughout this period, the willow warbler remained the most abundant species (Table 7.6), though the meadow pipit had declined considerably since 1966, perhaps because the scrub had spread. Among the most numerous birds were the ubiquitous blackbird and song thrush; the relatively high incidence of the reed bunting was unusual for a chalk site. Habitat preferences were as expected (Fig. 7.13), though possibly influenced by the fact that because some areas had reached saturation point certain pairs were forced to nest in suboptimal sites. A slight increase in woodland birds such as goldcrest and coal tit was cautiously noted by the author, again reflecting the advancing maturity of the scrub.

Smaller numbers of birds bred in the area of Salisbury Plain observed by Morgan (1975), as might be expected considering the floristic poorness and general sparseness of the scrub, and the uniformity of the surrounding soils

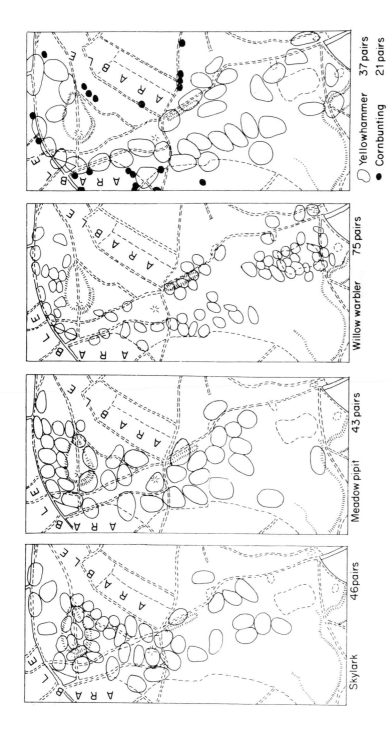

Fig. 7.12. The distribution of breeding territories of skylark, meadow pipit, willow warbler and yellowhammer, and nesting sites of cornbunting, in chalk grassland and scrub at Ivinghoe and Steps Hills, 1966. From Williamson (1967).

Skylark 46 pairs   Meadow pipit 43 pairs   Willow warbler 75 pairs

◯ Yellowhammer 37 pairs
● Cornbunting 21 pairs

Table 7.6

Territories of breeding pairs of birds mapped in a 24 ha observation plot of chalk grassland and scrub at Steps Hill, Buckinghamshire, from 1968 to 1974 (excluding 1972) [a]

| | | Average number of breeding territories per year | |
	Species	(a) On the 24 ha plot	(b) Per km^2
1. Species breeding 6 years out of 6			
Willow warbler	*Phylloscopus trochilus*	30·1	124·0
Linnet	*Acanthis cannabina*	20·8	86·0
Dunnock	*Prunella modularis*	19·5	80·0
Blackbird	*Turdus merula*	18·3	75·0
Song thrush	*T. philomelos*	10·9	45·0
Yellowhammer	*Emberiza citrinella*	10·3	42·5
Robin	*Erithacus rubecula*	8·7	36·0
Chaffinch	*Fringilla coelebs*	8·2	33·6
Wren	*Troglodytes troglodytes*	7·8	32·4
Wood-pigeon	*Columba palumbus*	7·5	31·0
Meadow pipit	*Anthus pratensis*	6·5	26·7
Bullfinch	*Pyrrhula pyrrhula*	5·2	21·2
Turtle dove	*Streptopelia turtur*	4·8	19·8
Whitethroat	*Sylvia communis*	4·7	19·3
Blue tit	*Parus caeruleus*	4·2	17·1
Skylark	*Alauda arvensis*	4·0	16·6
Great tit	*Parus major*	3·8	15·4
Garden warbler	*Sylvia borin*	3·7	15·1
Blackcap	*S. atricapilla*	2·2	8·9
Chiff-chaff	*Phylloscopus collybita*	2·0	8·3
Long-tailed tit	*Aegithalos caudatus*	1·8	7·6
2. Species breeding 5 years out of 6			
Lesser redpoll	*Acanthis flammea*	2·8	11·9
Tree sparrow	*Passer montanus*	2·0	8·2
Magpie	*Pica pica*	1·7	6·9
Willow tit	*Parus montanus*	0·9	3·8
Tawny owl	*Strix aluco*	0·8	3·5
Tree pipit	*Anthus trivialis*	0·8	3·5
3. Species breeding 4 years out of 6			
Grasshopper warbler	*Locustella naevia*	2·4	10·0
Reed bunting	*Emberiza schoeniculus*	2·0	8·2
Lesser whitethroat	*Sylvia curruca*	1·4	5·9

4. Species breeding 3 years or less out of 6 (in decreasing order of incidence)

Pheasant (*Phasianus colchicus*), coal tit (*Parus ater*), cuckoo (*Cuculus canorus*), goldcrest (*Regulus regulus*), jay (*Garrulus glandarius*), marsh tit (*Parus palustris*).

[a] From Williamson (1975).

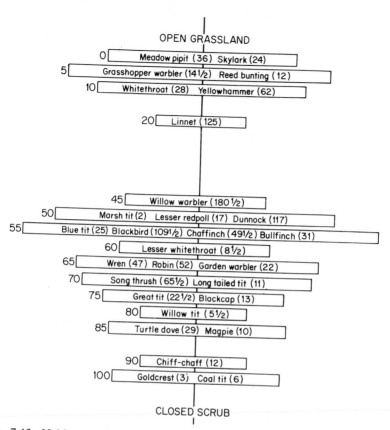

Fig. 7.13. Habitat preferences of 29 bird species of chalk scrub at Steps Hill, expressed as a percentage of territories established in closed scrub during 1968–1971, 1973 and 1974. The number of territories recorded for each species is shown in parentheses. After Williamson (1975).

Table 7.7

Area, number of breeding species and territories, and density of breeding pairs of birds in four kinds of chalkland habitat at Porton, Wiltshire, in 1973 [a]

Habitat	Census area (ha)	No. of breeding species	No. of territories	No. of territories km⁻² ("density")
Chalk grassland [b]	32	13	28	87·4
Hawthorn scrub	60	33	127	210·0
Juniper scrub	14	26	69	492·7
Pinus sylvestris plantation	10	16	34	340·0

[a] From Morgan (1975).

[b] Including fine-leaved fescue, *Helictotrichon pubescens/H. pratense*, and lichen-dominated types (see Chapter 5).

and land-use. Hawthorn scrub supported a larger number of species, but juniper scrub contained most birds (Table 7.7). This area has special significance, however, in supporting birds with large territorial demands such as the buzzard (*Buteo buteo*) and the much rarer hobby (*Falco subbuteo*) and stone curlew (Section 6.3.5). The role of yew scrub and woodland as a habitat for birds is considered in Chapter 8.

7.8 Hedges

7.8.1 The origin of hedges

Surely no feature of our rural landscape has come to be taken so much for granted as the hedge. Indeed, specific interest in the subject has sprung up only in the last decade or so, with increasing public concern over the zest with which some landowners have seen fit to remove hedges in the interests of agricultural efficiency. This is no place to deal in detail with the history of hedges and their place in the rural economy, a job which in any case has been admirably done by Pollard *et al.* (1974). Economic aspects of hedges are touched on in Chapter 9 and the controversy over hedgerow removal versus conservation in Chapter 10.

Hedges are man-made. The majority were planted with the express purpose of containing (or excluding) livestock, usually in connection with the conversion of former open fields in tenantry into larger units to which improved methods of husbandry could be applied. Until recently, the popular image of this procedure has been one of an all-encompassing eighteenth-century hedge-planting bonanza, with documentary evidence to support the idea in the form of the innumerable Parliamentary Acts and Awards of Enclosure of that period. But although it remains true that enclosure reached a peak between 1760 and 1820 (Pollard *et al.*), it is becoming increasingly clear that the process was a much more patchy and protracted business than formerly portrayed, and by no means necessarily ratified by Parliament. In some places, enclosure was already well under way by the late 1500s. In others, it has hardly occurred at all, either because of local customs and quirks of social history, or because it was physically impossible, or at least unprofitable, to parcel up the land into small fields (Jones, 1967; Kerridge, 1967).

The more remote chalk country undoubtedly comes into the latter category, and it seems safe to suggest that here there has never been more than a minimal framework of boundary and trackway hedges, and that the scenery has always been open and rolling, whether under sheep or corn. This was the champion country of the historical writers. A corollary of this, of course, is that most chalkland hedges are old, some very old indeed, and the older a hedge is, the greater is its ecological richness (p. 51). Moreover, the poorer the surrounding countryside is in shrubs and trees, whether it is sheep-grazed downland or intensively managed arable land, the greater the importance of the hedge as a habitat for wildlife, not only for the establishment of permanent plant and animal inhabitants but also to provide a corridor along which all kinds of animals, and plants for that matter, can migrate from one place to another. The herbaceous grass-verge habitat serves an exactly comparable function, and the importance of both hedge and verge (considered together as the hedgerow) in conservation is discussed in Chapter 10.

7.8.2 Hedge ecology

It is hard enough to find information of any kind on hedge ecology, let alone on those of the Chalk, but one example is Elton's survey of 1938–1939 of a double stretch of rather neglected hawthorn hedge bordering a grassy track known as Sheep Drove high on the Hampshire Chalk near Leckford (Elton, 1966, pp. 174–175 and Plate 24). In fact, Venables walked this area in his bird survey mentioned earlier in this chapter. No less than 20 woody species were recorded by Elton, suggesting, as might be expected, that the hedge was very old, but because no indication was given of how these species were distributed along the length of the hedge it is not possible to estimate its age. The track crossed a wide expanse of arable land supporting "cereals, roots, flax and other crops", and was connected at one point by another double hedge, at right angles to it, to a young plantation of beech and other trees. An important feature noted by Elton's keen eye was the way the track curved slightly so that the double hedgerow provided "particularly quiet shelter", and resembled "a large glade within walls of scrub".

It is possible here to touch only briefly on what was a very comprehensive exercise, and for more details the reader should dip into Chapter 9 of Elton's superb book. He writes (p.174):

> The chalk scrub flanking parts of these long hedges on the inner side was rich and well representative of this kind of association on pure chalk. There were blackthorn, guelder rose, wayfaring tree, wild rose, dogwood, elder, hazel, common buckthorn, privet, spindle and sallow (*Salix capraea*). . . . Also several climbers: ivy, honeysuckle, black bryony and white bryony. Of trees there were very few indeed: one common maple (*Acer campestre*), one wild pear (*Pyrus communis*), one crab-apple (*Malus sylvestris*), one holly, one young oak, one or two young yews, and several grown ash trees with some spread of young saplings from them.

The field layer which flanked the hedges contained "a great deal of bramble" in places.

On the fauna, Elton continues (p.175):

Apart from birds, the most conspicuous animals in summer were butterflies of nine species. Only one of these had its larvae on scrub—the brimstone (*Gonepteryx rhamni*) whose eggs were found on leaves of the buckthorn. The others fell into four ecological groups. The peacock (*Nymphalis* [*Inachis*] *io*) depends on nettles; the orange tip (*Euchloe* [*Anthocaris*] *cardamines*), the green-veined white (*Pieris napi*) and the small white (*P. rapae*), on cruciferous plants; the meadow brown (*Maniola jurtina*), gatekeeper (*M.* [*Pyronia*] *tithonus*) and large skipper (*Ochlodes* or *Augiodes venata*) on grasses; and the common blue (*Polyommatus icarus*) on birdsfoot trefoil, *Lotus corniculatus*. How far all of these lived and bred in Sheep Drove itself was not ascertained.

An interesting feature of Sheep Drove which illustrates the point made in the previous section about the way "alien" invertebrates may find shelter in chalk scrub (to some extent already demonstrated by some of the butterflies, although these must all have fed on the flowers of the scrub plants) was the abundance of the blackflies *Simulium equinum* and *S. ornatum*, which breed in "rather slow, weedy rivers". These were frequently encountered in swarms in the shelter of the hedges, and they could only have come from the River Test some 1·5 km distant. The mayfly *Baetis rhodani* was also noted. During the winter, the hedgerow hawthorns (and presumably other members of the woody flora) provided important refuges for many hibernating insects from the surrounding farmland. These insects, which dispersed again during spring and summer, included both allies and enemies of the farmer: a central argument in the interminable wrangles over the pros and cons of hedge removal (see Chapters 9 and 10).

8
Woodlands on the Chalk

8.1 Introduction

A chapter on woodlands logically follows one on scrub, though in fact only a
small proportion of our woods has evolved thus, many, perhaps a majority,
having been deliberately planted within the last 200 years or so (see Chapter
9). Even where documentary and other evidence indicates a much greater
age than this (the Chiltern beechwoods and parts of Grovely Wood, Wiltshire, are examples on the Chalk), the influence of deliberate management,
particularly coppicing, by generations of silviculturists can hardly be underestimated. This should be borne in mind throughout the following account
of the natural history of these woodlands.

8.2 Succession and Climax

8.2.1 Pioneer woodland

The actual transition of advanced scrub to pioneer woodland is mainly to be
seen on neglected marginal land flanking existing woods (Fig. 8.1). It is no
accident that most of what we know comes from the surveys of large
downland and Chiltern estates conducted during a period of severe agricultural depression (Adamson, 1921; Watt, 1923, 1924, 1925, 1934). Indeed,
remarkably little has been added to our knowledge of the ecology of chalk
woodlands since that work was carried out.

As with other transitional communities (ecotones), pioneeer woodland is
notable for its relatively large numbers of species, some declining as time
passes, others increasing either temporarily (the true woodland pioneers) or
more permanently. Regarding species which decline, it is a fact of observation that herbaceous vegetation typical of grassland and open scrub gives

way to more shade-tolerant species, or even bare ground, within perhaps a decade of the woodland canopy closing over, though, as was noted before, documentary evidence to illustrate this is extremely sparse. More is said about the herbaceous flora later.

Fig. 8.1. Ash (*Fraxinus excelsior*) and gean (*Prunus avium*) invade chalk grassland and scrub from adjacent woodland near Naphill, Buckinghamshire. The ready supply of these tree seeds and their efficient dispersal and establishment herald a short-lived scrub phase, though at what stage, if at all, beech enters remains to be seen. (C.J.S.)

Of the scrub species themselves, juniper succumbs relatively rapidly to shading despite its evergreen habit, though because of the durability of juniper wood the dead bushes may persist for decades (Fig. 7.10): a feature which helped Watt formulate his seral hypothesis (Section 7.2.1). Blackthorn is another species to die out quite quickly. Elder and privet may persist, drawn and unflowering, under the developing tree canopy as well as common hawthorn (*Crataegus monogyna*). Midland hawthorn (*C. oxyacanthoides*) is able to flower and set fruit in quite heavy shade, though this is rather local on the Chalk. Hazel may make a substantial contribution to the shrub or understorey layer, particularly where it has formerly been coppiced (Section 9.4.1). The evergreen spurge laurel (*Daphne laureola*) is a late invader of calcicolous scrub which forms a characteristic component of the woodland shrub layer.

The most typical trees of pioneer woodland on chalk, often coming in at a very early stage of scrub development and locally forming pure stands, include wind-sown ash (*Fraxinus excelsior*) and sycamore (*Acer pseudoplatanus*), and bird-sown whitebeam (*Sorbus aria*) and yew (*Taxus baccata*). Others include the willows (e.g. *Salix capraea, S. cinerea*), birch (typically *Betula pendula*), field maple (*Acer campestre*), holly (*Ilex aquifolium*), holm oak (*Quercus ilex*), rowan (*Sorbus aucuparia*), crab-apple (*Malus sylvestris*), guelder rose (*Viburnum opulus*) and gean (*Prunus avium*). The bird-cherry (*Prunus padus*) is typical of more northerly situations. An interesting point is that seedlings of oak (usually *Quercus robur*— but see Section 8.5) may be quite numerous on chalk soils—often more so than beech itself—but these seem unable, as a rule, to survive for longer than about two years (Adamson, 1921). This is presumably due largely to rabbits and other mammals, and is certainly not a phenomenon confined to the Chalk. Controversy has long raged, however, over how well those that do survive are able to thrive on chalk soils. Certainly, many remain stunted and display symptoms of chlorosis, yet sometimes oaks can be seen growing quite satisfactorily, if not optimally, on thin chalk rendzinas, as at Box Hill. These contrasts must reflect subtle differences in the composition of the chalk, drift and solifluxion deposits from which the soils are derived, but the matter requires investigation.

Less common are hornbeam (*Carpinus betulus*), of which the small-leaved var. *parvifolia* has been reported as occurring locally on chalk soils in the mid-Chilterns (Christy, 1924), the apparently naturalised walnut (*Juglans regia*) and Norway maple (*Acer platanoides*), the rather rare wild pear (*Pyrus communis*) and the occasional lime and elm. Among the limes, the large-leaved *Tilia platyphyllos* has a somewhat fragmented distribution, but both *T. cordata* and the hybrid between these two, *T. × europaea*, are more widely distributed and can be found as escapes from parkland where they have been commonly planted. More is said about *Tilia* below. Among the taxonomically complicated elms, *Ulmus glabra* (the wych elm) is the species most likely to be encountered in developing woodland on chalk. Finally, certain economic conifers, particularly *Pinus sylvestris*, commonly invade and contribute to pioneer woodland where a nearby plantation provides a seed source, but not, apparently the larches (*Larix* spp.), which seem singularly inefficient at regenerating from seed, at least on chalk.

It seems to be taken for granted that, at least in south-east England, beech (*Fagus sylvatica*) will ultimately emerge from this seral plethora of shrubs and trees, overwhelming them with its deep shade, and taking its place as the climax species (see below) as it appears to have done since it spread back to its present natural bounds in Anglo-Saxon times (Chapter 1). Yet, it is not at all common to see beech establishing itself in this way now, and although this may be due as much to a lack of suitable sites to colonise, or to rabbits and squirrels, as to any failings in beech itself, it is a point to be borne in mind. It has even been suggested (Streeter, 1965) that the British climate is no longer suitable for the natural regeneration of beech. We return to this point later in this chapter, and again in Chapters 9 (with regard to forestry) and 10 (conservation).

8.2.2 Mature woodland

Using the terminology of Watt (1947), the young wood next enters its *building phase*, during which the rate of biomass accumulation per unit area increases to a peak, the canopy closes over and becomes very dense, and both inter- and intraspecific competition are intense. Most of the pioneers and many potentially longer- lived species "become suppressed and die at an early age while their more successful neighbours expand their crowns above" (Cousens, 1974). In course of time, growth slows down, competition lessens and the canopy becomes more open, and it is during this, the *mature phase*, that the characteristic herbaceous plants of the woodland floor become fully established. The *degenerate phase* eventually follows as individual trees begin to senesce and die, though because in Britain these woods are usually managed commercially, this phase is not commonly seen, an important point in woodland conservation. The stand cycle is completed by the reinvasion of the gaps created by the death of the older trees: the *regeneration phase* in its strict sense.

Where the botanical composition of the trees being recruited to the community is not appreciably different from that of the mature stand as a whole, a climax woodland is indicated, though ecologists differ in their interpretation of this term, according to the time-scale used (Burges, 1960; Shimwell, 1971a), and to the degree to which anthropogenic influence is recognised. Thus beechwood on chalk may be regarded as an edaphic climax in the sense that oakwood could be expected to develop but for the calcareous soil (Cousens, 1974), or as a biotic climax in which originally mixed woodland has evolved into beechwood through the activities of early man (Ratcliffe, 1977). Outside the immediate influence of beech, ash is the usual constituent of climax woodlands, but in places these too are now seen as anthropogenic, and seral to communities in which other species such as oak, elm and lime may eventually make their appearance (see Section 8.4). More locally yew and box may form climax communities, while the ability of hawthorn to do this has already been mentioned in Chapter 7.

8.3 Beechwoods

8.3.1 Trees, shrubs and climbers

Beechwoods are seen most characteristically in the Chilterns, as well as in parts of the North, South and Berkshire Downs, as beech hangers, crowning the upper slopes of the escarpments and valley sides (Fig. 8.2), their undulating lower boundaries sometimes marking outcrops or accumulations of flints. The structure and composition of these woods varies, of course, according to the nature of the soil. Three main types (associations) of beechwood are recognised (Tansley, 1939): Fagetum calcicolum, found on

Fig. 8.2. A classic Chiltern beech hanger near Bradenham, Buckinghamshire. (C.J.S.)

thin rendzina soils of the Chalk and on Oolitic limestone in the Cotswolds; Fagetum rubosum, typical of the Clay-with-flints and similar deposits of the Chiltern and downland plateaux (and also of Old Red Sandstone in eastern Scotland and south-east Wales), in which a dense ground flora of brambles is characteristic; and Fagetum ericetosum, containing ericoids and other calcifuges (notably the grass *Deschampsia flexuosa*), of the acid sands and gravels of the Hampshire and London Basins, as in the New Forest, and at Burnham Beeches and Epping Forest. These contrasts emphasise the fact that beech is in no way a calcicole: indeed it frequently exhibits chlorosis on chalk (Section 9.4.2.5). In fact it is tolerant of a wide range of soils, provided they are freely drained.

The notes which follow relate mainly to beechwoods referable to Fagetum calcicolum, but of course intermediate types transitional to the other associations will usually be encountered (Fig. 8.3). This is hardly surprising, having seen (in Chapters 1 and 3) the variations in depth of the superficial deposits overlying the Chalk, and in the composition of the underlying rocks on the slopes—here pure chalk, there an intimate mixture of chalk, clay and flint. There is no more vivid demonstration of this fact than the aftermath of a severe winter gale when perhaps a dozen or more mature trees may be

Fig. 8.3. An unusually clear-cut boundary between herbaceous beechwood communities in Stocking Wood, Naphill. Brambles (*Rubus fruticosus*, knee-deep on the floor of the Fagetum rubosum woodland on Clay-with-flints just to the left of the picture) display symptoms of chlorosis as they encroach on to chalky soil supporting dog's mercury (*Mercurialis perennis*), characteristic of Fagetum calcicolum woods. (C.J.S.)

thrown within an area of a hectare or so, all too common a sight when earlier felling has opened up a wind-gap. Look at the soil adhering to the roots and at the profiles exposed where these have been wrenched from the ground and try to find two examples which are exactly alike. And note, too, how extremely thin a layer of acid surface soil is necessary to permit the establishment of strongly calcifuge species, despite the proximity of the solid chalk beneath.

A common feature of beechwoods on chalk is an almost complete absence of any other plants whatsoever (Fig. 8.4), but in places the odd ash, white-beam, birch, sycamore and gean may reach the canopy and persist for many decades, the last betraying its presence by its spectacular but brief display of

Fig. 8.4. The interior of a typical chalk-scarp beechwood at Whiteleaf Hill, Bucking-hamshire, virtually devoid of any other woody or herbaceous plants, and with little sign of regenerating beeches. (C.J.S.)

white blossom early in spring but in addition readily recognised by its bright red autumn tints and deep red bark which peels away in horizontal strips. Pines or larches in the mature canopy are a sure indication that the wood has been planted (Brown, 1953).

Most species likely to be found in the shrub layer, where this occurs, have been mentioned already. Those which appear to flourish, rather than simply hold their own, include yew, box, holly and spurge laurel. Closely related to the last, the deciduous mezereon (*Daphne mezereum*), long grown as a garden shrub for its fragrant pink blossom produced in early spring before its leaves open, is a native of beechwoods on chalk soils, but is now very rare indeed in the wild (Table 10.8). Although more characteristic of the plateau beechwoods, occasional brambles occur, including the dewberry (*Rubus caesius*). C. D. Pigott (personal communication) regards *R. vestitus* and some members of sect. *Corylifolius* as most characteristic of calcareous beechwoods. However, few of these flower in the deepest shade, which also has the effect of altering the nature of the prickles (see Heslop-Harrison, 1953, Fig. 1), adding further to the difficulties of accurately identifying this notorious group. Here too, the leaves of the older holly bushes (which often exhibit chlorosis on chalk soils) acquire uncharacteristic shapes, some losing all trace of lobes and spines. Among the climbers, *Clematis vitalba* may persist from the scrub stage where the canopy is not too dense. Ivy (*Hedera helix*) may scramble across the floor of the wood, but only where it climbs into adequate illumination does it produce its greenish flowers, so attractive to wasps and flies, between September and November. The berries ripen in the following spring. Honeysuckle (*Lonicera periclymenum*) climbs by twining and is much more of a threat than ivy to the trees which support it because of its well-known ability to strangle and distort the branches.

8.3.2 The herbaceous vegetation (including lower plants)

8.3.2.1 Plants of the shade

It is not difficult to draw up a list of the plant species which typically make up the herbaceous vegetation of "calcicolum" beechwoods, but the distribution of these from place to place is usually very patchy. Some parts of the wood, especially where the beech canopy is very dense or where there is an understorey of holly or yew, or where the floor is open to a long fetch of wind, are entirely devoid of herbaceous plants of any description and may be shin-deep in fallen leaves (whatever the time of year) or so bare of litter as to give the impression of having been swept by a broom. Here, only mosses such as *Ctenidium molluscum*, *Anomodon viticulosus*, the less calcicolous *Fissidens taxifolius* and *Brachythecium rutabulum* (Adamson, 1921) and pleurococcoid algae provide a touch of green. The algae and some mosses (notably *Hypnum cupressiforme* var. *filiforme*) grow epiphytically, the former on the smooth beech trunks and the latter on the humps of the tree roots, while some, such as *Orthotrichon affine*, even colonise flints. Watson (1968) provides a useful guide to the habitat preferences of these beechwood bryophytes. All these simple autotrophs become most conspicuous in winter when the crowns of the trees are leafless and the gaunt boles become suffused with what is almost a green irridescence.

Early summer may see the emergence from the leaf litter of the curious heterotrophic angiosperms, devoid of root or green leaf and mostly very local in their distribution. Examples are the yellow birds nest (*Monotropa hypopytis*, actually an aggregate species about which very little is known), the quite unrelated birds-nest orchid (*Neottia nidus-avis*), and one of the rarest species in the whole of the British flora, the ghost orchid (*Epipogium aphyllum*). Often quoted as saprophytes, these plants are actually parasites, depending entirely on their mycorrhizal fungal associates for sustenance. A valuable account of the biology of these elusive orchids is provided by Summerhayes (1951). Later in the year the saprophytic fungi make their appearance, though, as is well known, the display in any one place is likely to fluctuate widely from year to year. A list of these can be found in Sankey (1966) (see also Lange and Hora, 1965).

The more light-demanding herbs and grasses avoid the deepest shade, or rather fail to establish there. Some germinate and grow for as long as their food reserves last, but these weak, straggling plants rarely reach the flowering stage and usually succumb well before this. Others fail even to germinate (though their seeds may be present in enormous numbers), either because the light is simply too dim, or because it contains too large a proportion of infra-red to visible short-wave radiation (see Fig. 2.16c) for germination to be promoted. In a recent demonstration of the latter phenomenon, King (1975) showed that a double layer of *Tilia* × *europaea* leaves was sufficient to inhibit germination of seeds of *Arenaria serpyllifolia*, *Veronica arvensis* and *Cerastium holosteoides*, species of disturbed grassland (see Chapter 6). Either mechanism may explain the familiar phenomenon of seeds of the common mullein (*Verbascum thapsus*), or the foxglove in woods on acid soils, lying dormant for many years under woodland shade, only to germinate in profusion the moment the canopy is opened, as by wind-throw or deliberate felling. This, though, is dealt with later.

As with some of the shrubs, however, many herbaceous green plants are able to grow in surprisingly deep shade. Some, such as bugle (*Ajuga reptans*) and stinging nettle (*Urtica dioica*) grow vegetatively but do not flower; others, such as the common enchanter's nightshade (*Circaea lutetiana*), may flower, but then fail to set seed (Salisbury, 1976). It has been suggested that male and female clones of dioecious dog's mercury (*Mercurialis perennis*) may be differentially influenced by shading, but more of this in a moment. An interesting example of adaptive morphological plasticity is shown by the wall lettuce (*Mycelis muralis*) which occurs, as its name suggests, in open, dry, rubbly sites such as old walls yet which, curiously, is no less characteristic of deep shade in beechwoods on chalk (Fig. 8.5). This plant is readily identified by its large open panicles in which only a few yellow heads are in flower at any one time, others being either already at the fruiting stage or still

Fig. 8.5. Wall lettuce, *Mycelis muralis*.

in bud. On open ground this species grown vigorously and its smooth foliage often takes on a deep red colour. In beechwoods it retains its greenness, and forms a most delicate plant, its leaves as thin as paper, their beautiful outline accentuated, and looking just as if they had been pressed. The result is to increase the leaf area ratio (LAR) so as to intercept as much incident solar radiation as possible. C. D. Pigott (personal communication) has recorded exceptionally high LAR values in *Mycelis*. Considerable attention has been devoted to this phenomenon in an unrelated but ecologically similar species, *Impatiens parviflora* (Hughes, 1959), in which the thinning of the leaves was found to be due to a redistribution of the leaf cells as shown in Fig. 8.6. Later work (see, for example, Hughes, 1965) showed, however, that both environmental fluctuations in woodland shade and the morphogenetic response in *Impatiens* were extremely complex. Similar studies on *Oxalis acetosella* have recently been reported by Packham and Willis (1977).

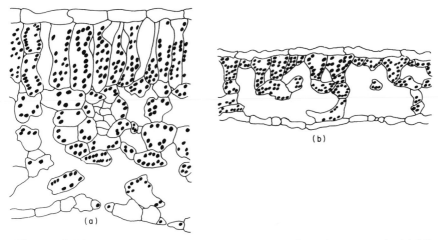

Fig. 8.6. Differences in leaf anatomy in *Impatiens parviflora* plants grown in (a) full daylight and (b) 7% daylight. The leaf in (a) was 270 μm thick, that in (b) 100 μm. From Hughes (1959).

Among the grasses, *Brachypodium sylvaticum* is very common in (though not exclusive to) chalk beechwoods but, like *Mycelis*, is also to be found in the open. This has led to the suggestion that *Brachypodium* makes a useful indicator of sites cleared of woodland in the recent past, but care should be taken in adopting too rigid a viewpoint on this matter, for it can sometimes be found invading chalk grassland and scrub of long standing from adjacent woods and hedgerows. When it occurs in open grassland, this grass is sometimes mistaken for tor grass (*B. pinnatum*) (see Chapters 5 and 6), but as reference to Hubbard (1968) will confirm, the foliage of *B. sylvaticum* is at least moderately and usually very hairy (*B. pinnatum* is normally quite hairless), and the spikelets are much more conspicuously awned in *B. sylvaticum* (awns up to 12 mm long) than they are in *B. pinnatum* (1–5 mm).

Several other grasses grow and flower under the summer tree canopy. The coarse tussock grass (*Deschampsia caespitosa*), sometimes occurring as the ssp. *parviflora*, is common and apparently increasing in chalkland beech-woods, especially where thinning is practised (see Chapter 9), but it is just as characteristic of wet and acid woodland and grassland plant communities as of beechwoods. Those more exclusive to woodland, though still of wide edaphic tolerance (and usually more abundant in the plateau woods than in those directly on chalk) include wood millet (*Milium effusum*), sometimes becoming much mildewed as the season progresses, the distinctive and often very abundant wood melick (*Melica uniflora*), the delicate *Poa nemoralis* and, flowering rather later in the year, *Festuca gigantea* and *Bromus ramosus* (the latter known in older papers as *Bromus asper*).

More restricted to calcareous soils are the rarer wood barley (*Hordelymus europaeus*) and *Bromus benekenii*. Referring again to the older literature, the first appears under its former name of *Hordeum silvaticum*, while the latter seems to have been overlooked on account of its close similarity to *Bromus ramosus*. These two species of *Bromus* are distinguishable in a number of subtle ways, but *B. benekenii* tends to be slightly less hairy about the upper leaf sheaths, and the tiny scale found at the base of some of the lower branches of the panicle in both species is hairy in *B. ramosus*, but glabrous in *B. benekenii*. A very rare grass which is naturalised in some Thames-side beechwoods is *Dactylis polygama* (Hubbard, 1968).

Of the other graminoids, the hairy woodrush (*Luzula pilosa*) is commonly found. Sedges are not as well represented in the chalk woods as they are on the more acid clays of the plateaux, although an interesting exception is the wood sedge (*Carex sylvatica*) which, though typical of "heavy, often wet soils in woods", is sometimes found "on chalky soils with little clay" (Jermy and Tutin, 1968). *Carex flacca* is a common associate of sanicle in some beechwoods, though it rarely flowers there. It is convenient, if phylogene-tically misplaced, to note here that two ferns occur in beechwoods on chalk: the male and common buckler ferns (*Dryopteris filix-mas* and *D. dilatata*), both evergreens.

Shade-tolerant broadleaved herbs of the beechwoods include sanicle (*Sanicula europaea*), herbs Robert (*Geranium robertianum*) and Bennett (*Geum urbanum*) and, already noted, *Mycelis muralis* and *Circaea lutetiana*, the last another victim of mildew in some summers. More local are the columbine (*Aquilegia vulgaris*), nettle-leaved bellflower (*Campanula trachelium*), and the curious herb Paris (*Paris quadrifolia*). Though by no means calcifuge, none of these is restricted to chalk woodlands and most are as common on clay as on calcareous strata: their main edaphic requirement seems to be a high base status. Species which *are* more confined to chalk woodlands include a number of orchids of which the white helleborine (*Cephalanthera damasonium*) is very characteristic: sometimes it is the sole herbaceous species present, and it is used by phytosociologists to distinguish beechwoods of the thin-soiled chalk scarps as the alliance Cephalanthero-Fagion (see Stott, 1970). Others are the much rarer long-leaved and red helleborines (*C. longifolia* and *C. rubra*).

8.3.2.2 The light phase

One way of coping with the dearth of summer light in deciduous woodland is, of course, for plants to vegetate and flower in early spring before the trees expand their leafy canopy (Fig. 2.16d). We touched on this in the reference to the green algal bloom of mid- to late-winter. For herbaceous plants with this strategy, growth (as distinct from the maintenance of winter greenness) must often begin as early as December in conditions of, at best, cool, short days of weak light intensity, necessitating the mobilisation of food reserves from storage tissues by enzymes adapted to low temperatures in order to supplement what little food the plant is able to fix for itself by photosynthesis.

The most familiar example of this geophytic strategy is seen in the bluebell (*Endymion non-scriptus*). Its bulbs begin to sprout beneath the leaf litter of the woods early in the winter period so that the first leaves are visible above the ground by the turn of the year. Less well known is the fact that, simultaneously, the black seeds of this species germinate in abundance to augment the carpet of green shoots with a large crop of seedlings each year, although these plants do not flower in their first year. Peak growth coincides with the increasingly strong light intensity and often surprisingly high temperatures of the light phase of the wood (see Chapter 2), and by the time the trees have opened up their leaves fully the mature bluebell plants have formed the following year's bulbs and seeds, the foliage dies and the population becomes dormant for the summer (aestivates), although the seeds may be retained for many weeks in the erect, dehisced fruits.

The ecology of the bluebell was the subject of a classic series of experiments by Blackman and Rutter (1946–1950), in which measurements of woodland light climates were combined with the then fairly new science of growth analysis for the first time in a non-agricultural study. Here it was shown that bluebells do, in fact, grow more strongly in full sunlight than in shade, but that they are normally unable to thrive in the open due to the growth of more vigorous competitors, especially grasses, and to trampling by livestock. Bluebells are thus forced to take refuge, so to speak, in the suboptimal environment of the woods; indeed, they occur much more sparsely, if at all, in the shade of evergreen yew or holly, and even brambles may suppress them.

Actually, the bluebell is not the most characteristic herb of the chalk beechwoods, and occurs in these woods, like the red campion (*Silene dioica*) and the stitchworts (*Stellaria graminea, S. holostea*) with which it is a common associate, only where there is some accumulation of leaf litter, alluvium, hillwash or more substantial superficial deposits. Thus, in a study of the woodlands of north-east Kent (Wilson, 1911), bluebells were found only "very sparingly" where the thickness of soil above the chalk was less than about 23 cm. Primrose (*Primula vulgaris*) and lords-and-ladies (*Arum maculatum*) showed a very similar distribution, except that the latter was more shade tolerant. Adamson (1921) recorded even bracken (*Pteridium aquilinum*) from the deeper chalk soils of the lower slopes of the Ditcham Park beechwoods. An interesting catena of calcicolous and eutrophic woodland herbs in ashwoods on limestone in the Derbyshire Dales was found by Pigott and Taylor (1964) to reflect soil depth, and was largely due to differences in available phosphate content of the soil (Fig. 8.12).

Wherever *Arum* is absent from the thinnest soils, however, it seems likely that physical rather than chemical factors are involved, for its tubers normally work their way down to a depth of 20–30 cm (Sowter, 1949), and a bed of hard chalk near the surface may preclude this. The species is certainly tolerant of high pH. It is common to find the young spathe of *Arum* bitten clean off, revealing the fertile lower part of the spadix which soon withers (Fig. 8.7). There is usually no sign either of the removed portion of the spathe or of the agent of its destruction, though Sowter has found torn-off spathes in the branches of trees, from which he deduces that birds may be responsible, feeding on slugs and snails which seek the shelter of the spathe. Both spathe and the sterile portion of the spadix are essential for pollination, the latter giving off a foetid smell and actually radiating heat, and enticing in midges of the genus *Psychoda*—up to 4000 have been found in a single spathe—which then act as pollinators (Prime, 1960; Sowter, 1949). It could, of course, be these which the birds are after.

Very typical of the chalk beechwood floor of early spring is dog's mercury (*Mercurialis perennis*), though this species is confined neither to chalk nor to beechwoods, occurring, like many of the woodland herbs already reviewed, in any reasonably moist soils of high base status, but calcicolous in the sense that it seems to demand a relatively large concentration of Ca^{2+} ions (de Silva, 1934). Like the bluebell, *Mercurialis* initiates its vegetative growth and comes into flower in the light phase of spring, though flowering is much earlier than in the bluebell. Indeed, the green flowers, each with three sepaloid perianth segments, are among the first to be seen as the new season opens when, despite their anemophilous appearance, they are visited by insects.

Fig. 8.7. Lords-and-ladies, *Arum maculatum*, showing damaged spathes (see text).

The perennating organ of *Mercurialis* is a creeping rhizome, so that quite large clones can develop from one plant. Moreover, dog's mercury is a dioecious species so that clones are usually either male or female, rarely both. Many years ago, Mukerjee (1936) presented evidence to suggest that only females flower and set seed in the deepest shade. Mukerjee's death even before his work was published was all the more tragic for he brought to light many tantalising facts about the autecology of this species which seem never to have been followed up. Another contrast with the truly vernal species is that the aerial shoots of mercury persist right through the summer, despite a tendency to wilt in hot, dry weather, and they may still be standing even at the end of the following winter if the season is an open one. It is interesting to observe that *Mercurialis perennis*, easily taken for granted or overlooked for the more showy herbs, plays a key role in the ecology of the woodland field layer, and has been recognised as an important species in the classification of woodland types. We return to this in a moment.

Other beechwood herbs of which growth and flowering are more or less confined to the spring light phase include wood sorrel (*Oxalis acetosella*), wood anemone (*Anemone nemorosa*), woodruff (*Galium odoratum*), ramsons (*Allium ursinum*), wood spurge (*Euphorbia amygdaloides*), the common wood violet (*Viola riviniana* ssp. *riviniana*), the beautiful yellow archangel (*Galeobdolon luteum*) with its golden yellow flowers and dark green foliage, and the curious goldilocks (*Ranunculus auricomus*), a woodland buttercup which has lost the ability to develop a complete set of petals, always looking in consequence as though the birds had been at it. An interesting rarity is the coral-wort (*Dentaria bulbifera*), actually a crucifer closely related to the familiar *Cardamine pratensis* of damp meadows, but possessing a fleshy-scaled white rhizome, and of particular interest for its production of axillary bulbils by which it is able to reproduce, having apparently lost the ability to set seed, at least in Britain. Vernal species more restricted to calcareous soils include the pale wood violet (*Viola reichenbachiana*) and the much rarer hellebores (not to be confused with the totally unrelated helleborines), *Helleborus foetidus* and *H. viridis*, the latter strictly ssp. *occidentalis*.

8.3.2.3 Limiting factors other than light

It is very easy to regard shading as the sole agency in beechwoods which restricts or prevents the growth of a vigorous ground flora. Certainly the light climate must be the major factor, but as we noted in Chapter 2 the woodland floor can at times get very dry, at least in places, and where water does not reach the ground by leaf- and twig-drip and stem-flow, it is not only mainly channelled around the very regions which cast most shade, but may well concentrate aerial pollutants which the canopy so effectively intercepts, as well as exudates from the trees themselves. The ominous black streaks apparently devoid of algal growth which run down many a beech trunk must surely have some ecological significance, however localised. There is also the highly potent factor of root competition, and long ago Watt and Fraser (1933) showed what a boost the herbaceous plants could derive simply by severing the roots of the trees with which the former had been competing for water and nutrients.

8.3.2.4 Mercury woods and sanicle woods

A widely quoted concept which came out of the early surveys of the English woodlands was that there are two main kinds of beechwood which occur directly over chalk, distinguished by whether *Mercurialis perennis* or *Sanicula europaea* consitute the most abundant herbaceous species of the woodland floor. "Mercury woods" were found to be associated with deep and relatively base-rich and moisture-retentive soils, very similar (apart from the layer of beech-leaf litter at the surface) to those found under hawthorn scrub. "Sanicle woods" on the other hand developed on thinner soils over harder chalk, as on the steeper slopes and ridges. This much had been recognised in the Ditcham Park woods as early as 1915 by Adamson (1921), but Watt went further and in his celebrated paper on the Chiltern beechwoods (Watt, 1934), he proposed the seral hypothesis already mentioned in Chapter 7.

This states that mercury beechwoods develop from hawthorn scrub, and sanicle woods from juniper scrub. We noted in Chapter 7 that despite the plausibility of this idea much of the evidence concerning the vegetation and its seral development is circumstantial, and we still need detailed records of actual succession over a period of time to demonstrate conclusively the seral concept, although with the continued demise of the juniper and the unpredictable regeneration of the beech, this becomes ever more impractical. Nevertheless, with well over 40 years' experience since its publication, Watt stands by his hypothesis, only warning against indiscriminate comparisions between, for example, beechwoods of the Chilterns and those of the South Downs. Thus, because of the greater atmospheric humidity in summer in the South Downs, mercury can grow on soils which in the drier Chilterns would be too shallow for it, and more likely to support sanicle. With this reservation in mind, the following characteristics of both woody and herbaceous vegetation were thought to distinguish the two types of woodland.

In the mature mercury beechwoods of both Downs and Chilterns, a number of ash trees invariably attained the height of the main beech canopy, although Adamson noted that these ashes were usually "very much drawn out, slender, and not flourishing". The beeches themselves were relatively good for chalk, ranging from 23 m to 28 m in height, with a mean of 25 m , and attaining a girth at 1 m above ground level of more than 2 m. But no other tree species were normally found in the main canopy, and the shrub layer was generally poorly represented. Even sapling ashes were largely lacking. The mature beech canopy of the sanicle woods, on the other hand, tended to be of lower stature (18–26 m high, mean 22 m), and the trees of smaller girth (about 1·4 m) and less uniform in conformation than in the mercury woods, and whitebeam as well as ash attained the main canopy. More shrubs and ash saplings were evident than in the mercury woods, and a characteristic species of the sanicle woods was ivy, though more as ground cover than as a tree-climber. From more recent surveys, Brown (1953) makes the interesting observation that "commonly . . . the mercury beechwood is a quality-class [see Section 9.4.2.5] higher than the sanicle beechwood".

Regarding the two key herbaceous species themselves (*Mercurialis* and *Sanicula*), it should be stressed that they are not mutually exclusive, and occur to some degree in "each other's" woods, though with mercury more often in sanicle woods than the other way about. Observations by Adamson and Watt, as well as more recent work by Wardle (1959), Wilson (1968) and Hutchings and Barkham (1976), indicate that wherever the soil is relatively moist and base-rich, mercury forms extensive and highly competitive stands of large leaf-area-duration (LAD), among which only taller-growing, base-demanding plants (such as *Campanula trachelium*, *Circaea lutetiana* and occasionally the ferns *Dryopteris filix-mas* and *D. dilatata*), tolerant of shading from the tree canopy, are normally able to maintain themselves. Watt adds *Urtica dioica*, *Scrophularia nodosa* and *Arctium minus* as associates of mercury, though these become much more numerous following thinning or felling (see Chapter 4 and later sections in this chapter). Mosses such as *Eurhynchium striatum* and *Brachythecium rutabulum* occur among the mercury shoots, probably making the bulk of their growth in late winter when the latter eventually die down.

Sanicle woods are, in effect, simply those in which soil conditions are inadequate for mercury to exert its full potential. The much more varied herbaceous and bryophytic flora of this type of beechwood (of which typical examples of the former, as well as sanicle itself, are *Primula vulgaris*, *Viola reichenbachiana*, *Fragaria vesca*, *Anemone nemorosa*, *Geranium robertianum* and *Galium odoratum*) occur there not so much because the shallower soils suit them better, but because they meet with less competition from mercury (Table 8.1). Here is another example of one of the fundamental laws of *ecological tolerance*, met with earlier in connection with the distribution of the bluebell, in which competition restricts a species to a suboptimal environment. This interpretation is supported in that where

Table 8.1

Indications of contrasts in vigour between populations of Mercurialis perennis *in mercury and in sanicle woods, and in transitional types* [a]

	Mercury woods	Mercury/sanicle woods	Sanicle woods
"Gregariousness" [b]	5	3	1
Colour of leaves	"Dark green"	"Paler green"	—
State of flowering	Flowering	Flowering	Not flowering
Height of shoots (cm)			
Mean	37·5	—	15·0
Range	22·5–51·0	22·5–30·0	—

[a] From Watt (1934).

[b] A phytosociological measure in which a score of 5 represents a completely continuous cover, and 1 scattered, isolated individuals; 3, naturally, represents a situation somewhere between the two. (See also Wardle's descriptive data, Section 8.4.)

sanicle associates such as *Geranium robertianum* and *Galium odoratum* do occur in mercury woods, they tend to grow more robustly than they do in sanicle woods.

8.3.3 Wood-edge, glade and clearing

8.3.3.1 The wood-edge

The microclimate of the wood-edge (Chapter 2) causes the trees to develop a predominance of branches (and roots) on the side facing away from the wood; they are quite distinct on the one hand from specimen (pioneer) trees growing relatively free from competition in scrub or hedgerow, and on the other from the often drawn-up, spindly individuals of the densely packed woodland interior, favoured on the whole by the forester for their straightness and lack of branches. As noted before, the edge is botanically the richest part of the wood, including as it does a whole range of seral and scrub species which give the profile a fully clothed appearance from the tree canopy to ground level. Sometimes browsing by livestock in adjacent pastures may trim this canopy to a level base some 2 m or so above the ground but, as we shall see, a "natural" wood-edge, whether browsed or not, is of very great significance in animal ecology, especially for birds (Section 8.8.2). Paths and rides provide an important network of wood-edge communities, as well as linear glades.

8.3.3.2 Gaps in the canopy: the woodland glade

An important feature of natural woodlands is the creation of gaps in the canopy as trees die of old age, sustained attack by pests and diseases, drought, fire or lightning strike. Where a very large tree falls, others may be brought down too, so that quite an extensive glade can be formed. Sometimes wind-gaps may be created in this way, whereupon more trees, previously well sheltered and quite stable, follow suit, their roots no longer providing an adequate anchorage. Deliberate clear-felling can, of course, produce exactly the same results and, where large areas of mature woodland are cleared, the exposed trunks of the remaining trees may suffer from extremes of insolation and radiative cooling, which can be so severe as to split open a whole expanse of bark from top to bottom on the side facing the clearing. The thinness of its bark renders the beech especially susceptible to this phenomenon.

In the beechwoods of south-east England a common cause of gaps in the canopy is the condition known as beech-snap, in which the tree breaks clean off, usually about 4–5 m above the ground (Fig. 8.8). It has long been known that fungal attack in general on beech is more common both on the wetter flushes in the chalk as where a marly seam outcrops (see Bourne, 1931;

Tansley, 1939, Plate 46), and on the shallowest rendzinas over hard chalk (see Chapter 9). Beech-snap appears to be caused by the combined activities of the beech coccus insect *Cryptococcus fagi*, which damages the bark as it probes for sap, several canker fungi of the genus *Nectria*, and probably also the bracket fungus *Polyporus adustus* (see Brown, 1953). Gnawing by mammals, particularly the grey squirrel (*Sciurus carolinensis*), and other

Fig. 8.8. Ash saplings grow up in profusion around the stump and fallen trunk of an ancient beech brought down by beech-snap on the steep chalk scarp above Peters-field, Hampshire. Mosses grow on the trunk, and a buttercup (*Ranunculus*) has established itself on the stump. Yews form the background. Sycamore is absent here, but is found increasingly commonly in this situation. (C.J.S.)

damage such as that sustained from careless trimming or even gun-shot wounds, may exacerbate the condition. Other fungi may invade the damaged tissue, and an oblique line of fruiting bodies (toadstools) down the trunk betrays the imminent demise of the tree; in this state the rotten wood inside can sometimes be smelled from several metres away. Anyone encoun-tering a beech tree exhibiting these symptoms, especially in windy weather, is advised to make a detour of at least the equivalent of the height of the crown from the affected part of the trunk! With no root anchorage to restrain the fall as when a whole tree is thrown, the entire upper part of the tree crashes abruptly to the ground with disastrous consequences for anyone or anything beneath. Large branches may be shed in a similar way, although this happens as often in very hot dry weather as when it is windy.

Gaps in the canopy are also caused by the death of trees of relatively short life span which have grown up among the beeches. The gean (*Prunus avium*) is a common example, birch another. Both seem to be susceptible to all manner of cankers and rots. Of course, relatively open canopies with extensive glades may be of entirely edaphic origin, as where soils are too shallow or flinty to maintain the normal climax vegetation. This is one possible explanation of early man's choice of chalkland sites for settlement (Chapter 1).

8.3.3.3 The herbaceous flora of the gaps and glades

Of course almost any plant which happens to be in the right place when the canopy opens up cannot help but respond by greatly increasing its rate of growth, not only because of the improvement in the light climate, but also because of the increase in available water and nutrients in the surface layers of the soil. The most spectacular examples of this great burst of vegetative growth following tree-felling were seen in the traditional management of woodland under the coppicing system (see Chapter 9), but any clear-felled area demonstrates perfectly adequately what happens. Indeed, the tangle of undergrowth may proliferate so vigorously as to greatly hinder replanting operations, or for that matter any natural re-establishment of trees. Much depends on the size of the gap (Fig. 2.16e).

Nevertheless, a distinct succession of colonising plants can usually be observed. Especially where the soil is disturbed, the classic coloniser of the glades of chalk beechwoods is *Verbascum thapsus* (Fig. 8.9), the calcicolous counterpart, as it were, of the foxglove (*Digitalis purpurea*), though unlike the foxglove *Verbascum* usually appears in profusion for one year only, unless the ground is disturbed again. Other opportunists include figworts, burdocks, thistles and occasionally comfreys (*Symphytum* spp.). More persistent are the perennial willow herbs (especially *Chamaenerion angustifolium* and *Epilobium montanum*), and the grass *Deschampsia caespitosa*. Some of these have already received mention in Chapter 4.

The natural course of events in the glades is for trees ultimately to re-establish themselves, although of course regeneration may be impeded either, as suggested previously, by unfavourable edaphic conditons, by grazing animals, especially rabbits, or by the proliferation of dense patches of scrub such as brambles and dogwood. It surely cannot be ruled out, moreover, that grazed glades of this description enabled some of our chalk grassland and other pasture species to survive from Postglacial to Neolithic times, even though it is generally accepted that chalk cliffs and steep banks were the most significant refugia in this respect (see Chapter 4).

Fig. 8.9. The spectacular mullein (*Verbascum thapsus*) is a familiar species of woodland glades on chalk, especially where the soil has been disturbed, as by rabbits. The plants shown here were photographed at Lodge Hill, Buckinghamshire. (C.J.S.)

8.3.3.4 Recolonisation by trees

The classic woody pioneer of the glades and clearings is, of course, the ash (Fig. 8.8). The usual pattern is, at least in small gaps, for a cluster of young ash saplings (poles) to grow up towards the circle of sky exposed by the missing beech crown or crowns. As we have seen, the odd ash tree may actually reach the height of the main canopy here and there, but normally the crowns of the beeches expand and merge once again before this can

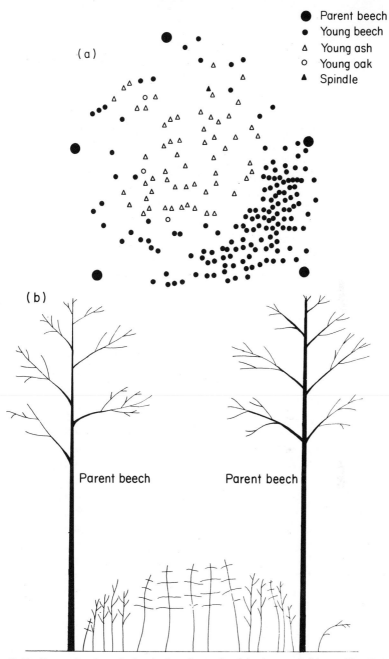

Fig. 8.10. Reproduction circles in beechwoods, (a) plan and (b) profile. Beech seedlings establish themselves only at the periphery of the gap, since the nuts mostly fall vertically downwards from the parent trees. The centre of the gap becomes filled with young ash which is later shaded out by the growing beech. In (b), the parent trees are about 13 m apart. Young beeches are shown with alternate branching, young ashes with opposite. Note how the latter bend into the glade, away from the shade of the parent beeches. From Watt (1925).

happen, and the spindly young ashes die off and eventually fall over. Evidence for this phenomenon in all its stages can regularly be seen in beechwoods, and it was placed on record more than 50 years ago by Watt (1925) from whose paper Fig. 8.10 is reproduced.

In larger clearings, seral ashwood may establish and persist for as much as several decades though, theoretically at least, this must eventually be overcome by climax beech which, although regenerating more slowly, eventually overtops the ash with its heavily shading canopy. We shall see in a moment, however, that beech regeneration is not nearly so reliable as this model situation suggests, and the position is also complicated by the sycamore (*Acer pseudoplatanus*). Dealing with the latter point first, the place of the ash is now frequently taken by this increasingly abundant tree, which was originally introduced into Britain in the fifteenth to sixteenth century but is now completely naturalised. Not only is sycamore so invasive, but it casts a far denser shade than ash and, locally at least, threatens to replace beech as the dominant species. The problem was foreseen by Watt as long ago as 1925 when he concluded that sycamore "competes with ash for vacant ground and with beech for a place in the canopy".

8.3.4 The regeneration of beech

The question of regeneration of beech itself in woodlands of which it forms the dominant species has long dogged woodland ecologists and foresters (see Bourne, 1931, 1942; Brown 1953). Beech may begin to flower and fruit at 40 years of age, but it is usually nearer 80 years old before this happens, and even then it is an irregular, unpredictable process, primarily correlated, apparently, with weather conditions during the preceding summer when the flower primordia are laid down. This process is greatly stimulated by warm sunshine. (On the Continent, good masts are alleged to follow vintage champagne years!) When it occurs, flowering takes place in spring, usually passing unnoticed at least until the spent male tassels begin to litter the floor of the wood well after the leaves have expanded. By this time, if conditions have been right (frost, for example, destroys beech pollen and squirrels eat the flowers), the fruit (mast) will have set, though it is not until autumn that the familiar, bristly-husked fruits turn brown and split open to reveal whether or not the sharply triangular, shining brown beechnuts (one or two to a fruit) have in fact developed.

But not all of these seeds germinate, of course; some never even reach the ground. In his painstaking and informative study of 1923, Watt found that caterpillars of the moth *Cydia fagiglandana*, which hollow out the seeds leaving a tell-tale hole in the husk as they emerge, consumed or damaged about 2% of the crop while still on the tree. Much greater depredations were made by canopy-foraging squirrels (still the red, *Sciurus vulgaris*, at that time, even in Sussex, though since its decline this niche has been filled all too effectively by the introduced grey squirrel, *S. carolinensis*—see Chapter 9). Simms (1971) notes that bramblings (*Fringilla montifringilla*), which have a strong preference for beechnuts, feed directly from the trees "by hovering or reaching out from the twigs".

On the ground, squirrels again, as well as wood-mice (*Apodemus sylvaticus*), were found to take the beechnuts, and huge numbers of nuts were consumed by great flocks of wood-pigeons (*Columba palumbus*), which moved in from surrounding arable land and systematically scoured the woodland floor. Other foraging vertebrates mentioned by Watt include voles, dormice, hares, rabbits, badgers and roe deer, as well as blackbirds, thrushes, fieldfares, pheasants, partridges, rooks and collared doves, the last now very much more abundant than they were at the time of Watt's survey. Simms (1971) adds great, blue, coal and marsh tits and nuthatch, of which the last three have actually been seen to hide beechnuts. Within their very limited range in Britain (Section 8.8.1), edible dormice (*Glis glis*) have been seen eating beechnuts. As for more or less direct anthropogenic agencies, neither the running of pigs in beechwoods for the mast (pannage) nor the collection of the nuts for either human or livestock consumption can be regarded any longer as significant factors in the non-regeneration of beech, assuming that they ever were in times past (see also below).

As for germination itself, no special conditions seem to be required for the process to be initiated; indeed, the total lack of dormancy in the beechnut means that there is no reserve to bridge the gaps between good mast years. The successful completion of germination is undoubtedly enhanced by burial, which helps to protect the seeds from the hordes of seed-eaters just mentioned, from desiccation and especially from frost, to which the extruded radicle is particularly sensitive. Falling leaves cover the seeds but can easily smother them altogether, or blow away again at the crucial moment. Earthworms may cast atop the seeds or drag them into their burrows and squirrels and mice may bury and then abandon their caches. But there can be little doubt that much the most effective agency in enhancing regeneration in beechwoods is trampling by animals: "pigging"— penning pigs in woodland compartments for several weeks before or after felling—has been used with considerable success on the West Wycombe Estate (F. Dashwood, personal communication). This topic is raised again later in connection with forestry and conservation.

Following a good mast year, it is well known that, notwithstanding all the inroads on the viable seed crop and problems in germination, carpets of the familiar seedlings (Fig. 8.11) are to be found covering the ground in huge numbers. In places these may be so numerous as to overlap each other, and

Watt cites Forbes' claim in 1901 of beech seedling densities reaching 2000 per square yard (equivalent to 1 every 2 cm² or almost 24 million per hectare!); a more realistic average is in the order of 80 000 plants per hectare (Edlin, 1955). But what happens to them all? Obviously, intraspecific competition between individuals in such huge populations must result in a large measure of density-dependent mortality, especially in the dry spring period, and there is no mystery in the tremendous toll of seedlings during this stage, despite their satisfactory mycorrhizal infection (Harley, 1937, 1949), and their ability to survive for at least a year in a light intensity one-hundredth that of full daylight and to grow at an intensity of one-sixtieth (Watt, 1923).

Fig. 8.11. The unmistakable seedlings of beech, showing their fleshy, rounded epigeal cotyledons.

An obvious scapegoat is the grey squirrel again which, along with other rodents including the wood-mouse (Brown, 1959) is known to consume the succulent seedlings in large quantities. However it is enlightening to look back at Watt's careful analysis of seedling mortality in his observation plots at Goodwood, Sussex (Table 8.2), in which the demise of all except 12 (which disappeared without trace) of an original stand of 119 beech seedlings between April 11, 1921 and April 28, 1922 was attributable to damage by invertebrates. Damage was either direct ("catastrophic"), as when a seedling was bitten clean off, or indirect ("crippling"), not killing the

Table 8.2

Seedling mortality in beech and its possible causes, Goodwood 1921–1922 [a]

| | Degree of protection | | | | |
	"Complete"	Against rabbits	None	Total	Percentage mortality
Initial number of seedlings in observation plot, April 11, 1921	57	24	38	119	—
Seedlings dead by July 26, 1921 [b]					
(a) With radicle or hypocotyl cut through [c]	20	9	19	48	40·3
(b) With cotyledons and leaves eaten to various degrees [d]	36	14	14	64	53·8
(c) With aphids (Phyllaphis fagi) present on leaves	10	2	2	14	11·8
(d) With leaf-hoppers (Typhlocyba cruenta) on leaves	9	2	2	13	10·9
(e) No trace of seedling	1	1	5	7	5·8
Number of seedlings still alive on July 26, 1921	5	0	0	5	4·2
Number of seedlings still alive on April 28, 1922	0	0	0	0	0

[a] From Watt (1923).

[b] Some seedlings suffered from more than one set of symptoms, and not in every case could the apparent causal agent be positively deemed to have killed the seedling, directly or indirectly. Thus, traces of slug activity were frequently found but (at least at Goodwood) never the slugs themselves.

[c] Positively identified agents of this damage included caterpillars of Noctua pronuba and Hepialus lupulinus.

[d] For example by caterpillars of Erannis defoliaria, Operophtera brumata, Pandemis corylana and a Tortrix sp., as well as by these beetles: Agriotes sp. (tunnelled into radicles and germinating nuts); Strophosomus coryli (ate radicles, cotyledons and leaves and gnawed hypocotyls); and the weevil, Orchestes fagi (ate circular holes in primary leaves, which were also sometimes mined by the larvae of this species).

seedling outright, but impairing its ability to recover and maintain growth in the deep shade. In fact, by far the greatest proportion of seedlings was already dead by July 26, 1921: only five remained, and these had gone by the following spring. Details are shown in Table 8.2, but particular attention was directed to the "crippling" effect of the leaf-hopper *Typhlocyba cruenta* which, through sucking the cells of the lower surfaces of the leaves, led to dechlorophyllisation which was confirmed experimentally to greatly impair photosynthesis. This insect, as well as the beech aphis (Table 8.2), is not confined to seedlings, and both species affect mature beech trees as well, often in enormous numbers (see also Brown, 1953).

The situation is exacerbated by the way in which beech seedlings are found only on the periphery of the clearings (Fig. 8.10), and this must result from two major causes, both physical. Firstly, beech fruits seem, surprisingly, to depend for their spread mainly on wind: Watt advances several plausible reasons why biotic agencies of spread may on the whole be dismissed (though see King, 1974). Certainly, it is obvious enough that horizontal dispersal is extremely limited. Fruits may be found as much as 20 m downwind of an isolated tree, but in the shelter of the wood virtually every fruit drops vertically to the ground beneath the crown of the tree which produced it. Secondly, any seedlings which do arise in the centre of the clearing are, during early spring, exposed to the risk of both frost and drought (see Chapter 2), to which they are extremely susceptible (see continental references cited by Watt, 1923).

8.3.5 Degeneration

The concept of a degenerate beechwood is not new: Adamson (1921) used it to describe beechwoods on part of the Ditcham Park estate to which rabbits, sheep and cattle had access, so preventing regeneration of the woodland flora in a very obvious way. The beechwood flora was replaced by scrub and chalk grassland (in much the same way as must have begun to happen 6000 years before—Chapter 1), and in places by yew-wood. The term retrogression is sometimes applied to this process. At the present time there are signs of degeneration *en masse* in the ailing beechwoods of the Chilterns and elsewhere, and left to themselves these might well display retrogression (see Chapters 9 and 10).

8.4 Ashwoods on Chalk

Ash flowers and seeds much sooner in its life span than beech, usually beginning at 15–20 years of age, and is among the most regular and prolific seeders of the chalkland tree species. Even regularly coppiced ash can set seed between fellings. Its winged fruits (keys), each containing one seed, are readily dispersed, resulting in the abundance of ash seedlings and saplings already mentioned. But ash has a very open canopy (Fig. 2.16a) and has a shorter growing season than any other comparable tree of similar stature. Not until June does its canopy of compound leaves—Hopkins' "fringe and fray of greenery"—fully expand, yet these begin to drop, without any appreciable change in colour, usually before October is half-way through.

Within the natural geographical range of beech, ash usually behaves as a seral species on chalk soils, forming stands of relatively short-term duration, though it is more persistent in regions of relatively high rainfall and atmospheric humidity or where the soils are deeper, richer in bases and more retentive of water. Thus ash is of much greater significance in the woodlands of the western South Downs than it is either east of the Arun River in the same range of hills, or in the Chilterns (see Watt, 1924, 1925, 1934). On a more local scale, Adamson (1921) noted the tendency for ash to occupy deeper soils, especially those derived from Lower Chalk, lower down the scarp slopes and in the valley bottoms at Ditcham. Similarly the woodlands at Wain Hill near Bledlow on the Buckinghamshire–Oxfordshire border, as well as those further west at Aston Hill, still (1980) retain the predominance of ash first recorded by Watt in his survey of 1934, and the distinct wetness of this stretch of the Chiltern escarpment was also noted in Chapter 2. Bourne (1931) makes the interesting observation that a "wet beechwood" on chalk is easily converted to ashwood.

Away from the influence of beech, ash occurs as a more permanent species on calcareous soils, where it may form climax woodlands (Fig. 1.24). These are particularly well seen on chalk in the relatively warm and humid climates of the Isle of Wight and southern Wessex; elsewhere, as in northern Wiltshire and the Lincolnshire and Yorkshire Wolds, ash tends to occur in mixture with other species (see next section). It is now evident (Pigott, 1969; Merton, 1970) that the celebrated ashwoods of the Derbyshire Dales are derived from woods in which limes (*Tilia cordata* and *T. platyphyllos*) were formerly of greater importance, though whether this could apply to the nearby Wolds is not known.

Like beech, ash is very vulnerable in its seed, seedling and sapling stages to a plethora of pests and pathogens (Wardle, 1959), the effects of which are augmented in the Chilterns by the greater tendency for spring frosts and lower summer humidity. Thus, even when still on the tree, ash fruits can be consumed by caterpillars of *Pseudargyotyza conwayana*. Once shed, many are eaten by rodents during their full season's dormancy, while others may sustain damage and subsequent invasion by the fungus *Amerosporium chaetostroma* until, as Wardle puts it, "only seed coats and pycnidia remain". Intact seeds which survive their first full season germinate freely, but growth analysis and measurement of the light profile showed that when the resulting seedlings emerge under a dense canopy of *Mercurialis perennis* the light intensity is well below their compensation point, quoted as about 7% of full daylight. The poor growth resulting from this heavy shading renders the ash seedlings particularly prone to damping off by *Pythium* spp. and attack by caterpillars of *Gracillaria syringella*.

Even after establishment, ash can still succumb to frost damage (dramatically demonstrated by the great frost of May 31, 1975), and grazing by rabbits, squirrels, deer and hares. Watt noted stem infection by *Chionaspis salicis* and the canker *Nectria galligena*. Forking of the main stem and lateral branches follows damage to the terminal buds eaten by caterpillars of the ash-bud moth (*Prays fraxinella*). Pawsey (1973) adds the cryptic ash decline or dieback, the cause of which is so far unknown.

Yet it is a curious fact that, once established as a mature tree, ash supports relatively few phytophagous invertebrates. Wardle (1961) lists some 22 species of mites and insects associated with the foliage and twigs of ash but, as Elton (1966) puts it, "there is none of the rich variety of the oak, nor the aphid biomass of the sycamore", despite the apparent tenderness of its leaves. Elton goes on to suggest that the leaves may simply be too delicate:

> The extremely unstable and rather small leaflets of the ash . . . are neither firm platforms for most invertebrates, nor easy for small birds to explore. The chief user of the ash canopy for food, the coal tit, *Parus ater*, is one of the best acrobats among titmice.

Even the ubiquitous grey squirrel seems averse to nesting in ash trees.

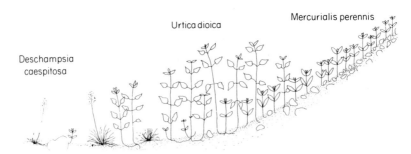

Fig. 8.12. Distribution of woodland herbaceous communities in the bottom of Monk's Dale on Carboniferous Limestone in Derbyshire (Pigott and Taylor, 1964).

On the whole, the shrub, field and ground flora of the ashwoods resembles in kind that of the beechwoods, differing only in the much greater profusion and vigour of its component species on account of the higher light intensity inside the wood. In the herbaceous flora, there is a greater tendency for eutrophic species such as *Urtica dioica* to occur (especially towards the bottoms of slopes—Fig. 8.12), due partly to the ash's "choice" of better soils, and also, it must be presumed, to the much quicker turnover of the far more easily decomposable leaf litter of the ash. Some species characteristic of the woodlands of the southern Chalk peter out northwards, such as *Euonymus europaeus* and *Daphne laureola* among the shrubs, and *Euphorbia amygdaloides*, *Campanula trachelium*, orchids such as *Cephalanthera damasonium*, and to a lesser extent *Galeobdolon luteum*, among the herbs. A few species increase northwards, such as *Campanula latifolia*, and there is a general increase in ferns and bryophytes.

8.5 Other Deciduous Woodlands

Of course, not all deciduous woodlands on chalk soils fall into the two major categories of beech and ashwoods just described. Apart from plantations of relatively recent origin (see Chapter 9), we have the old formerly coppiced woodlands of Sussex, Hampshire, Wiltshire and Dorset in which hazel predominates, with various combinations of ash, oak, gean, maple, sweet chestnut, hornbeam and others (Section 9.4.1). Further north, mixed woods of ash, sycamore and wych elm take on more significance. There are, too, occasional mixed woods of apparently semi-natural but obscure origin: large-leaved lime (*Tilia platyphyllos*) grows among the beeches at Box Hill, Surrey; woods of pedunculate oak (*Quercus robur*), ash and field maple with other deciduous species (but not beech) are found on deep chalk soils on the North Downs near Dover and on the Pewsey Vale scarp in north Wiltshire (Ratcliffe, 1977). On the edge of Darenth Wood in north Kent, sessile oak (*Quercus petraea*) has spread onto the chalk from adjacent Thanet Beds, and is continuing to regenerate freely among calcicolous grassland and scrub with little evidence of chlorosis.

8.6 The Age and Origin of the Deciduous Woods

It is not difficult to recognise a planted wood of relatively recent origin from any of several attributes: the geometry of its boundaries, the regimentation of its trees and the presence of exotic conifers are examples. None of these is infallible, but documentary records usually exist from which the history of the wood can be confirmed (see Barber, 1976; Rackham, 1976). But what about the older woods? Just how old are they? Are they all secondary (deliberately planted or naturally regenerated following a period of clearance and cultivation or grassland farming)? Or could some be primary woodlands, directly descended from the primeval forest? And is there any way of telling? Long experience in the field may lead to informed guesses. For example, A. S. Watt (personal communication) is of the opinion that the beech hangers of the Chiltern escarpment arose subspontaneously from those of the plateau, where the primary forest cover survived into mediaeval times (see Chapter 9). And on the basis that woods dominated by one species reflect anthropogenic modification, it is suggested in Ratcliffe (1977) that the oak–ash–maple woods mentioned in the previous section could represent a type pre-dating the beechwoods.

Unfortunately, on the steeper slopes where the woods are most likely to be very ancient, there is no tendency for the surface layers of the chalky soils to exhibit symptoms of leaching as might be expected on long-undisturbed flatter sites. However, recent work by Peterken (1974) in the English Midlands has indicated that certain species are restricted there to woodlands known to be ancient, and it is interesting to note that some of these are characteristic of many chalkland beechwoods. Thus *Milium effusum, Oxalis acetosella, Campanula trachelium* and the rarer *Aquilegia vulgaris, Paris quadrifolia, Vicia sylvatica* and *Neottia nidus-avis* (as well as other species, including sedges, of wetter, more acid soils) were found only in woods known to have ancient origins, while species "almost confined" to these primary woodlands included *Melica uniflora, Luzula pilosa, Galeobdolon luteum, Galium odoratum, Anemone nemorosa, Ranunculus auricomus* and *Viola reichenbachiana.*

Much depends, however, on the degree of isolation of a wood from its neighbours (an important point taken up in Chapter 10 in relation to the establishment of nature reserves). How long a woodland species can persist either as a thriving population or as dormant seed between periods of woodland occupation is also important. Less significance may be attached to at least some of these key species in areas which are well wooded, than where individual woods resemble islands in a sea of arable land largely devoid even of hedges; Peterken's work derives, so far, from the latter situation, and it would be naive indeed to blindly extrapolate his findings to regions with quite different histories. Nevertheless the very presence of these plants, particularly several of them together, is strongly suggestive of a long association with woodland conditions, and this fascinating subject warrants further study. Certain lower plants, as well as invertebrates, appear to have similar potential as indicators of primary woodland (Barber, 1976).

8.7 Evergreen Woodlands

8.7.1 Plantations

Much planting of conifers has taken place on the Chalk over the past 200 years, mainly of introduced pines and larches, of which the latter are, of course, deciduous. All these, however, are dealt with under the heading of forestry in Chapter 9. Here, the intention is to describe the natural (or semi-natural) evergreen woodlands of the Chalk, which include yew and box, though it is a matter of opinion whether box should strictly be considered as a scrub rather than a woodland species. As we have already noted, this too has been widely planted.

8.7.2 Yew-woods

Yew is a common tree of chalk scrub and pioneer woodland, typically associated with juniper, whitebeam and, especially in the South Downs, ash. It also occurs as an understorey species in mature beechwoods on chalk and clay alike. In places, however, especially on the slopes of chalk coombes, the trees grow up in such profusion as to form virtually pure yew-woods. It has been suggested (Bourne, 1931) that these woods indicate locations which are too dry for beech, but this now seems unlikely, since the latter has been established successfully on sites of former yew-woods (D.F. Fourt, personal communication). In his classic surveys at Kingley Vale, Sussex, Watt (1926) traced the successional development of these woods, though of course the exact sequence could only be inferred. The pioneer yews, frequently nursed by junipers which are subsequently overshadowed (Section 7.4.1), grow into specimen trees, sometimes multi-stemmed and often of great girth, and producing copious crops of their attractive red "berries", each actually a much-modified female cone in which the solitary naked seed is encircled by a fleshy aril.

Many birds, including robin, song thrush, blackbird, mistle thrush, fieldfare, redwing and pheasant feed avidly on these fruits (see Chapter 7).

> And do they feast! [wrote Hudson in 1900]. It is worse than a feast, it is a perfect orgy. When a bird, with incredible greediness, has gorged to repletion he flies down to a spot where there is a nice green turf and disgorges, then relieved, he goes back . . . to gorge again, and then again. The result is that every patch or strip of green turf among or near the trees is thickly sprinkled over with little masses or blobs of disgorged fruit, bright pinky red in colour, looking like strawberries scattered about the ground and crushed by passing feet.

A. S. Watt has observed robins passing droppings with a similar appearance. Badgers and foxes also take the fruits, and their droppings, too, may contain numerous seeds. More than 120 have been found in a single fox stool (R.M. Tittensor, personal communication). Yew seeds thus find their way to the surface of the ground in huge numbers, so that even though many are cracked open or eaten altogether by birds and rodents, notably bullfinches (*Pyrrhula pyrrhula*), hawfinches (*Coccothraustes coccothraustes*) and various mice, plenty remain to germinate where conditions permit.

Yew seedlings consequently spring up in abundance in the vicinity of the parent trees. Moreover, while seedlings in the shade develop a cluster of glossy dark green leaves in their first season, those in the open remain stunted, and their leaves, which barely grow at all, turn a light orangey-brown colour. Indeed it is often difficult to tell whether they are alive or dead. Grazing by hares or rabbits further inhibits growth. Watt found one yew at Kingley Vale 18 cm high which proved to be 55 years old. In consequence, the pattern of development of the yew-wood is for secondary trees, usually of a more straggly conformation, to establish around the fringes of the pioneers so that the centres of invasion gradually expand and coalesce. During this time, the grassland flora and most of the scrub is steadily eclipsed, except sometimes for a few individuals of whitebeam which may persist in the yew canopy for many years, their silvery-green foliage contrasting attractively with the deep green of the yews (Fig. 8.13a).

There is essentially no understorey flora, herbaceous or otherwise, in the shade of the yews (Fig. 8.13b). Except for the odd herb or straggly elder the woodland floor is usually utterly devoid of plants, giving the yew-wood its undoubtedly mystic feel: Hudson's "sacred dim interior" among the "dark religious trees". This dearth of herbaceous plants and bryophytes must result largely from the extremely dense shade cast by the evergreen yews, together with intense competition from their roots. From time to time the possibility is raised that toxic leaf drip and stem flow enhance these effects, but until someone does a "Lee and Monsi" (Section 2.3.3.5) this must remain an open question.

Great claims are often made for the age which yews can attain. Edlin (1955) asserts that some giants could be 2000 years old, and makes the interesting point that some yews may actually pre-date the churches in whose sacred environs they are assumed to have been planted. Tansley (1939) set the ages of the oldest Kingley Vale yews as in excess of 500 years, and calculations based on ring counts by A. S. Watt indicate that the largest single-stemmed trees there could be between 800 and 1100 years old. A recent survey by Tittensor (1976), however, suggests that most are very much younger than this, and that, apart from those planted along parish boundaries, yews were relatively rare in the South Downs until about 200 years ago. Indeed, most of the Kingley Vale yews appear to date from about 1870–1880, when changes in sheep husbandry (Chapter 9) allowed seedlings from the boundary trees to invade the former sheepwalks. Little regeneration now appears to be taking place within these woods, lending support to the traditional view that yew-woods form a successional dead end, rather like a gigantic heather clump or fescue tussock with a greatly extended time-scale. Tansley considered that once the oldest trees began to degenerate, only ash and elder seemed generally able to colonise the gaps: another

Fig. 8.13. The yew-wood at Watlington Hill, Oxfordshire. (a, top) View of the eastern edge of the wood with whitebeam (*Sorbus aria*) visible in the canopy, and the ex-arable sward pictured in Fig. 5.4 in the foreground. (b, bottom) The interior of the yew-wood. (C.J.S.)

example of retrogression (see Section 8.3.5). Yet, as he took pains to stress, this was as much a result of rabbit as of yew ecology, and we still know very little indeed about the long-term behaviour of yew-woods in the absence of heavy rabbit grazing.

8.7.3 Box-woods

Very local, and strikingly confined to the steepest chalk slopes and screes, box (*Buxus sempervirens*) must have evoked more discussion than any comparable species about whether it is indigenous or introduced (see Pigott and Walters, 1953). There is no doubt that this species, familiar in gardens as much from its distinct smell as from its appearance, has been widely planted, for example as pheasant cover, and that it occurs in many places as an escape from cultivation. But the perennial argument is whether the well-known box coombes at Box Hill, Surrey, and Ellesborough and Kimble Warrens, Buckinghamshire (Fig. 8.14) are genuinely native or not. In a characteristic reflection, Tansley (1939) wrote: "The late Count Solms-Laubach, who was

Fig. 8.14. Box (*Buxus sempervirens*) growing on the extremely steep and curiously scree-like south-facing slope of Kimble Warren, Buckinghamshire. Other shrubs here include privet (*Ligustrum vulgare*), wayfaring tree (*Viburnum lantana*) and *Cotoneaster horizontalis*. The box grows more luxuriantly among the trees below, which include walnut (*Juglans regia*). The whole area now lies within BBONT's Chequers Reserve to which access is strictly controlled. (C.J.S.)

a connoisseur of such things, was convinced that the Boxhill wood is a fragment of old native forest, and that is certainly the impression it gives''. One feels obliged to add "and there's an end on't", and at least for these Surrey and Buckinghamshire localities present opinion seems to favour the idea (Staples, 1970; Ratcliffe, 1977). But elsewhere, the argument has to remain an open one.

Almost nothing seems to have been written on the ecology of box. It seems reasonable to regard it as especially tolerant of exposure to heat and drought, for it is usually the only woody species on the most precipitous banks, though yew and beech accompany it where the gradient is less steep. Box is able to flourish in the shade of beech, but, as with yew, virtually nothing grows under its canopy. Where the stand is open, herbaceous species characteristic of bare chalk can be found among the bushes, notably *Echium vulgare*. The rare cut-leaved germander (*Teucrium botrys*) occurs at Box Hill.

8.8 The Animal Life of the Woods

8.8.1 The mammals

Any of the ubiquitous mammals—rabbit, hare, field and bank voles, common and pygmy shrews (*Sorex araneus* and *S. minutus*), hedgehog (*Erinaceus europaeus*), stoat (*Mustela erminea*) and weasel (*M. nivalis*), as well as bats—may be encountered in the woods, especially along the margins. Conversely, even the most characteristic woodland mammals are fairly mobile, not only moving about freely from place to place within the wood, but frequently leaving it altogether for more open country. Much will depend on the extent and continuity of the woods, and on the nature of the intervening areas. Thus, fox and badger, and more locally the introduced muntjac (*Muntiacus reevesi*) and fallow deer (*Dama dama*), are obvious examples of animals which may dwell within, but forage (usually nocturnally) well away from the wood. Where they occur, roe deer (*Capreolus capreolus*) tend, by contrast, to stay within the woodland bounds. A very thorough study of the roe deer of Cranborne Chase has been made by Prior (1968).

Of the smaller mammals, the red and grey squirrel (*Sciurus vulgaris* and *S. carolinensis*), and the wood-mouse (*Apodemus sylvaticus*) are more restricted to woods, and have already received mention in connection with the regeneration of beech from seed. The red squirrel has now been displaced throughout the south-eastern chalk country by the grey (see Shorten, 1962), though the red persists in parts of East Anglia and on the Isle of Wight. The yellow-necked mouse (*Apodemus flavicollis*) occurs in woods, but is more local than *A. sylvaticus*. The dormouse (*Muscardinus avellanarius*) is perhaps more exclusively a woodland species, and this applies even more to its close relative the edible dormouse (*Glis glis*), well-known as a Roman delicacy, and reintroduced to Britain in 1890 by Lord Rothschild on his estate near Tring. This surprisingly large animal (Fig. 8.15) has spread very little from its point of release. It is occasionally to be found in scrub, as on the Chalk scarp at Whipsnade (Middleton, 1937), and it is common in the woods of the Chiltern escarpment on the borders of Hertfordshire and Buckinghamshire. However, because it is both arboreal (preferring, moreover, the concealment of a coniferous canopy) and nocturnal, it is seen rarely enough in summer, and not at all in winter when, like the common dormouse, it hibernates. Chiltern foresters, nevertheless, are all too aware of its presence, for it causes considerable damage (see Chapter 9).

Fig. 8.15. The edible dormouse (*Glis glis*), confined in Britain to a small area of Chiltern woodlands in the vicinity of the locality to which it was re-introduced in 1890. About one-fifth life-size.

It follows from the foregoing remarks that to start talking about the significance of geological boundaries and contrasting soil types in the present context would be absurd. None of the mammals referred to in this section (nor the birds or the majority of invertebrates in the next) can be regarded as exclusively chalkland species: they simply inhabit, or are associated with, lowland woodland country as a whole.

8.8.2 Woodland birds

The influence of the structural features of the woodland community is particularly well demonstrated by the diverse and easily observed bird fauna. It is not difficult to draw up a list of the most characteristic woodland bird species, each of which "during all or part of the year can fulfil all demands of its life within a woodland habitat", to adopt Campbell's definition quoted in Simms (1971).

Regarding the beechwoods, we have already noted how, following a good mast year, the beechnuts attract large foraging parties of finches, tits, pigeons and other birds, both native and migrant. But these woods tend to support a poorer breeding bird fauna than the structurally more complex and floristically richer oak and mixed woods of the deeper and heavier soils, and this paucity of species is especially true of the beechwoods of chalk soils which are essentially all crown and bare trunk, with relatively little understorey or ground flora, if any at all (Fig. 8.4).

Here, the most characteristic species, readily discernible in spring and early summer to anyone with an ear for birdsong, include the ubiquitous blackbird, invariably to be found scuffling about for food among the leaf litter; and robin, chaffinch and tits (blue, great and coal), all these nesting in holes in the trunks and branches and feeding on the invertebrate crop of the canopy. The tiny goldcrest is more frequent than seems to be realised, but though its high-pitched song is clear enough, it is hard to see among the very tops of the crowns. Bark-feeders such as tree-creeper, great spotted woodpecker and nuthatch are occasionally to be seen, though the thin bark of the beeches can provide little food for the first two, or crevices in which to lodge seeds for the last. One relatively uncommon species which does appear to find mature beechwoods especially favourable, however, is the wood warbler (*Phylloscopus sibilatrix*), virtually restricted to this habitat in south-east England. It is a rewarding experience to come across this delightful little bird, uttering its distinctive trill as it flies from one branch to another at the bottom of the main beech canopy. Among the larger birds, woodpigeons seem ever-present, flapping noisily about the tree-tops, while as Henry Williamson's readers will recall, beechwood rookeries are not unknown. Towards nightfall the tawny owl may reveal its presence, and though strictly a wetland bird, the woodcock (*Scolopax rusticola*) is commonly seen (and heard) on its roding flights over any sufficiently extensive tract of beechwood.

Wherever there is a greater mixture of tree species, and particularly a more varied canopy structure, so the bird population is likely to be more diverse (see, for example, Elton, 1966; Simms, 1971; Ratcliffe, 1977; and also Fig. 10.5). A good tangle of scrub and undergrowth will attract wren, jay, blackcap, garden warbler and many others—even the elusive nightingale—while an open glade encourages redstart and flycatcher and is the classic habitat for the diminishing nightjar. Conifers are likely to attract the crossbill (*Loxia curvirostra*) though, as with beech, uniform, densely planted stands can be relatively destitute of bird life.

As a component of scrub and along the edge of woods, yew is associated with a rich bird fauna, and we have already noted the importance of its berries as food, especially for thrushes. But in pure yew-woods the spectrum is very much reduced, and again largely confined to hole-nesters such as robins and tits, though, as always, the exact composition will vary from place to place and from year to year. Thus, Lack and Venables (1939) found tits especially abundant in the yew-woods they covered, and Williamson and Williamson (1973) recorded coal and great tits as making up between 10% and 20% of the breeding bird population of the Kingley Vale yew-woods during 1967–1971. Yet, in another survey recorded by Simms (1971), tits were "almost entirely absent" from the yew-woods of Butser Hill. Williamson and Williamson's survey was directed in particular to the relationship of bird ecology to the successional status of the yew-woods. They found that the smallest number of species occurred in relatively young stands, where the canopy was densest, but that species richness increased again in the vicinity of trees of greater age.

8.8.3 The invertebrates

We come now to animals which, at least for part of their life-cycle, tend to stay put, and can be treated more like plants in the sense that it is possible to avoid many of the complicated conditional phrases which accompany the account of the birds. Indeed, among the more stenophagous invertebrates, it is possible, if not particularly exciting, to refer once again to chalkland species where these are found only on plants which themselves are edaphically (or geographically) restricted to chalk soils, though in fact relatively few woodland invertebrates are known which show this restricted distribution. This is for the simple reason that, as we have seen, the overwhelming bulk of the trees and herbs of the chalkland woods occur widely elsewhere (more so than those of grassland and scrub). A number of invertebrates have recently been recognised as possible indicators of primary woodland, though as yet this work is in its early stages.

Even such obviously mobile species as the butterflies often stay very strictly within the bounds of their acknowledged habitat. A striking example of this is seen in the speckled wood (*Pararge aegeria*), as religiously restricted to dappled shade as its close relative the wall (*P. megera*) is to walls and warm, bare soil.

The wood-edge and glade provides much the richest assortment in the wood of spiders and insects in relation to the area (or more correctly, volume) sampled, and a quiet hour spent just watching the traffic in a woodland glade on a sunny afternoon in early summer is an enlightening experience, especially for those whose natural instinct is either to count plants or gaze at everything through binoculars. The exquisite lacewings (Planipennia), the scorpion-flies (Mecoptera) and robber-flies (Asilidae and Empididae), the dancing columns of gnats (Culicidae) and long-horn moths (Incurvariidae), the bewildering variety of variously striped solitary wasps

(Sphecoidea) and the dipterous hoverflies (Syrphidae) that mimic them, spiders large and small, the long-legged harvestmen (Opiliones): any of these can be seen in or near, for example, a clump of stinging nettles without recourse to any beating. On the other hand, the woodland interior can be singularly devoid of these invertebrates—a point already made in connection with the habits of insectivorous birds.

A much more thorough and systematic analysis is necessary, of course, to obtain a complete picture of all the invertebrate species present, their relative abundance and importance, and their distribution throughout the woodland profile from the top of the canopy to the litter layer of the woodland floor, but a detailed analysis is not proposed here for three reasons. Firstly, the phytophagous invertebrate faunas of two representative tree species, beech (this chapter) and yew (Chapter 7) have already received adequate mention for the present purpose. Secondly, the ecology of this vast fauna is a subject in itself (the monograph on the litter fauna of Danish beechwood soils by van der Drift (1950) runs to 168 pages), and very much a job for the ultra-specialist: in a preliminary survey of the protozoan fauna of the Hampden beechwoods, Buckinghamshire, Stout (1963) listed 51 species from the mineral horizon of a chalk rendzina under Fagetum calcicolum, and 36 from the overlying leaf litter. The third reason for brevity is that as comprehensive and readable an account of the whole woodland fauna as is ever likely to appear in print is already available in Charles Elton's magnificent "Pattern of Animal Communities", to which constant reference has been made throughout this book. Elton has distilled the essence of his long experience in Wytham Woods, and much, if not all, of what he writes about the invertebrates of those celebrated woods (for all that there are relatively few beeches) is applicable to the chalkland woods, on which in any case very little work has been done. A few examples of characteristic species are illustrated in Fig. 8.16.

8.9 The Woodland Ecosystem

It is not difficult to combine descriptive information of the kind contained in this chapter—essentially woodland natural history—with the now well-established concept of energy flow through successive trophic levels, and to obtain at least a qualitative impression of how the woodland ecosystem works. Examples of food webs are shown in Fig. 8.16, though of course these are grossly oversimplified and may vary in time, particularly seasonally. The decomposition phase, about which relatively little is known, is not included.

Quantitative ecosystem ecology is even more in its infancy, and studies so far have largely been confined to managed, rather than natural com-

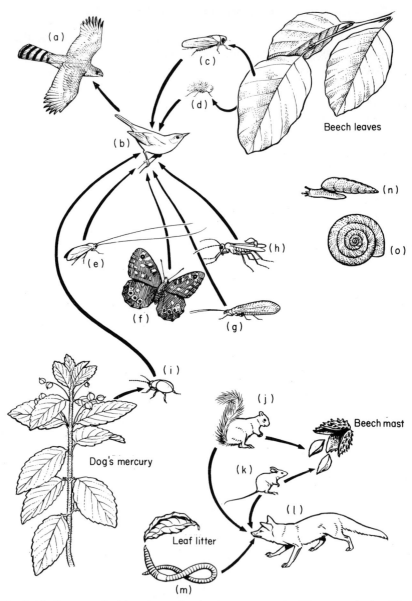

Fig. 8.16. Some typical beechwood animals, not to scale. These much-simplified food chains include (a) sparrowhawk, (b) wood warbler, (c) the leaf-hopper *Typhlocyba cruenta*, (d) the aphid *Phyllaphis fagi*, (e) a long-horn moth, (f) the speckled wood butterfly *Pararge aegeria*, (g) a lacewing (*Chrysopa* sp.), (h) a male scorpion fly (*Panorpa* sp.), (i) a flea-beetle, (j) grey squirrel, (k) wood-mouse, (l) fox and (m) earthworm. Also shown are two characteristic beechwood snails, (n) *Cochlodina laminata* and (o) *Discus rotundatus*.

munities. The International Biological Programme (IBP) was set up in order to redress this deficiency. Useful accounts of what needs to be done and how to go about it are given by Cousens (1974) and Chapman (1976b). In the meantime, we continue to refer intuitively to "climax beechwood", and cling to the idea, which certainly seems to hold true and make sense, that the more intricate the pattern of food chains and webs (i.e. the more diverse the ecosystem) the more stable that ecosystem is, and able to resist or accommodate environmental change. But until we have reliable estimates of biomass accumulation and turnover, and energy and nutrient budgets, we cannot say with certainty whether or not a particular piece of woodland is truly in dynamic equilibrium with its environment (which is how a climax ecosystem is defined), nor whether any particular plant or animal has any special significance or status. Further reference is made to quantitative woodland ecology under the heading of forestry in Chapter 9, and see also p.493.

9
Economic Aspects of the Chalklands

9.1 Resources

Any excursion into economics demands first of all a consideration of resources, which may be non-renewable, as in the case of mineral deposits, or renewable. In the latter case, renewal may be achieved relatively rapidly as in farming and forestry, or only slowly as in the regeneration of natural and semi-natural plant and animal communities of the kind reviewed in the foregoing chapters. It is extremely difficult to put an economic value on resources of this latter kind, and discussion of the role of natural ecosystems in this sense is deferred until Chapter 10. Soil, of course, is a resource, slowly renewable with adequate care, as is the vast store of underground water contained in the Chalk formation, but the management of these resources is more conveniently discussed in the context of agriculture and forestry, to which subjects the bulk of this chapter is devoted, and to which we turn shortly. But first a brief word about chalk itself as a resource.

9.2 Chalk as a Resource

Chalk has been quarried at least since Roman times (see Chapter 1), when it was spread on adjacent clayland to "sweeten" it. The great protagonist of chalking in historic times was William Ellis, who farmed in the Hertfordshire Chilterns at Berkhamsted, and whose book, "Chiltern and Vale Farming Explained" (e.g. Ellis, 1745) contains a wealth of information on this, as well as numerous other matters of comparative agronomy. "Chalk", declares Ellis, ". . . will shorten and crumble the clay before the plough, to that degree as to make one ploughing go as far as two or three without it", provided it is of the "fat, soft kind, with a yellow coat", and that it is applied,

Fig. 9.1. One of the last gangs of chalk drawers, photographed with their equipment at Harpenden in 1913. From Gardner (1967). (H. B. Hutchinson, via the Royal Agricultural Society of England)

in sufficient quantity, during autumn and winter when the lumps are saturated and so most effectively pulverised by frost.

Defoe (e.g. 1724) records in great detail the traffic in chalk from Thames-side Kent to the claylands of East Anglia:

> From these chalky cliffs on the river side, the rubbish of the chalk, which crumbles away when they dig the larger chalk for lime . . . is bought and fetch'd away by lighters and hoys, and carry'd to all the ports and creeks in the opposite county of Essex, and even to Suffolk and Norfolk, and sold there to the country farmers to lay upon their land, and that in prodigious quantities; . . . . thus the barren soil of Kent, for such the chalky grounds are esteem'd, make the Essex lands rich and fruitful, and the mixture of earth forms a composition, which out of two barren extreames, makes one prolifick medium.

Chalk gangs (Fig. 9.1) were still digging their bell-pits in Hertfordshire in 1913 (Gardner, 1967), and Russell and Keen (1921) confirmed at Rothamsted the relative ease with which a plough could be drawn across heavy clayland after a dressing of chalk had been applied (Table 9.1), though yields of cereals and potatoes were not necessarily increased. Chalking neutral and acid soils raises the pH, of course, and the effects of sustained applications of chalk to a formerly acid permanent pasture on Greensand soil at Abinger, Surrey (Hope-Simpson, 1938) have already been described in Chapter 5.

Table 9.1

The effect of a dressing of chalk on a clay soil [a]

	Without chalk	With chalk
Average speed of ploughing (km hr^{-1}) [b]		
with Cockshutt plough	3·49	3·57
with Ransome plough	3·17	3·54
Average drawbar pull (kg) [b]		
with Cockshutt plough	698	616
with Ransome plough	731	647

[a] From Russell and Keen (1921).

[b] The "average of all results" was "a saving of 180 lb [82 kg] drawbar pull and an additional mile of ploughing in every 9 hours of work [1 km in about 5½ hr]".

The quarrying of various hard bands in the Chalk for building was touched on in Chapter 1. Hard or "hurlocky" chalk of any kind, usually to be found in the Middle or Lower Chalk, was used, as freestone or *clunch*, mainly for interior work, where it was least likely to weather and so could be ornately carved. A particularly fine chalk freestone quarried from Amberley, Sussex, was used in the construction of vaulted ceilings in Arundel Castle and Chichester Cathedral. The Clunch Rock of Cambridgeshire was used in many of the early Cambridge colleges and in parts of Ely Cathedral. Some of the named hardgrounds were known and used well beyond their source: Totternhoe Stone, at one time actually tunnelled from the Bedfordshire scarp, was used in parts of Windsor Castle (as well as chalk freestone from Bisham, near Marlow), while the celebrated Beer Stone from the Middle Chalk of Devon was used as far afield as Norwich Cathedral. Many buildings can be found, however, in which chalk is incorporated into outer walls, often as single blocks or cornerstones, sometimes in combination with flint, itself skilfully cleaved (knapped) by craftsmen. Occasionally entire buildings (save the roof) were constructed of chalk, such as the old school at Medmenham, Buckinghamshire. In places, cottages were built of *cob,* a mixture of clay and pounded chalk, and sometimes bricks were made by compounding chalk dust with a range of diverse substances (North, 1930; Edmonds *et al.*, 1969; see also Jukes-Browne and Hill, 1904; Howe, 1910). Happily, interest in these, as well as more fundamentally agricultural, aspects of past rural economies has been revived through institutions such as Reading University's Museum of English Rural Life and the Weald and Downland Museum at Singleton. The Chiltern Open Air Museum is an imaginative venture recently launched at Newland Park (Smith, 1980).

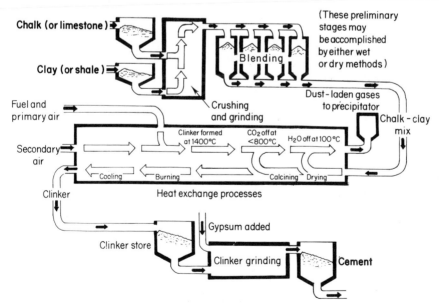

Fig. 9.2. Schematic diagram of the manufacture of cement.

Of course, not all chalk-derived industries are confined to museums. Chalk is still burned to provide mortar and agricultural lime, and is consumed on an altogether more massive scale in the manufacture of cement. The latter process, outlined in Fig. 9.2, is typically carried out in large modern works of the kind shown in Fig. 9.3. Cement manufacture is not confined to the Chalk, but this formation provides well over half of the total tonnage of Portland cement produced annually in the United Kingdom. Moreover, the highly predictable chemical composition of Upper, Middle and Lower Chalks makes these materials admirably suitable for the production of high-class cement. Indeed, "good" chalk is very much at a premium in some localities, and one company actually pumps chalk slurry along a 92 km pipeline from a quarry in Bedfordshire to its factory in the Midlands—a journey which takes it two days to complete. An informative review of current trends in chalk quarrying can be found in Blunden (1975).

9.3 Farming on the Chalk

9.3.1 Introduction

It is rather an artificial procedure to turn to agriculture as though it were something quite separate and new. Everything that has been said so far on the ecology of chalk downland, for example, is intimately associated with man's farming activities, and particularly with sheep husbandry. Note, however, that the term chalk downland tends to be used in a broader sense by agriculturists than ecologists, and often encompasses land under Clay-with-flints and other superficial deposits of the upper slopes and plateaux, as well as those parts which are situated strictly upon the Chalk itself. Moreover, it is common practice to continue to refer to downland as such, even after it has been ploughed for arable crops, or for that matter planted with trees.

We pick up the story here at the point where we left it in Chapter 1, and look first at the role of sheep and cattle in the downland economy from mediaeval times to the present day. Many of the more astute contemporary agricultural writers have described farming on chalk, often in considerable detail. Indeed, quotations taken directly from some of these sources are cited throughout this book. However, an immense amount of information resides in old Parliamentary, parochial and legal records and documents as well as private diaries and papers: information which is open to those blessed with the knowledge, skill and perseverance to track them down, to master them, and to winnow the grain from the chaff of their contents. In this respect we owe a considerable debt of gratitude to Kerridge (1967) who,

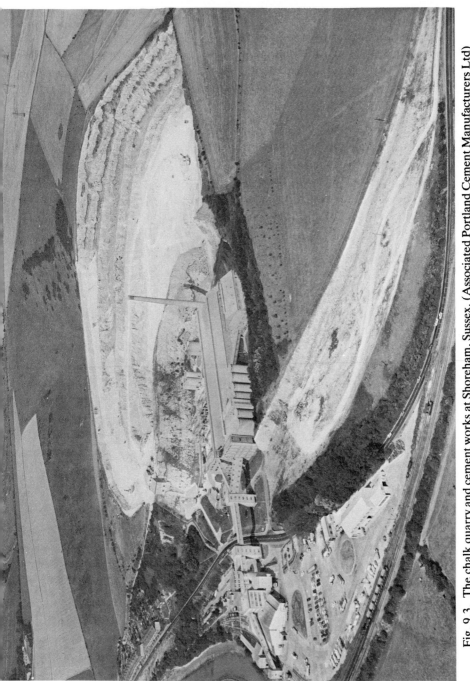

Fig. 9.3. The chalk quarry and cement works at Shoreham, Sussex. (Associated Portland Cement Manufacturers Ltd)

through his masterly handling of a wealth of historical manuscripts and publications, has thrown much light on the state of farming over the two centuries traditionally regarded as preceding the agricultural revolution. The account which follows draws heavily on this work, and it is helpful to use Kerridge's terminology for the various agricultural regions (farming countries) of the chalklands. These are shown in Fig. 9.4.

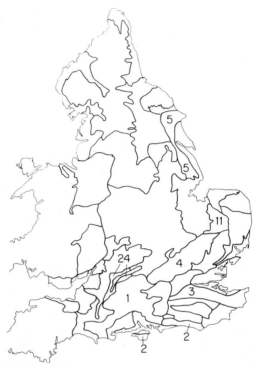

Fig. 9.4. English farming countries of early modern times mentioned in the text (see Kerridge, 1967, 1973). 1, Chalk Country; 2, Southdown Country; 3, Northdown Country; 4, Chiltern Country; 5, Northwold Country; 11, Breckland; 24, Cheese Country. Note that Chalk Country in ths context, with capital letters, has, like the Chalk Rock in geology, a narrower and more specific meaning than when lower case letters are used throughout.

9.3.2 Historical background

9.3.2.1 The sheep-fold
Frequent reference has been made to the traditional system of running large sheep flocks on the downland sheepwalks, but it is essential to realise that these flocks in no way resembled those of the moors of the northern and

western uplands. Those mountain and moorland flocks were of true hill sheep, black of face, coarse in the wool, quite untamed, and neither enclosed nor shepherded at any time other than lambing. By contrast, sheep of the downland flocks were much larger animals, tended constantly by their shepherd who moved them about the farm virtually every single day of the year (see Hudson, 1910).

The breed common to all the chalklands but the Southdown and Northwold Countries (Fig. 9.4) was variously known as the Western, the Old Wiltshire or the Old Hampshire.

> This was a big, hardy animal [writes Kerridge]. It had a large head, big eyes, roman nose, long arched face, wide nostrils, horns (in both sexes) falling back behind the ears, a wide, deep chest, straight back and long, sturdy legs. Both face and legs were white. With a long but light frame, the sheep was naturally active and agile, a good walker and climber. It could pass with ease up and down the hills, often abrupt, that separated the nightly fold from the daily pasture. A greedy feeder, it was slow to fatten, but sometimes attained great weight and its mutton was good. It bore a fleece of about two pounds of good carding wool, but the underparts, from sympathy with the warm, dry soils grew only a short silvery hair. For long bred on open downs, this sheep was a strong, healthy, intelligent animal, well suited to its habitat and function. It was held in the highest esteem for its folding quality, for its propensity to leave its droppings on the arable at night, and for its ability as a walking dung-cart, robbing the downs for the sake of the tillage, but maintaining the down pastures by feeding them closely.

Its polled counterparts in the Sussex Downs (the black-faced Old Southdown) and the Lincolnshire and Yorkshire Wolds (the Old Northwold) provided short, fine wool suitable for clothing, but these, too, were essentially folding sheep (see Kerridge, 1972).

In fact, the chalkland sheep was just as much an agent of arable as of grassland management, and the technique of folding the flock in pens on the arable land by day (Fig. 9.5), and drifting them on the down by night can be traced back at least as far as 1570. In classic chalk country, the long, narrow parishes established before the Norman conquest spread typically from the top to the bottom and beyond of down, escarpment or valley side (Fig. 9.6). "Down pastures, arable fields and bottom meadows all formed part of each farm and all were used together in a single farming system." By the same token, cattle would sometimes be put to graze on the downs. Most villages and townships had their cow-down, and there was a period when Highland cattle were kept on the higher Wolds. Cattle were also allowed to graze newly restored grassland: sheep were considered harmful until the third year. But on the whole cattle were beasts of the lowland meadows, and were but temporary and supplementary grazers on the chalklands as long as the sheep reigned supreme. A lack of water (dewponds—see Figs 1.30 and 9.7—were barely adequate for sheep, let alone cattle) and the need for

Fig. 9.5. Sheep folded onto arable land at Ipsden, Oxfordshire, probably in the early 1940s. The crop appears to consist of alternate bands of kale and turnips. (Museum of English Rural Life)

supplementary feeding were the major limiting factors, though these were to be overcome as more intensive farming methods were introduced.

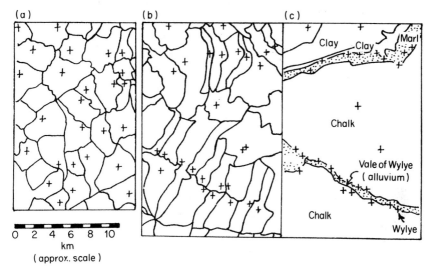

Fig. 9.6. (a) The relative uniformity of parish shapes in typical clayland country, compared with (b) the long, narrow parishes characteristic of the chalklands, here shown for the Wylye area of Wiltshire of which the underlying geology is depicted in (c). Each parish church is shown as a plus sign. From Stamp (1969a), Figs 15 and 16, where all the parish names are given.

By the latter half of the seventeenth century, folded sheep were beginning to be offered more than just the pickings of the stubbles and fallows, for new crops had been introduced, of which dwarf rape (*Brassica napus*) and Cotswold sainfoin (*Onobrychis viciifolia*) were most generally typical of the chalklands. Turnips (*Brassica rapa*) were rarely successful on the thinnest chalk soils for, in the words of William Ellis, "they failed to apple out". Weld (*Reseda luteola*), grown as a biennial for its yellow dye and folded by sheep in its first year, was a speciality of the Chiltern Country, while lucerne (*Medicago sativa*) was favoured by Northdown farmers. Pure stands or simple mixtures of indigenous and imported white and red clovers (*Trifolium repens* and *T. pratense*), trefoil (*Lotus corniculatus*), melilots (*Melilotus* spp.) and other herbs were similarly introduced into the arable rotation. So were "collected grasses" such as perennial ryegrass (*Lolium perenne*) and timothy (*Phleum pratense*), although ley farming had long been practised—as up-and-down husbandry—simply by allowing natural regeneration of the barely buried turf as soon as the last cereal crop was gathered.

9.3.2.2 Ploughing the downs

Thus, the arable soils of the lower slopes were steadily improved by the combined effects of sown grasses and legumes, and the dunging of the sheep. In addition, organic residues of every description were applied: composted straw bedding and thatching, bracken, thistles, old clothes, even offal pilchards and sticklebacks! But such treatments were rarely, if ever, meted

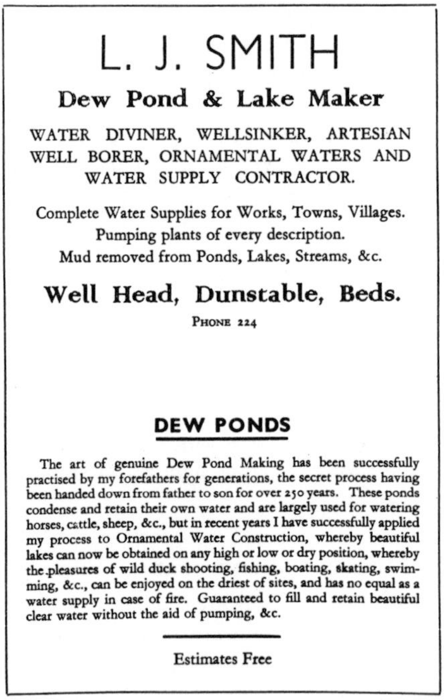

L. J. SMITH
Dew Pond & Lake Maker

WATER DIVINER, WELLSINKER, ARTESIAN
WELL BORER, ORNAMENTAL WATERS AND
WATER SUPPLY CONTRACTOR.

Complete Water Supplies for Works, Towns, Villages.
Pumping plants of every description.
Mud removed from Ponds, Lakes, Streams, &c.

Well Head, Dunstable, Beds.
PHONE 224

DEW PONDS

The art of genuine Dew Pond Making has been successfully practised by my forefathers for generations, the secret process having been handed down from father to son for over 250 years. These ponds condense and retain their own water and are largely used for watering horses, cattle, sheep, &c., but in recent years I have successfully applied my process to Ornamental Water Construction, whereby beautiful lakes can now be obtained on any high or low or dry position, whereby the pleasures of wild duck shooting, fishing, boating, skating, swimming, &c., can be enjoyed on the driest of sites, and has no equal as a water supply in case of fire. Guaranteed to fill and retain beautiful clear water without the aid of pumping, &c.

Estimates Free

Fig. 9.7. A trade-card of the late 1930s which speaks for itself. L. J. Smith was a descendant of the celebrated Smith family of Market Lavington on the northern edge of Salisbury Plain, who were responsible for most of the dewponds in Wiltshire and Berkshire. From Pugsley (1939).

out to the downland sward. Although great ingenuity had long been exercised in flooding the lower-lying water meadows, culminating in the sophisticated floating technique (see Kerridge, 1967, pp. 254–267), the permanent pastures of the high downs had almost 200 years to wait before their improvement by artificial means was to be contemplated. The only way in which the sward was altered—and then drastically—was when the need for more cultivated land, or the financial incentive to expand, pushed the arable tide-line further up the slope to take in virgin downland.

Travelling through Wiltshire in the late 1600s, Defoe witnessed with some surprise

> how a great part of these downs comes by a new method of husbandry, to be not only made arable, which they never were in former days, but to bear excellent wheat, and great crops too, tho' otherwise poor barren land, and never known to our ancestors to be capable of any such thing; nay, they would perhaps have laughed at anyone that would have gone about to plough up the wild downs and hills, where the sheep were wont to go. But experience has made the present age wiser, and more skilful in husbandry; for by only folding the sheep upon the plow'd lands, those lands which otherwise are barren, and where the plough goes within three or four inches of the solid rock of chalk, are made fruitful, and bear very good wheat, as well as rye and barley.

But without the fold, crops were poor and frequently failed altogether, more so on the blacklands than the redlands or "bake" (Chapter 3), and especially where hard (hurlocky) chalk formed the bedrock, or the climate was that much drier. Thus Defoe refers to his own "barren soil of Kent, for such the chalky grounds are esteemed . . . and indifferently fruitful", while Cobbett writes thus (in 1823) of the tract of downs including Longwood Warren to the south-east of Winchester:

> These hills are amongst the most barren of the Downs of England; yet a part of them was broken up during the rage for improvements. . . . The Down itself was poor; what then must it be as cornland! . . . Instead of grass it will now, for twenty years to come, bear nothing but that species of weeds which is hardy enough to grow where the grass will not grow.

Of course, this must partly have been the result of warrening itself, "the most fatal policy of all", according to Hudson (1910).

> How hideous they are—those great stretches of downland, enclosed in big wire fences and rabbit netting, with little but wiry weeds, moss and lichen growing on them, the earth dug up everywhere by the disorderly little beasts! For a while there is a profit—"It will serve me my time", the owner says—but the end is utter barrenness.

Many ingenious techniques were devised in the hope of improving the chances of successful establishment of the first crop—usually dwarf rape. One method involved stripping off the turf, piling it into heaps, and setting light to it to smoulder slowly to ashes. This was known more generally as paring and burning, but locally it went under various names, such as *denshiring* and *burnbaking*. It was practised as early as 1638 in southern Wiltshire on the redlands, but only when corn prices soared in the eighteenth and nineteenth centuries on the blacklands, and here the results were invariably disappointing if not disastrous. An alternative to burnbaking was *raftering*, in which alternate strips of turf were lifted and inverted on to the strips left *in situ*, to provide a double sod into which the corn was then sown. On the redlands, five years of crops might be obtained before the soil was ready for improvement again, usually through folding a tumbledown fallow, perhaps supplemented with sainfoin or hayseeds. The blacklands could produce at most two successive corn crops before they were exhausted, and Kerridge describes in some detail the methods and consequences of tilling these soils.

> Once the grass was broken on "blackland" soils nothing but burnbaking could induce a crop and nothing but famine prices justify it. The usual field-course in "blackland" was (1) turnips or rape sown in the ashes and fed off by sheep for (2) oats undersown with rye grass. At the first burnbaking good crops were obtained but these merely exhausted the soil. The grass grown under the oats soon gave out and what took its place was not sheep down grass but black couch and bent [*Agrostis* spp., see Section 5.4.3]. A young tender-mouthed flock of sheep would sooner starve than eat this, so the farmer was forced to put down his ewes in favour of wethers. These fed badly and provided poor fold. Yet without folding or mucking the land would grow nothing. Hence it had to be either folded with sheep fed elsewhere, to the detriment of the other arable, or abandoned altogether . . . .

Kerridge continues:

> The "blackland" was natural sheep down, and its plough-up deprived the flocks of their best feed, so that many had to be put down and the sheep that remained contracted goggles, while the staple of the wool was prejudiced. In an attempt to overcome these troubles, the Horn sheep were crossed with the new Southdowns to produce the Hampshire Down. But these were not such good folding sheep, and were less suited to close cropping the down pastures, which were allowed slowly to deteriorate. By this and the wasting of the "blacklands", the sheep fold itself was jeopardised. As soon as grain prices fell from their wartime peak, great stretches of downland went out of cultivation altogether, and even the best of the swards started to become coarse and tufty, so that when much of the land was later taken over for military training, it was already semi-derelict.

Little wonder that these tracts of downland reflect to the present day such marked differences in botanical composition (Section 5.4.3).

9.3.2.3 The Great Depression: the return to grass

Kerridge's reference to falling prices relates to the period of economic depression immediately following the Napoleonic Wars, but the agricultural industry recovered to enjoy half a century of prosperity—the Golden Age— before bad times returned in earnest in the Great Depression of the late 1800s during which time all manner of drastic changes befell the British farmer, particularly in lowland England, the chalklands included. The Depression was ushered in by a run of bad seasons in the early 1870s, in which low temperatures and incessant rain plagued grassland and arable farmers alike, especially in the south and east. But this was no more than a prelude to the prolonged recession which was only broken by the 1914–1918 War, and which lasted to all intents and purposes, at least in the chalk country, until the outbreak of the 1939–1945 War. The cause was a complicated mixture of economics and politics, both national and international, which even now it is not easy to disentangle. A valuable analysis has recently been provided by Perry (1974), who notes that the closing decades of the nineteenth century saw a marked general decline in the national importance of British agriculture: the industry provided 20% of the Gross National Product (GNP) and employed one-fifth of the population in the 1850s, but produced only 6% of the GNP, and employed less than one-tenth of the population by the turn of the century.

As a consequence of the Depression, the labour-intensive high-farming methods of the preceding Golden Age were substituted by low-cost systems which led to the grassing-down, either deliberately or through natural succession from tumbledown fallows, of much arable land. Again quoting from Perry, tillage crops (excluding temporary grass) accounted for 45% of the acreage of rural Britain in 1866–1875, but only 32% in 1896–1905, while permanent grassland (in the broad sense, including rough grazing and even fallow land) rose from 44% to 56%. Over the same period, the wheat acreage fell from more than 3 million acres to "rarely over 2 million . . . and occasionally less than 1½ million", though in places oats, and especially barley, were grown instead. The number of sheep fell from 28 to 26 million, though the marked decline in folding sheep on the Chalk was masked to some extent by the substitution of "grass sheep" (such as the Cheviots and Blackfaces which were brought to Salisbury Plain), and especially by increases in sheep in the Highland Zone, in which a relatively stable economic climate was maintained throughout the depression. Much of the marginal land on the Chalk went out of agricultural production altogether and was abandoned to the rabbits.

9.3.2.4 Grassland improvement: some famous pioneers

This was not to say that no improvements in grassland husbandry were made, or novel enterprises tried. Basic slag, a by-product of the steel industry rich in phosphate, became available towards the end of the nineteenth century, and its effects on permanent pasture had been convincingly demonstrated by Gilchrist and Somerville at Cockle Park on the Northumbrian Boulder Clay. It seems less widely known that following the success of the Cockle Park trials, Somerville experimented with basic slag on other types of permanent pasture, including chalk downland. Actually the downland site at Sevington, near Alresford, Hampshire, where the effects of slagging were followed most closely (Ashcroft, 1910) was on Clay-with-flints (Ashcroft, 1901).

Early accounts of the effects of slag on chalk pastures in the strict sense are provided by Somerville (1911, 1918). During a visit to the Applesham Estate near Shoreham, Sussex, he had been particularly impressed by the response of clovers in the treated sward, and he wrote (in 1911):

> The improvement effected on this poor Downland is one of the most striking object lessons that it has ever been my good fortune to see . . . and where untreated pasture is available for comparison, as also where portions have been accidentally missed by the sowing machine, the line between the slagged and unslagged is a very sharp one . . . on the one side of the fence one sees poor, weedy, thin herbage with flints and chalk hardly obscured, while on the other side one has a tangled mass of clover that would do credit to high class land.

Somerville's paper of 1918 records similar spectacular responses to slagging in legumes such as *Lotus corniculatus, Hippocrepis comosa* and particularly white clover again, on his experimental farm, Poverty Bottom, near Newhaven, although in both cases clover in these quantities would suggest earlier improvement by reseeding. Describing the experimental plots established on the Wiltshire Downs by the University of Bristol in 1929, Warne (1934) found that the more "natural" fine-leaved fescue swards there responded to phosphate (and some other fertilisers) by becoming altogether more grassy, at the expense of broad-leaved herbs such as *Plantago lanceolata* and *Poterium sanguisorba*. We look more closely at other work along these lines in the next section.

During the 1920s, an innovation on the Wiltshire Chalk, at least in terms of scale, was dairying, an enterprise greatly stimulated by the milking bail (Fig. 9.8) developed by the celebrated Hosiers (see Hosier and Hosier, 1951). Milk could be produced extremely cheaply from cows kept out on the downs all the year round, though the system still demanded adequate water and suitably productive pastures. Laying on water was no problem to the inventive Hosiers (whose earliest projects were aided by a government job-creation scheme for the unemployed of those difficult times), while the pastures were improved simply by stocking heavily with cows fed on hay grown elsewhere on the farm, supplemented with concentrates. The soil was enriched by the accumulated dung, urine, spilt food and washings from the bail, which was moved on daily to fresh ground: indeed its effect was analogous to the sheep-fold, except for the copious quantities of water involved, and that the treatment was applied not to the arable land but to the long-starved downland turf itself. A range of chemical fertilisers— "artificials"—was available by now, but this system did not require them.

Of course, none of this would have been possible without adequate outlets. The Hosiers pioneered direct delivery of fresh milk to London by road using glass-lined tankers, but throughout most of the chalk country diversification of this kind was severely limited by the lack of markets close by, or transport to more distant ones. In the western Chilterns, for example, dairying could be sustained by the renowned Reading biscuit industry, but further east milk production was largely confined to the valleys, not so much because of climate and soil but because of rail links to London and other markets (Coppock, 1961).

Fig. 9.8. A milking bail still in regular use at Stoney Green, Buckinghamshire. (C.J.S.)

Not until the 1930s was any nationally concerted effort made to reclaim downland (as well as other) permanent pastures, and the main emphasis was on grubbing out scrub, putting in the plough, and switching to ley farming (Stapledon and Davies, 1940; Hosier and Hosier, 1951; Thomas, 1951). The outbreak of the 1939–1945 War greatly boosted this campaign. Newly available varieties of grasses and legumes greatly improved in yield and digestibility—the famous Aberystwyth "S" strains among them—alternated with cereals. Perennial ryegrass, cocksfoot, meadow fescue (*Festuca pratensis*), timothy and white clover were most commonly sown, the choice of exact variety depending on the proposed duration of the ley. Soil type was almost entirely disregarded although lucerne (*Medicago sativa*) was advocated for thin chalk soils (e.g. Jones and Dermott, 1951). It is convenient to make the point here that at the present time improved pastures of this nature, composed largely of a new generation of selected cultivars, form the basis of milk and meat production in the chalklands, as elsewhere in Britain. With appropriate grazing management and the application of heavy dressings of fertilisers, including up to 400 kg ha^{-1} of N suitably split to allow for the seasonal pattern of herbage growth (Fig. 6.1), yields of up to 15 000 kg ha^{-1} of dry matter can now be obtained in a season from a grass ley (Cowling, 1966).

9.3.3 More recent experiments on the improvement of chalk downland pastures

9.3.3.1 Permanent grassland

Strictly speaking, the establishment of leys comes under the heading of arable farming, which is dealt with later. For the moment we stay with the subject of permanent grassland in the strict sense, and look more closely at the various ways in which the downland sward can be improved without recourse to outright ploughing. We shall consider modern work on the use of fertilisers; cultural treatments involving partial or complete destruction of the turf mechanically, or by burning or treading, or by the application of herbicides; and reseeding, either by surface broadcasting or sod-seeding. Regarding fertilisers, it might be noted here that many articles in the farming press indicate, contrary to expectation, a marked response of downland pastures to liming. It may be that locally, as we noted in Chapter 3, soils on level ground long undisturbed by the plough become leached free of calcium carbonate in their surface layers, but in general these observations seem more likely to apply to chalk downland in the broader sense (Section 9.3.1) where superficial deposits, not leaching, depress pH. This aspect is not pursued here.

9.3.3.2 The effects of potash on chalk grassland

Potash alone usually has little effect on either the yield or botanical composition of chalk grassland, but where deficiencies of phosphate and nitrogen have been corrected, responses to K have been observed. These are discussed in the two following sections on the effects of P and N.

9.3.3.3 The effects of phosphate and potassium fertilisers

Norman (1956) investigated the effects of applying contrasting rates and frequencies of superphosphate, with or without potash, to chalk grassland at Huriey, Berkshire. The sward was not typical of grazed downland, for it had been ploughed and cropped during 1940–1942. *Agrostis stolonifera* shared dominance with *Festuca rubra*, while *Arrhenatherum elatius* was "locally abundant", and there was a large proportion of dicotyledons—in fact it was a typical ex-arable sward (Section 5.4.3). Each treatment combination (Table 9.2) was applied to duplicate plots 0·03 ha in area, each separately

Table 9.2

Fertiliser applications to chalk grassland plots at the Grassland Research Institute, Hurley, 1952–1954 [a,b]

Code [c]	1952 March	1952 June	1952 Sept	1953 March	1953 June	1953 Sept	1954 March	1954 June	1954 Sept	Referred to in Table 9.3 and text as
k_0p_0	—	—	—	—	—	—	—	—	—	No K or P
k_1p_0	62 K	—	—	62 K	—	—	62 K	—	—	K only
k_0p_1 B	10 P	—	—	10 P	—	—	10 P	—	—	Low P without K
k_1p_1 A	30 P 62 K	—	—	62 K	—	—	62 K	—	—	Low P with K
k_1p_1 B	10 P 62 K	—	—	10 P 62 K	—	—	10 P 62 K	—	—	Low P with K
k_1p_1 C	3 P 62 K	3 P	3 P	3 P 62 K	3 P	3 P	3 P 62 K	3 P	3 P	Low P with K
k_1p_2 A	89 P 62 K	—	—	62 K	—	—	62 K	—	—	High P with K
k_1p_2 B	30 P 62 K	—	—	30 P 62 K	—	—	30 P 62 K	—	—	High P with K
k_1p_2 C	10 P 62 K	10 P	10 P	10 P 62 K	10 P	10 P	10 P 62 K	10 P	10 P	High P with K

[a] From Norman (1956).

[b] K was applied as muriate of potash (KCl, 49·8% K) and P as superphosphate (7·9% P). Application rates have been recalculated on the basis of kg elemental K and P ha^{-1}, and rounded off to the nearest whole kilogram. All plots received in addition 39 kg ha^{-1} N as nitrochalk in April and August 1952, 1953 and 1954.

[c] The letters A, B, and C refer to the degree to which the P applications were split: in A, all the P was applied at the start; in B, it was split into three annual spring-time dressings; in C, it was applied three times a year for three years. The effects of these different applications were small and are not discussed in the text.

fenced, and grazed by sheep for three to four days at a time, four to five times a year through the three growing seasons of 1952–1954. The quantity and feeding value of the herbage, and its botanical composition, were assessed immediately before each grazing period. In the first year, the effects of K and P alone on dry-matter yield were negligible, but in combination these elements raised yields by 14·8% and 25·0%, depending on the rate of phosphate applied (Table 9.3). The effects of these treatments progressively increased over the following two years, apparently through the accumulation of the residues of the nutrients applied: for example, by the end of the third year, the higher rate of P, in combination with K, increased the yield by over 70%. This was partly explained, however, by smaller yields from the unfertilised plots in the second and third years. Details are provided in the original paper of the effects of splitting the superphosphate dressings, but this is not pursued here.

The effects on botanical composition took longer to show up, but by the end of the three-year period, distinct differences were evident in some of the plots (Table 9.4). The total cover of most species was increased by K, but only very slightly, even though all plots received two top-dressings of nitro-chalk each year. Similarly, P had little effect on its own, although it substantially increased the cover of *Holcus lanatus*, *Lolium perenne*, *Poa pratensis* and *P. trivialis*, and the legumes *Medicago lupulina* and *Trifolium repens*. The overall effect was that plots receiving the phosphate dressing alone were more grassy, but no less herby, than the control plots (cf. Warne's observations, Section 9.3.2.4). The grazing animals themselves, of course, must have influenced the overall results.

Where K and P were applied together, the increased contribution of the grasses was enhanced, though there was nothing like the complete domination by *Festuca rubra* which in other experiments has been found to follow nitrogen application (see below). Indeed, this species played little part in the response to phosphate, which was much more marked in *Arrhenatherum elatius*, *Dactylis glomerata*, *Holcus lanatus*, *Lolium perenne*, *Poa pratensis* and *P. trivialis*. The persistence of *Arrhenatherum* under three years' sheep grazing is at variance with the well-documented generalisation about the inability of this species to stand up to grazing, although, of course, the grazing regime was not continuous (and see Section 4.2.2.4). Only *Agrostis stolonifera* decreased with the heavier rates of P fertiliser. Norman attributed this to its poor ability to compete with the other grasses, rather than to any direct effect of the P. Of the non-grasses, only *Trifolium repens* showed a marked positive response to P (see Somerville's observations, above). Among the rest, some responded slightly to P, but their growth varied from season to season: to borrow a phrase from Brenchley (1924), they "took turns in being the most prominent miscellaneous species". The herbs in the

Hurley plots showed a decline with increasing P, though not so marked as the increase in the grasses.

Table 9.3

Herbage dry-matter yields as affected by phosphate application, with or without potash [a,b]

Year	No K or P		K only		Low P without K		Low P with K		High P with K	
	DM	%	DM	%	DM	%	DM	%	DM	%
1952	2870	100	2836	98·8	2982	103·9	3408	114·8	3587	125·0
1953	2309	100	2634	114·1	2724	118·0	3083	133·5	3531	152·9
1954	2421	100	2993	123·6	3016	124·5	3643	150·5	4159	171·8

[a] From Norman (1956).

[b] DM = dry-matter yield (kg ha^{-1}); % = percentage of the "no K or P" yield for that year. See Table 9.2 for details.

9.3.3.4 The effects of nitrogen fertilisers

Earlier work (such as that of Warne (Section 9.3.2.4) and Green, Norman and others at the Grassland Research Institute) had established the response of chalk grassland to nitrogen, but more detailed information comes from the series of experiments begun by Elston on *Festuca rubra* grassland at Swyncombe, Oxfordshire, touched on already in Chapters 3 and 5 (Elston, 1963; Mirghani, 1965; Bunting and Elston, 1966). Plots 0·03 ha in area were top-dressed in May 1960 with N and K fertilisers and were either mown later that summer and twice annually thereafter or left uncut. These treatments were applied in factorial combination, replicated four times, and in addition all plots received a top-dressing of P (Table 9.5). The yield of herbage dry matter from the mown plots was measured annually for ten years (Fig. 9.9), and in June 1970 a detailed assessment was made of the botanical composition of the plots (Smith *et al.*, 1971). Both N and K, applied alone, increased yields, but the effect of N was always much greater than that of K, which, as in Norman's experience at Hurley, never attained statistical significance. There was, however, a significant N × K interaction (Table 9.6).

The treatments led to marked contrasts in botanical composition which help to explain the yield differences. Plots which had received only N (in addition to the overall P application) supported a virtually pure sward of *Festuca rubra* (Fig. 9.10 and Table 9.7). Where N and K were added in combination, the *F. rubra* was much taller and accompanied by large herbs, such as *Cirsium vulgare, Senecio jacobaea, Taraxacum officinale, Pastinaca*

Table 9.4

Botanical composition of the Hurley plots as affected by three years' applications of the treatments detailed in Table 9.2 [a,b]

Species	No K or P	K only	Treatment Low P without K	Low P with K	High P with K
Festuca rubra	10·0	11·0	11·1	12·0	12·7
Agrostis stolonifera	9·1	8·5	8·2	6·2	6·4
Arrhenatherum elatius	3·6	3·9	2·9	5·5	5·7
Dactylis glomerata	2·4	2·0	2·2	4·0	4·8
Poa pratensis	2·2	2·2	4·2	4·2	4·7
P. trivialis	0·7	0·9	3·9	3·9	4·9
Holcus lanatus	0·9	1·2	2·3	3·3	3·0
Lolium perenne	0·9	1·6	3·3	3·1	4·4
Total grasses	29·8	31·3	38·1	42·2	46·6
Medicago lupulina	2·3	2·9	4·6	3·0	3·2
Trifolium repens	0·5	0·5	2·8	4·7	5·3
Total legumes	2·8	3·4	7·4	7·7	8·5
Ranunculus bulbosus	2·5	2·5	2·5	3·0	2·9
Prunella vulgaris	3·0	3·8	2·9	1·7	1·7
Plantago lanceolata	6·2	7·0	5·8	5·8	6·1
Veronica chamaedrys	1·7	1·9	2·6	2·3	2·3
Crepis capillaris	2·6	2·9	4·0	3·3	3·3
Leontodon spp.	11·9	10·4	9·4	9·0	7·7
Total herbs	27·9	28·5	27·2	25·1	24·0

[a] From Norman (1956).
[b] Values represent the square root of "total cover" (see original paper for definition), determined by placing a ten-pin frame of points 30 times in each plot.

Table 9.5

Fertiliser and mowing treatments applied to the Swyncombe Down plots [a]

(a) In factorial combination
  (i) Fertiliser N: none, or 53 kg ha^{-1} N as nitrochalk (21% N), applied after each cut (see below), and in 1960 and from 1968 onwards as a spring top-dressing as well.
  (ii) Fertiliser K: none, or 126 kg ha^{-1} K as muriate of potash (KCl, 49·8% K), applied at the same times as for nitrochalk.
  (iii) Cutting: uncut, or cut (usually) twice a year with a motorscythe, the herbage being removed from the plots for weighing and sampling, and then discarded.

(b) All plots received an annual spring top-dressing of 19 kg ha^{-1} P as superphosphate (7·9% P).

[a] From Smith et al. (1971).

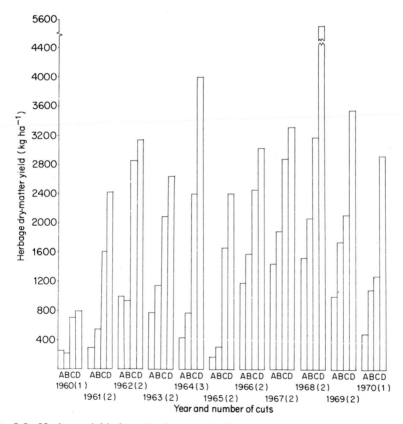

Fig. 9.9. Herbage yields from the Swyncombe Down plots between 1960 and 1970. Treatments were: A, no N or K; B, K only; C, N only; D, N and K together. See text and Table 9.5.

Table 9.6

Mean annual herbage yields, Swyncombe Down, 1960–1970 (kg dry matter ha⁻¹ year⁻¹) [a]

	No N	N	Effect of N
No K	730	1967	+1237 [c]
K	1013	2794	+1781 [c]
Effect of K	+283 (NS)	+827 [c]	—
N × K interaction	—	—	+544 [b]

[a] From Smith *et al.* (1971).
[b] Significance $P < 0.01$.
[c] Significance $P < 0.001$.

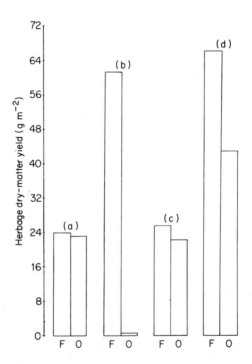

Fig. 9.10. Botanical composition of the Swyncombe Down plots in 1970. F = *Festuca rubra*, O = other species. (a) No N or K, (b) N only, (c) K only and (d) N and K together.

Table 9.7

Summary of data on botanical composition of the Swyncombe Down plots, based on dry-matter production in 1970 [a]

		Treatment		
	No N or K	N alone	K alone	N and K
Total weight of herbage (g m^{-2})	47·3	61·4	48·3	109·2
Festuca rubra (g m^{-2})	24·2	61·1	25·9	66·2
Other species (g m^{-2})	23·1	0·3	22·3	43·0
F. rubra (% of total)	51·2	99·5	53·6	60·6
Number of species	28	12	21	14

[a] From Smith *et al.* (1971).

sativa, Filipendula vulgaris and *Tragopogon pratensis*. It is interesting to note (Table 9.7) that with N and K in combination, the yield of *F. rubra* was only very slightly greater than with N alone. In contrast, the yield of the other species was greatly increased by N and K together. Thus, while the effect of N on yield was accounted for by *F. rubra* alone, the positive N × K interaction resulted from the growth of other grasses and herbs.

Without N, the number of species was larger and more typical of close-cropped chalk grassland. Where neither N nor K were applied, a characteristic—indeed exclusive—species was *Scabiosa columbaria*. With K alone, the legumes *Medicago lupulina* and *Lotus corniculatus* came in. Indeed, there was a suggestion in the data (Fig. 9.9) that the effect of K was beginning to reach significance, at least up to 1969, and it seems quite likely that the legumes were responsible for this increase, both directly by their own contribution to the herbage, and indirectly through their obviously stimulating effect on the surrounding turf. Remember, however, that all the plots had received an overall dressing of phosphate (Table 9.5).

In the unmown plots, botanical composition was estimated visually, and so the results lack both the detail and precision of those from the mown plots. Many of the smaller herbs, sedges and grasses were absent, and the herbage had become extremely hummocky and matted (Fig. 6.17), but in general the fertilisers had similar effects on the unmown plots as on the mown ones. Thus, N encouraged the growth of *F. rubra* at the expense of even the larger herbs such as *Poterium sanguisorba* and *Filipendula vulgaris* which virtually disappeared. By 1970, several of the plots had been invaded by creeping herbaceous perennials, such as *Glechoma hederacea*, *Urtica dioica*, *Cirsium arvense* and *Sonchus arvensis*, and during the period 1971–1973 these had a dramatic effect upon the previously strongly persistent mounds of fescue herbage and litter (Section 6.5.3.1). A few bushes (mainly buckthorn and hawthorn) and trees (mainly birch) became established, but woody invasion was not nearly as extensive or rapid as might have been expected (see Chapter 7).

The virtual dominance of the turf by *Festuca rubra* in both mown and unmown plots receiving N but no K is striking, but agrees with findings from earlier work. Warne (1934) reported it from his fertiliser experiments on the Wiltshire Downs, while during routine sampling at sites in the Chilterns and Berkshire Downs (C. J. Smith, unpublished), soils from under turf in which *F. rubra* was the dominant grass were consistently found to contain less total and available K than soils in which taller grasses such as *Arrhenatherum elatius, Bromus erectus, Dactylis glomerata, Helictotrichon pratense* or *H. pubescens* were predominant (Table 9.8). As we saw in Chapter 5, Perring (1959) found *F. rubra* more typical of the plateaux and gentle slopes facing

Table 9.8

Total and available K content of soils from fescue and "non-fescue" sites in the Chilterns and Berkshire Downs (see text)

	Total K (% oven-dry soil)	Available K (ppm oven-dry soil)
Fescue sites (mean of 8)	0·09±0·02	4·85±1·36
"Non-fescue" sites (mean of 7)	0·16±0·05	6·12±2·15

between NW. and E., and it was on NE. slopes that he found the smallest exchangeable K concentrations in the soil (Chapter 3). In an extensive review of the fertilising of pastures, Castle and Holmes (1960) report that if N is given where K is limiting, fine-leaved fescues invariably become predominant.

Using the techniques described in Chapters 3 and 5, Elston went on to show how these visual differences in the Swyncombe turf resulting from the application of nitrogen fertilisers were paralleled by differences in water use (Table 9.9) which were explained by the deeper penetration of roots to a level in the subsoil where more water was available (Fig. 9.11). Later work by Mirghani (1965), however, suggested that Elston's results were very much influenced by the relatively heavy rainfall of the year in question (1964), and that in the following year much less rain fell, and the response in water use to added N was not nearly so marked.

9.3.3.5 Liquid manures and aerial top-dressings

Not surprisingly, the most spectacular responses to added nutrients are seen when these are applied in liquid form (Section 9.3.2.4). It was for a time a

Table 9.9

Actual transpiration from untreated and nitrogen-manured weighed transpirometers at Swyncombe Down, 1962 [a]

Week beginning	Transpiration (mm) No N	+N
July 16	15·5	17·0
July 23	24·1	24·1
August 13	23·1	24·6
August 20	15·5	18·0
August 27	16·3	21·1
September 3	9·9	11·7
September 10	25·6	26·4
September 17	7·9	10·2
October 8	7·1	7·4
October 15	7·4	8·9
Total	152·4	169·4

[a] From Elston (1963).

regular practice in certain rural areas for the contents of domestic cesspools to be discharged from tanker lorries on to local pasture land. A similar but more sophisticated service to farmers now under the auspices of the regional Water Authorities provides liquid digested sewage with far less risk to health, although there has been controversy in some industrial areas where routine sampling has indicated high levels of industrial pollutants. Slurry from intensive livestock enterprises, usually applied after a period of maturation in forbidding "lagoons", similarly provides liquid organic fertiliser. Until recently, the steepest tracts of land (and those small areas in which sarsens bar access to wheeled vehicles) have escaped most, if not all, of these diverse, enriching substances, and where these areas have been continually grazed they retain their original species richness and provide some of our most valuable sites for conservation (see Chapter 10). But now even these are threatened (where they are not formally safeguarded) by the application of fertilisers from aircraft.

9.3.3.6 Other nutrients

Attention has focused in the foregoing paragraphs on the major nutrients N, P and K, and their interactions. The response of sown pastures to added sulphur on chalk soils in the Thames Valley was mentioned in Chapter 3, as was the possible deficiency, especially in lush spring growth of grass-dominated swards, of magnesium, leading to hypomagnesaemia in livestock. Animals on Icknield soils are liable to suffer from a shortage of

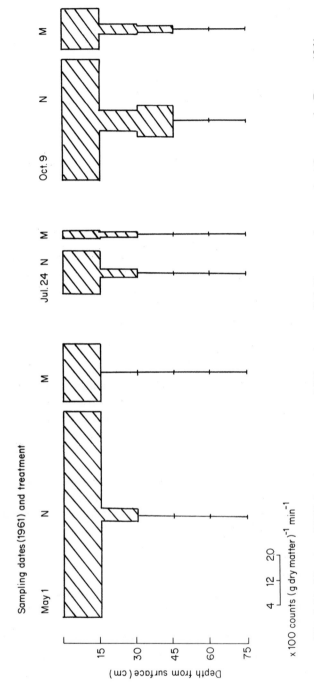

Fig. 9.11. Uptake of ⁸⁶Rb by nitrogen manured (N) and mown (M) *Festuca rubra* swards at Swyncombe Down, 1961 (Elston, 1963). The tracer was dispensed down the profile at 15 cm intervals one week before sampling. Nitrogen and cutting treatments were applied two and eight months before the first sampling.

copper, and on Andover soils from both copper and cobalt deficiencies (Cope, 1976). However, deficiencies of this kind can normally be readily taken care of by providing livestock with salt-licks and other dietary additives.

9.3.3.7 Other methods of permanent grassland improvement*

We saw in Chapter 6 that one of the consequences of lax grazing is for the coarser element of the sward to gain dominance. This may be an advantage where the species are relatively palatable and nutritious, such as tall oat, tall brome, cocksfoot and red fescue, though the first is easily killed out when close grazing is resumed, and cocksfoot quickly forms tough tussocks if it is not checked while fairly young. Where the problem is no more than several seasons' accumulation of rank growth, the feeding value of the herbage can usually be restored by reintroducing a more intensive grazing regime, especially if store cattle, or other hardy, coarse-feeding stock are available. But other species are less desirable, and may pose a serious threat to animal productivity, and even health. Ragwort is an obvious example of the latter: though harmless to sheep, it is poisonous to cattle. Despite the inroads made by cinnabar larvae, there seems little hope of really effective biological control of ragwort with this insect, and the traditional method of hand-pulling is still the most effective, provided that the rogued plants are not left lying about, for in the wilted state they are a fatal attraction to cattle. Thistles, particularly creeping (*Cirsium arvense*), spear (*C. vulgare*) and marsh (*C. palustre*), though not poisonous, are unsightly and useless to the grazier, but less easily removed. Goats and donkeys may be persuaded to graze them. Chemical control of these and other "weeds" is touched on below.

A problem in the dryer regions of the chalk country is tor grass (*Brachypodium pinnatum*), highly invasive in the absence of close grazing, and not only unpalatable to farm livestock, but apparently also to many small mammals and possibly some invertebrates of ecological interest (Chapter 6). Even by 1910 tor grass was invading the old Sussex sheepwalks and methods were being sought for its effective control (Hall and Russell, 1911; Hope-Simpson, 1940c). Downland farmers there had made some inroads by grazing it hard with cattle, sometimes sprinkling salt on it beforehand to enhance its palatability. Alternatively it could be burned and this, timed correctly, effected some degree of control, especially if it was followed by grazing, when the new green shoots stimulated by the fire were themselves at

---

* The reclamation of grassland from scrub, involving the removal of bushes and trees, is usually accompanied by ploughing and reseeding, considered in the next section. Scrub removal with minimal disturbance to the soil is normally practised as a conservation measure, and is dealt with in Chapter 10.

their most palatable and vulnerable. Any coarse pasture could be improved by feeding mature, seed-bearing hay to livestock on heavily trampled land (the Hosiers' method mentioned above), though of course better results could be obtained by broadcasting seed of selected grass and legume culti-vars. Pigs have been used very effectively in this way (Hope-Simpson, 1940c).

The renovation of downland pasture by surface cultivation and reseeding may be worthwhile if the sward is sufficiently open and unproductive, but where the turf contains relatively large quantities of *Festuca rubra*, the operation may be uneconomic. For one thing, though hardly in the same category as the leys mentioned earlier, well-fertilised *F. rubra* turf produces quite good yields of herbage (those shown in Table 9.6 resulted from a very modest application of fertilisers by modern standards), it stands up well to grazing, and maintains its digestibility well into the winter. Secondly, the bulky shoots, litter and roots of *F. rubra* are extremely difficult to eradicate effectively, resulting in an inhospitable medium for the germination of seeds, and readily regenerate to compete with any seedlings of the sown species which do manage to establish. Thus, Norman and Green (1957a) found that following seeding with a mixture consisting of Italian ryegrass (*Lolium multiflorum*), S.23 and S.24 perennial ryegrass, S.143 cocksfoot, S.48 and S.50 timothy (*Phleum pratense* and *P. bertolonii*), S.215 and S.53 meadow fescue (*Festuca pratensis*) and S.100 white clover, only the clover and the perennial ryegrasses established satisfactorily, despite four passes with a ripper-harrow, two with a spring-tine cultivator, and applications of nitrochalk, basic slag and muriate of potash fertilisers.

Pre-treatment by heavy grazing (and perhaps by burning—see Section 10.3.4.3)reduces the bulk of the fescue turf, and of course herbicides are an additional potent aid in its destruction. However, although further work by Norman and Green (1957b) showed that spraying the old turf with 2,4-D aided the establishment of S.24 perennial ryegrass and S.100 white clover, this was mainly by suppression of the dicotyledonous species which pre-dominated in the Hurley swards. *Festuca rubra* (and also *Agrostis stoloni-fera*) actually increased following spraying, so that within two years no differences could be seen between sprayed and unsprayed plots. More success can be expected from non-selective herbicides, of course, yet during experiments at Swyncombe Down (Section 9.3.3.4.), *Festuca rubra* turf was found to possess marked powers of recovery even from Paraquat. All this may well prove to be of purely academic interest, however, for land-owners are now more than ever likely to be encouraged to conserve, rather than to renovate, the dwindling examples of native chalk grassland which remain (Chapter 10).

9.3.4 Modern arable farming on the Chalk

9.3.4.1 The economic climate

Although the more productive chalk soils have remained essentially continuously under the plough throughout the ups and downs of the last 100 years, arable farming returned to the chalk country in earnest with the outbreak of the 1939–1945 War. This was precipitated, of course, by the demands of the wartime economy, and an incentive to farmers was the subsidy authorised in 1939 to support the ploughing-up of permanent grassland. Nevertheless by now sufficient developments were taking place to radically alter the prospects for chalkland arable farmers, particularly as prices continued to rise. These included increasing mechanisation and the availability of relatively cheap chemical fertilisers and herbicides, though neither the number of tractors and combine harvesters nor the range of agrochemicals at the beginning of this period were nearly as large as frequently seems to be assumed (Table 9.10). However, the early 1940s were a turning point. One consequence of the increasing battery of mechanical and chemical aids for farmers was to diminish the significance of soil

Table 9.10

Increase in combine harvesters in Britain between 1942 and 1965, and in corn driers between 1946 and 1966 [a]

Year	Combine harvesters	Corn driers
1942	940	—
1944	2400	—
1946	3253	1002
1948	4969	1133
1950	10 048	1309
1952	16 470	2400
1954	21 117	3545
1956	31 020	7690
1957	35 370	9790
1958	39 890	11 880
1959	48 370	14 460
1960	47 930	16 520
1961	50 190	17 360
1962	52 350	19 710
1963	55 060	—
1964	—	27 010
1965	57 950	—
1966	—	23 410

[a] From MAFF (1968).

type, so that information specifically on chalk soils in the modern, as distinct from the historical, literature is surprisingly scant. But as costs of labour, machinery, feedstuffs, fuel and chemicals continue to soar, attention is again turning to the needs of specific crops on specific soils, and it is to be hoped that information of this kind, of which a great deal exists in internal reports, archives and above all in the heads of farmers and advisors, will be made more widely available.

9.3.4.2 Regional contrasts

Obviously cereals are overwhelmingly the most important arable crop on the Chalk, with barley at the forefront, though the era of the chalkland barley barons which reached its peak in the 1960s is past, and a more diverse cropping system is emerging, as we shall see shortly. Regional differences can be accounted for partly by climate and partly by the location of markets, processing plants and other outlets. Local traditions and the successes of progressive individuals can also strongly influence choice of enterprise. First a few examples of climatic influences.

In the south-west, where the winters are open and the summers cool (and crops particularly vulnerable to spores blown across the English Channel), certain cereal diseases occur, such as black rust of wheat (*Puccinia graminis*), leaf-blotch of barley (*Rhynchosporium secalis*) and *Septoria* infections of both these crops, which are unknown or less prevalent in the east and north. Here, too, where potato growing is feasible, main-crop varieties demand protection from blight (*Phytophthora infestans*) as a matter of course, except in the very driest of seasons. In the far northern chalklands a significant feature of the climate is wind—mainly due to a combination of altitude and latitude (the Yorkshire Wolds rise to over 200 m above sea-level and extend north of 54° N.), as well as to proximity to the east coast. This windiness, combined as it is with relatively low rainfall, results in an unusual problem for cereal farmers: ripening corn crops tend to shed their grain as it matures (Sutherland, 1967). On the other hand, the cool, windy climate minimises the risk of transmission by aphids of virus diseases of potatoes, and an increasingly important enterprise in the Yorkshire Wolds is the raising of seed potatoes (Mundy and Roebuck, 1966).

Certain crops are restricted to the south and south-east on account of their need of a sufficiently long growing season. Grain maize is a good example, and it is now accepted that despite Cobbett's renowned reports of successful crops in the Tyne Valley, and notwithstanding the recent breeding of new varieties such as Maris Carmine specifically for British conditions, the crop can be profitably and reliably grown (for grain, note, as distinct from forage) no further north than a line from Bristol to the Wash (Gunn, 1968). Even within this area, the deeper valleys and more lofty summits and plateaux of the chalk country may prove too cool: delayed ripening even of temperate cereals has been reported in the shadowy coombes beneath the South Downs scarp, for example (Jesse, 1960). As noted in Chapter 3, the lighter-coloured chalk soils (such as the Upton series) can take longer to warm up in spring than their darker counterparts. Even in the greyer soils of the Chalk Marl, otherwise ideal for maize, germination and emergence can be seriously delayed, and many farmers can confirm Ludwig and Harper's (1958) experimental findings to this effect. In the Champagne Crayeuse of northern France, photodegradable plastic mulches have been used with great success to boost emergence and early growth of maize on chalk soils (Ballif and Dutil, 1975). Most of the few English vineyards are located on, or in the lee of, south-facing slopes in the chalk country of Kent, Sussex, Surrey, Hampshire and the Isle of Wight.

On the other hand, north-facing slopes have been used in the south-east for potatoes, mangolds and other root crops which might otherwise fail to bulk-up through shortage of water. Hall and Russell (1911) cite a classic example in the concentration of potato growing along the northern slope of the Hog's Back west of Guildford, where "the cool northerly exposure keeps the main crop varieties growing steadily throughout the summer and ensures a large yield of good quality". Much the same applied to mangolds, while further east a similar combination of topography and soil gave rise to the market gardens of north-west Kent, now much diminished through the expansion of the Wen. The unusually low and flat terrain of the Thanet chalk country is ideal arable land: specialist crops here include early potatoes and broccoli, again mainly for the London markets (Garrad, 1954; Duncan, 1972; McRae, 1973).

Hops, top fruit, rye and sugar beet are examples of crops found in certain areas of the chalk country due to causes other than climate. Thus, while most of the Kentish hops come from the Weald, in the east of the county some are grown on soils of the Lower Chalk and deeper Coombe Deposits of the scarp and valley bottoms. A similar situation prevails with top fruit, which also impinges on the Lower Chalk in Berkshire, though more extensively grown there on the adjacent Lower Greensand. Still grown mainly in East Anglia, rye has made a comeback in parts of the Wessex chalklands within range of the Ryvita factory at Poole, and sugar beet is grown on the Lincolnshire and Yorkshire Wolds and other parts of the eastern chalk country with access to British Sugar Corporation factories.

9.3.4.3 Topography
Apart from the effects on microclimate just mentioned, steeply undulating topography can discourage arable cropping—indeed any mechanised farming operation—by interfering with the efficient use of machinery. In the

"Land Use Capability Classification" of the Soil Surveys of England and Wales and of Scotland (Bibby and Mackney, 1969), six categories of slope are recognised (Table 9.11). Obviously the steepest grades are simply too dangerous to tackle (and may even be too steep to graze with cattle): Bibby and Mackney regard 25° as the limit to mechanised operations without specialised equipment, but 15° as the limit to normal arable farming. Slopes greater than 15° are costly to cultivate and unsuitable for normal rotations, remaining in grass for long periods. "Slopes greater than 20° are difficult to plough, lime and fertilise and even if these dangers are accepted costs are large and normal rotations impossible. Between 20° and 25° occasional tillage for pasture improvement is sometimes practised" (Fig. 9.12).

Table 9.11

Slope classes recognised in the Soil Survey's Land Use Capability Classification [a]

Gradient	Description
0–3°	Gently sloping
3–7°	Moderately sloping
7–11°	Strongly sloping
11–15°	Moderately steeply sloping
15–25°	Steeply sloping
>25°	Very steeply sloping

[a] From Bibby and Mackay (1969).

Even "moderately steeply sloping" ground creates problems, however. It can be ploughed, but uphill work is difficult, even with four-wheel drive and crawler tractors, and it is customary to plough downhill only, running or reversing light back to the top at the end of each run. Either method makes for heavy wear and fuel consumption, and in prolonged rainy spells in autumn and winter even chalkland may suffer from sheet erosion (E. W. Russell, personal communication). But even if these problems are accepted (or circumvented by adopting direct-drilling techniques), the crop still has to be harvested, and manoeuvring loaded trailers and especially combine harvesters without accident or spillage becomes increasingly difficult at grades steeper than 10° (though combines modified for cross-slope working are available). Bibby and Mackney note that on "moderately sloping" ground (3–7°), "difficulties may be experienced with gapping machines or mechanised weeders, precision seeders and some root crop harvesters". To these might be added the frustrations of accurate "tram-lining" (Section

9.3.4.8) on sloping ground. Most of the regularly cultivated chalk soils fall into the Soil Survey's Land Capability Class 2 (some Lower Chalk soils are Class 1), and where gradient is a limitation, the suffix g is added. An example of this category is shown in Fig. 9.13.

Fig. 9.12. Bembridge Down, Isle of Wight, during reseeding operations in the spring of 1976. (C.J.S.)

9.3.4.4 Stoniness

Another limiting factor in chalkland arable farming is the occurrence of flints and chalk stones which locally may be so thickly strewn as to virtually hide the soil surface (Fig. 9.14). In moderation, this can be an advantage, for stones aid drainage and cut down heating and water loss at the soil surface by reflecting short-wave solar radiation and by forming a mulch. But in large quantities stones have the important disadvantages of causing excessive drainage, of taking up valuable soil volume which could otherwise be occupied by roots, of causing costly wear and tear on tyres, ploughshares, tines, discs and cutter bars, and of blocking up seed drills and interfering

Fig. 9.13. Land-use capability categories applied to the chalklands. In the foreground, well-drained shallow chalky loams (Class 2s) allow arable cropping; in the background, moderately steep slopes (Class 4g) preclude regular cultivation and other operations. From Bibby and Mackney (1969). (J. M. Ragg, via Soil Survey of England and Wales)

with the harvesting of root crops, especially potatoes. Stoniness is a soil limitation grouped with shallowness in the Land Capability Classification and is signified by the suffix s.

Fig. 9.14. Flints on the surface of a cultivated Icknield soil. The rule is 30 cm long. From Jarvis (1973). (M. G. Jarvis, via Soil Survey of England and Wales)

9.3.4.5 Water relations

Farmers have long known well enough the difficulties of maintaining grass and growing crops on the thinner soils of the Chalk, particularly on the dry downs of Kent. Defoe's reference to the barren chalk soils of Kent has already been mentioned. Hall and Russell (1911) stress that "The great difficulty of East Kent farming upon the Chalk is to retain moisture enough in the soil, and all the cultivation operations must be directed to that end." Nevertheless, many of these early writers on farming, particularly Ellis (1745), Young (e.g. 1769, 1813) and Cobbett (1830), were emphatic that crops suffered from drought less on chalk than on many contrasting soils. "The chalk takes care of all" was one of Cobbett's favourite quotations. A modern equivalent of these older writings is provided by Bunting and Elston (1966). Stressing the high standard of arable farming on the chalk soils of south-east England, they quote the experience of a Blewbury farmer "who told us in 1958 that on his land crops wilted less frequently in dry summers on chalk soils than on the traditionally 'moisture retentive' clays in the vale below them". A. A. Hudson's comments to this effect were noted earlier. Clearly the key to this paradox must lie in the extent to which plants are able to exploit the large reserve of water so tantalisingly close to their roots (Chapter 3). Arthur Young (1769) was obviously aware of this when he commented during a tour of Salisbury Plain: "White earth is very dry, sound corn land, very shallow; for the pure chalk is at a few inches of depth, and the surface varies only from it in its dryness".

The key to successful establishment is to encourage deep-rooting (provided the crop has the genetic potential for this) so that in due course plants can draw on water and nutrient reserves at depth once the surface layers dry beyond the available range. This is plain enough, as is the most straightforward means of achieving it: autumn sowing. Winter crops rarely run into summer water stress on chalkland.

With spring-sown crops, unless of course irrigation is available, which is uncommon, the situation is more difficult. Effective consolidation by rolling is essential for good water relations, especially in the puffy rendzinas, though this can sometimes be exasperatingly difficult to achieve because a wave of soil builds up ahead of the roller (Cope, 1976). Wheat and oats are naturally deep-rooting, and this can be encouraged by the incorporation of fertilisers, but other crops, notably beans (which ironically are particularly well suited to calcareous soils) and maize, are inherently shallow-rooted and fail frequently on chalk soils in a dry season. Spring barley sometimes fares badly for the same reason. This is well illustrated in the opening remarks of a recent report of Bridgets Experimental Husbandry Farm on the Hampshire Chalk (Jones, 1976), in which the overall yield of grain from the 1975 spring barley crop was $5 \cdot 25$ t ha^{-1}, with the best fields producing over $6 \cdot 3$ t ha^{-1}.

> This was in a season when many heavy-land farmers were reporting disastrous yields due to late drilling and drought conditions in the early growing months. These high yields were typical of crops drilled in good time on chalk soils and indicate the value of timeliness for maximum production.

The situation is most critical for crops sown at the height of summer, such as ley grasses and kale. A recent innovation has been to mechanise the old gardener's trick of injecting water into the soil with the seed as it is drilled, and this has shown promising results on the dry soils of the Kentish Chalk.

9.3.4.6 Fertiliser responses of crops on chalk soils

We have already noted (Chapter 3) the relatively low levels of available N, P and K inherent in chalk soils, and the influence of water on the supply of these nutrients to plants. Usually, however, any major nutrient deficiencies can be corrected with appropriate fertilisers and, as suggested in the preceding paragraph, chalkland farmers can expect reasonably satisfactory yields of cereals in most years, at least on the deeper soils (Gardner, 1959; Jones, 1967, 1976; Carter, 1970). Cooke (1967) has suggested that the residual effect of sustained dressings of P has made further applications of this element unnecessary, at least for the time being, and this is certainly in accord with recent findings from spring barley trials at some 40 chalkland sites in southern England reported by Johnson et al. (1977): the value of the average response to P was "much less" than the cost of the fertiliser. Johnson's team also noted a response to K in Andover soils which could improve the efficiency of N utilisation, but that an unknown factor limited the response to fertiliser N on Icknield soils (see below).

Magnesium deficiency is sometimes reported in cereal seedlings on the non-humic and more chalky rendzinas, though symptoms are usually transient, and the application of Mg fertiliser has no significant effect on grain yield (Cope, 1976). As on other soils, applications of all the major nutrients, as well as magnesium and sodium, can improve the performance of sugar beet, but yields are generally rather poor on rendzinas and brown calcareous soils over chalk (Webster et al., 1977), and there is evidence to suggest that nematodes may be involved here (Section 9.3.4.8). A relatively large demand for sulphur in the brassicas, particularly oil-seed rape, has been reported from studies on chalk soils in northern France (Radet, 1958).

Micronutrient deficiencies and responses have been reported on chalk soils. For example, pears, apples, plums and soft fruits grown on Gore and Wantage soils on Lower Chalk on the southern fringe of the Greensand orchard belt of Berkshire–Oxfordshire have shown classic symptoms of iron deficiency, as might be anticipated (Jarvis, 1973), and Cope (1976) reports both iron and manganese deficiencies on Gore soils in Wiltshire. Manganese deficiency may be a factor in the incidence of potato scab. Current trials of oil-seed lupins confirm the strongly calcifuge nature of this crop on chalk, especially in cultivars of the narrow-leaved Lupinus angustifolius. The seeds germinate readily enough, but the plants make little growth once their food reserves are exhausted, and the foliage becomes grossly chlorotic. Maize has long been known to have a higher requirement than most crops for zinc, but although zinc deficiency is widely associated with calcareous soils in the U.S.A. (Olson and Lucas, 1966), this has yet to be conclusively demonstrated in Britain (B. Payne, personal communication).

Of course, the exact response of any crop depends greatly on such matters as season and previous cropping and fertiliser programmes, on the incidence of pests and diseases, and on local variations in soil—good and bad—which may or may not be easily explained, but which are known intimately to the husbandman. Perhaps the most frustrating feature of chalk soils is their unpredicatability: a bumper crop one year may be followed by a meagre one the next—or vice versa—for no obvious reason. Thus in a series of experiments in which kale was grown on an Icknield soil, T. Batey (personal communication) recorded a remarkably high dry-matter yield of $12·3$ t ha^{-1} in 1964, but only $7·9$ and $9·5$ t ha^{-1} from the same site in 1965 and 1966. Johnson et al. describe similar examples from their spring barley trials mentioned above. Unpublished results of experiments on chalkland sites by ICI and ADAS show that cereal crops may exhibit symptoms of K or P deficiency (the former particularly following a grass crop used largely for conservation) where soil analyses indicate adequate supplies of these nutrients, so that it may be extremely difficult to predict and advise on fertiliser needs. Such problems may well prove insoluble simply because, as in the investigation of natural communities, so many variables are involved. The sheer complexity of the situation facing farmers and advisors is well illustrated by the curious phenomenon of melanism in wheat crops on newly ploughed chalk downland, and the role of copper in its control.

It has been noted (Hooper and Davies, 1968) that wheat crops grown on Icknield-type humic rendzinas some seven to ten years after reclamation from chalk downland frequently give poor yields of grain, which cannot be explained by the incidence of pests or pathogens, nor by major nutrient deficiencies. Indeed the condition is exacerbated by additional fertiliser N, especially in mineral form, as also by a preceding crop of kale or other brassica. In the worst cases, the plants produce rat's-tail ears containing grossly shrivelled grain, if any at all, and exhibit symptoms of melanism: dark olive-green pigments develop on the upper part of the culm, especially between the nodes, and on the glumes. Control has been obtained by a single application of copper sulphate or copper oxychloride foliar spray, or by applying copper sulphate to the seedbed in larger amounts.

9.3.4.7 Monoculture, break-crops and the incidence of disease

Barley has always been the traditional chalkland cereal crop, particularly for malting. For this purpose grain of the best quality, containing less than about $1·5\%$ by weight of N, can be expected, in contrast to samples of higher N content which are acceptable as feeding barleys. But other reasons account for the notable rise of barley-growing on the Chalk in the 1950s and 1960s (Coppock, 1971). Firstly, its relatively low price encouraged the introduction of barley beef and an increase in pig production. In some systems pigs were (and still are) kept out-of-doors on leys which themselves grew in rotation with the barley (Fig. 9.15; see also Boddington, 1972). Secondly, traditional rotations were being widely abandoned for continuous cereal cropping, to which barley was found particularly suited on account of its much lower susceptibility to the stem-rotting fungal diseases take-all (*Gaeumannomyces*, formerly *Ophiobolus, graminis*) and eye-spot (*Cercosporella herpotrichoides*), compared with wheat (Jones, 1967).

Fig. 9.15. Pigs in the open on the Lower Chalk below Swyncombe Down, Oxfordshire. (C.J.S.)

The barley era is passing, however. The economic climate has changed again, and many disadvantages of continuous cropping have come to light: soil compaction and structural deterioration, and intractable pest, disease and weed problems (see below) have all dealt a timely reminder of the economic and ecological wisdom of diversity. Loss of soil structure occurs to a smaller extent on chalk than on heavier, less calcareous soils, but compaction is a problem whenever cultivation with heavy machinery has to be carried out in sticky conditions (MAFF, 1970). The Upton series is the most susceptible of the chalk rendzinas to compaction (Cope, 1976), but this condition has been noted even on a humic Icknield soil on the Bedfordshire Chalk by Batey and Davies (1971), where part of an experimental barley crop had to be sown during wet weather. Final grain yield in this plot was only $1 \cdot 5$ t ha^{-1}, compared with $3 \cdot 2$ t ha^{-1} from the rest of the field which had been drilled a week earlier in a spell of more favourable weather. Although there were differences in the exact timing and method of incorporation of fertiliser, the authors conclude that compaction was the main cause of this poor yield. The outcome of this complex combination of economic and agronomic factors is that wheat has begun to creep back as the main chalkland cereal crop, for example on the downs of Sussex (Wyatt, 1971) and Wiltshire (Cope, 1976), and everywhere, whatever the relative balance between barley and wheat, a partial return to rotations is seen in the quest for break-crops.

On mixed farms (in the sense that one or more livestock enterprises are run alongside the arable), the obvious choice is a grass-break, laying down short- or longer-term leys, sometimes by undersowing the last straw crop, for grazing and for conservation as hay and silage. Long ago, Somerville (1914, 1916) showed marked increases in oat yields following slagged pastures on thin chalk soils, and by now the effects of well-managed grass–clover leys in restoring soil structure and fertility are too well known to need elaboration (see Clement and Williams, 1964). Yet it was necessary to stress the role of the ley in the Agricultural Advisory Council's report, "Modern Farming and the Soil" (MAFF, 1970), as though this were a novelty. One or two crops of wheat typically follow restorative leys on chalk, though precautions have sometimes to be taken against leather-jackets and other soil-borne invertebrates (see below).

An important property of the break-crop on the all-arable farm is that it should be possible to sow, raise and harvest it without recourse to specialised machinery. Oats and rye are ideal in this respect, and have the advantage of low susceptibility to take-all and eye-spot, although in the chalk country oats are very susceptible to cereal root eelworm (see below) and should not be grown where this is known to be a problem. Crown rust of oats (*Puccinia coronata*) might be expected to occur on chalk in particular, since the alternate host of this fungus is buckthorn (*Rhamnus catharticus*), a common component of chalk scrub (Chapter 7), but the disease does not seem to be very common, and fortunately no one has yet suggested the eradication of buckthorn as an economic necessity. Arable-break grass crops are sometimes grown for seed, a practice which is especially popular in Hampshire. Perennial ryegrass, timothy and cocksfoot all yield well on chalk; the last two have the disadvantage of ripening at the same time as the bulk of the cereals, but this problem has been overcome by the use of desiccants (such as Diquat) which permit earlier harvesting (Dowse, 1967). Partridge-rearing can be satisfactorily accomplished in a herbage seed crop (FWAG, 1973).

The relatively small geographical area in which grain maize will do consistently well has already been mentioned, but, whether grown for grain or for forage, maize has one particularly valuable asset as a break-crop in that it possesses a very high resistance to the potent triazine herbicides such as Atrazine: it actually detoxifies them internally. This makes maize ideal for the control of couch-grass, which is susceptible to the triazines, although after a dry season the herbicide may persist in the soil, necessitating either a second maize crop, beans (which are moderately resistant), or a winter fallow. Maize has not been outstandingly successful on the thinner chalk soils, however. Low spring soil temperatures and the possibility of zinc deficiency have already been mooted, but poor performance of maize seems mainly due to the inability of its roots to reach adequate depth before the surface soil dries out.

Outside the Gramineae, the most widely grown break-crop on the chalk is now oil-seed rape; other oil crops include linseed and sunflower. Kale and other leafy brassicas are alternatives to leys where stock are kept, and can double as cover for game birds. Buckwheat (*Fagopyron esculentum*) is occasionally grown specifically for the latter purpose. Among the legumes, beans are edaphically ideally suited to chalk soils, though notoriously susceptible to a plethora of ills. Spring beans, for example, are very prone to drought as well as to the ubiquitous black bean aphid (*Aphis fabae*), while winter-sown varieties contract infection by the chocolate-spot fungus (*Botrytis fabae*) for which there is no effective control. Vining peas are sometimes grown on contract for canning and freezing. In race-horse country such as the Berkshire Downs, sainfoin and lucerne have long been grown for sale off the farm, and always do well on chalk.

Roots, which may locally constitute a major enterprise rather than just a break-crop, mainly include potatoes and, in the eastern counties, sugar beet. The classic plague of potatoes on chalk is common scab, caused by the actinomycete *Streptomyces scabies*. The traditional cure is to incorporate copious quantities of organic manures, while recent experiments on irrigating potatoes have demonstrated the importance of favourable water relations during tuber formation. These treatments appear to operate through a rather complicated interaction between antagonistic actinomycete and bacterial populations, in which the availability of soil manganese may also be involved (Rogers, 1969; Lewis, 1970).

9.3.4.8 Chalkland pests

Several invertebrate pests characteristic of (though not exclusive to) chalkland crops have already been touched on in the foregoing sections. The most serious pest of cereals is probably now the grain aphid (*Sitobion avenae*), though other aphids occur, such as the rose-grain (*Metopolophium dirhodum*), grass (*M. festucae*), blackberry (*Sitobion fragariae*) and, provided its alternate host *Prunus padus* grows in the locality (Section 8.2.1), the bird-cherry aphid (*Rhopalosiphum padi*). All of these damage cereal plants by feeding, and in addition the grain, rose-grain and bird-cherry aphids are known to transmit barley yellow dwarf virus (BYDV), which affects wheat and oats as well as barley (MAFF, 1973). Where the level of infestation warrants it, spraying, for example with Dimethoate, can be undertaken, either from aircraft or from tractor-mounted booms driven along "tram-lines"—gaps left in the crop for just such late-stage operations. Another classic chalkland pest, the cereal cyst eelworm (*Heterodera avenae*), has declined somewhat in the past few years, partly due to the introduction of resistant cultivars such as the barley Tyra, and partly because the nematode has fallen prey to fungal pathogens such as *Verticillium chlamydosporium* and species of *Entomophthora* which infect the cysts (Graham and Stone, 1975; L. E. W. Stone, personal communication). Cereals following a grass-break or downland ploughing are liable to damage by leather-jackets and wireworms (respectively the larvae of crane-flies— *Tipula* spp.—and click-beetles—*Agriotes* spp.), as well as by slugs.

As for pests of the other major arable crops, blossom beetles (*Meligethes* spp.) can be a nuisance in oil-seed rape and other brassicas grown for seed, by destroying the flowers in the bud stage. The larvae of seed weevils, such as *Ceutorhynchus assimilis*, hollow out the developing seeds and encourage the ingress of the brassica pod midge (*Dasyneura brassicae*) which causes premature ripening and seed-shedding. Control is essential, but more difficult than with the cereal pests because of the danger to bees which pollinate the crop, as well as to other beneficial insects. The potent insecticide Azinphos-methyl, highly effective against these brassica pests, is particularly poisonous to bees, and it is essential to treat the crop before the flowers open (MAFF, 1971).

Regarding sugar beet, wireworms have always been particularly troublesome on chalk soils (MAFF, 1972), but an interesting discovery of recent years is that the slow and stunted early growth of beet commonly seen on chalk rendzinas and brown calcareous soils in East Anglia appears to result from heavy infestations of the spiral nematode *Helicotylenchus vulgaris*, though further work is needed to confirm this (Dunning *et al.*, 1977).

Finally, we should not overlook the rabbit. Following its recovery from myxomatosis (Chapter 6), this animal continues to ravage crops (Fig. 9.16), especially in winter, when journeys of up to 1 km may be undertaken daily from the burrow. With fewer hedges and larger fields, rabbits may inflict relatively less damage than when they had a denser network of hedges from which to operate, but their ability to reduce the plant population of winter wheat by 80% and to delay ripening by 18–20 days (Gough and Dunnett, 1950) can hardly be overlooked. Under the Pests Act, 1954, every landowner or occupier is responsible for either destroying wild rabbits on his or her land, or for preventing them from doing damage (MAFF, 1975).

9.3.4.9 Weeds of chalkland agriculture

Whether in arable land or pasture, weeds represent no more than the first stages of colonisation of disturbed ground: they are pioneers, in this case of secondary succession (see Chapter 4). The appearance of weed species in established pasture is seen by the enlightened grassland farmer as signs of its deterioration, demanding improvement or replacement. Wells (1967b) traced the reversion of a sown ley on chalk at Strawberry Down, Kent, to an increasingly "natural" sward. Five species had been sown: Italian and perennial ryegrasses, cocksfoot, and white and red clovers. Predictably, the first had disappeared by the end of the second season, and the last contributed relatively little herbage to the ley; perennial ryegrass, cocksfoot and white clover made up most of the bulk of the sward over the eight years that records were kept, though by the end of this period only the perennial

Fig. 9.16. Rabbit damage to spring barley on the fringe of the old rifle range at Kimble, Buckinghamshire. (C.J.S.)

ryegrass was holding its own, the cocksfoot and particularly the white clover having declined appreciably. During this time, nearly 50 plant species appeared in the ley, among which *Agrostis stolonifera*, already noted as a characteristic invader of ploughed downland, increased most markedly. Typical arable weeds in the early years were *Poa annua*, *Cerastium holosteoides*, *Geranium molle*, *Polygonum aviculare* and *Stellaria media*, while species more characteristic of grassland were *Medicago lupulina*, *Plantago lanceolata*, *P. media*, *Potentilla reptans*, *Prunella vulgaris*, *Ranunculus bulbosus*, *Taraxacum officinale* and *Veronica chamaedrys*, as well as several grasses. A number of these had been among the 38 species recorded by A. S. Thomas in the original sward before it was ploughed, but this had been composed mainly of plants more exclusive to permanent downland, of which only nine reappeared in the ley (*Brachypodium pinnatum*, *Bromus erectus*,

Festuca rubra, Koeleria cristata, Poa compressa, Trisetum flavescens, Cirsium acaulon, Linum catharticum and *Thymus drucei*); and of these only *B. pinnatum* was showing signs of increasing. *Leontodon hispidus* invaded the ley apparently from outside the study area. Conversely, *Carex flacca, Lotus corniculatus* and *Poterium sanguisorba*, formerly abundant on Strawberry Down, had not appeared by the end of the eighth year.

Arable weeds predominate in newly established crops of all kinds though under the influence of grazing or cutting these normally disappear within the first year or two, and to all intents and purposes they can be regarded as being restricted to land which is regularly disturbed by cultivation. Of course, these have declined dramatically in extent in the last 50 years. When Brenchley conducted her classic surveys of the weed floras of contrasting soil types in Norfolk, Bedfordshire, Hertfordshire and Wiltshire (Brenchley, 1911, 1912, 1913), she recognised over 100 species as exclusive to, characteristic of, or common on chalk soils under arable cultivation. Individual species reflect this decline: the corncockle (*Agrostemma githago*) is a commonly quoted example, but others are the beautiful pheasant's eye (*Adonis annua*), and the crested cow-wheat (*Melampyrum arvense*). *Adonis* was noted by Salisbury (1952) as "once so common in the cornfields of Sussex and Surrey as to have been collected and sold in Covent Garden market". The semi-parasitic *Melampyrum*, formerly the scourge of southern arable farmers, whence its alternative common name of poverty weed, is now so rare as to warrant deliberate protection as an endangered species (see Chapter 10).

Improvements in marketing and seed-cleaning have been partly responsible for these changes, but the great breakthrough came with the development in the 1940s of the selectively toxic growth regulators, notably the chlorinated phenoxy-acetic acids MCPA and 2,4-D, to which the majority of broad-leaved weeds succumbed (Fryer and Chancellor, 1970). Now, especially in the more intensively cultivated chalklands, the spectrum of weeds has been narrowed to a handful of ubiquitous species, such as cleavers (*Galium aparine*), chickweed (*Stellaria media*), speedwells (*Veronica*) and mayweeds among the dicots, and wild oats (*Avena fatua* and *A. ludoviciana*), blackgrass (*Alopecurus myosuroides*) and couch (*Agropyron repens*) among the grasses. And even these are falling, one by one, to more and more specific (and costly) concoctions of the agrochemist: further phenoxy-acid derivatives such as Mecoprop and Dichlorprop; the carbamates Barban and Triallate; Ioxynil; the triazines already mentioned; and many more (see Fryer and Makepiece, 1977). Whether the total eradication of every last weed is really necessary is an open question, but the battle continues for all that.

Yet (dare one say fortunately?), this is not to say that some variety of chalkland weeds is no longer to be found, and they make an interesting study. Most characteristic, perhaps, are charlock (*Sinapis arvensis*) and poppies, of which the most common species in the south is *Papaver rhoeas*. These are liable to reappear even in land from which they have been excluded for many years because their seeds possess marked powers of longevity. Miss a strip in the routine application of herbicides, and either of these species is quick to take its chance, the familiar and nostalgic stripes of yellow and scarlet showing up as the season advances. Occasionally other combinations occur: a barley field on a steep chalk bank in the Chilterns exhibited in the summer of 1974 a display of white campion (*Silene alba*) and forget-me-not (*Myosotis arvensis*) which would have done credit to any herbaceous border! The practice of direct drilling appears to be encouraging certain perennial weeds (such as *Heracleum sphondylium*) not formerly associated with arable agriculture.

Field bindweed (*Convolvulus arvensis*), toadflaxes (e.g. *Chaenorhinum minus* and *Kickxia* spp.), the attractive hemp-nettle *Galeopsis angustifolia*, field pansy (*Viola arvensis*), fumitories (*Fumaria micrantha, F. officinalis, F. parviflora* and *F. vaillantii*—not easy to distinguish in the field) and wild mignonette (*Reseda lutea*) are still quite common on chalk soils, the last three exclusively so. Other weeds are associated with the heavier soils of the Chalk Marl, such as the corn buttercup (*Ranunculus arvensis*), remarkable for its grotesque fruits. An interesting relationship within the mayweeds has been noted on the Berkshire Chalk by Kay (1971a,b), where *Anthemis cotula* on heavy Wantage-type soil over Lower Chalk gives way to *Tripleurospermum maritimum* ssp. *inodorum* on the less marly soil of the Middle Chalk. *Anthemis arvensis* is another mayweed associated with the lighter chalk soils. These papers by Kay provide a useful summary of chalkland weed ecology. The arable flora of Breckland, of course, is a study in itself.

9.3.5 Concluding comments

Chalkland arable farming, like specialised arable farming everywhere in lowland Britain, still depends heavily on highly capital-intensive equipment and a pharmacopoeia of agrochemicals. The great purge of hedges (which, as we have seen, are less widespread in the chalk country than they are elsewhere) has subsided, but in very few cases indeed are any new ones being planted. With innovations in processing straw, economics may again favour its utilisation as a valuable resource, but there are few signs as yet of any lessening of the practice of burning. Even where livestock enterprises are reinstated, the tendency is for highly intensive methods of husbandry and housing to prevail. In short, there seems little reason to expect any great changes in the appearance of the landscape of the intensively farmed chalklands. However, on many farms and estates there is renewed interest in either leaving untended or planting-up odd corners and marginal banks in the interests of aesthetics, game and wildlife conservation, and even timber production. These issues are taken up in the following section on forestry and again in Chapter 10.

9.4 Forestry on the Chalk

9.4.1 The historical background

At first glance, accounts of the rural economy of lowland Britain in mediaeval times conjure up a picture of enviable harmony between man and nature, not least in the woodland communities, and serve as a reminder that conservation is not, as some might be forgiven for thinking, a creation of the 1970s. Extensive tracts of grassland, scrub and woodland were protected (afforested in the legal sense of the word) as royal hunting preserves, in which the rights of the population as a whole were strictly limited. Cranborne Chase was a notable example on the Dorset Chalk, extending at one time to well over 300 000 ha (Prior, 1968). Outside these royal forests the woodlands were put to more diverse use, usually based on a system of *coppicing* and, through the multitude of skills encompassed in the art of woodcraft, numerous household commodities were provided (see Edlin, 1949; Rackham, 1976).

Coppicing itself entailed cutting the trees back more or less to ground level and allowing a new crop of shoots (poles) to grow up from the stools. Parcels of woodland were felled in rotation, usually every 12–14 years, but ranging from as little as 7 to more than 20 years. In some localities the stumps were maintained at a greater height so that the regrowth was beyond the reach of browsing animals, and here the term *pollarding* was applied. Sometimes selected trees (standards) would be left intact through repeated cycles of coppicing, hence *coppice-with-standards*. On chalk soils, ash and hazel were the usual coppiced species (and see Section 8.5). Hazel coppice with hornbeam standards can be seen at the Weald and Downland Open Air Museum at Singleton, Sussex. Though less amenable to the practice, beech was formerly coppiced and pollarded in the Chilterns (Brown, 1953; Roden, 1968). Not surprisingly, coppicing led to marked fluctuations in the topoclimate and herbaceous vegetation of the woods, which can be gauged by reference to Chapters 2 and 8 (and see Wilson, 1911; Salisbury, 1916, 1918a; Adamson, 1921; Ash and Barkham, 1977).

As time went on, however, continuing inroads by the growing rural population, as well as increasing urban and industrial demands, began to place greater and greater strain on these woodland resources. Ironically, the increasing availability of coal, which might have been expected to relieve some of the pressures on the woods, frequently led to dereliction as enterprises which relied solely on supplying small wood and charcoal for fuel went out of business. With improvements in husbandry of the kind referred to earlier in this chapter, arable cropping became for some owners a more profitable alternative. Defoe (1724) wrote thus of the woodlanders of the north Kent Chalk:

> . . . since the taverns in London are come to make coal fires in their upper rooms . . . what an alteration it makes in the value of these woods in Kent, and how many more of them than usual are yearly stubb'd up, and the land made fit for the plow.

In places, notably the Chilterns (see below), woodlands alternated with arable farming as corn prices fluctuated.

Of course, in typical chalk downland country, where sheep and corn formed the principal enterprises, woodlands had long ceased to form a significant part of the landscape, if indeed this had ever been the case since the original Neolithic forest clearances. During the Middle Ages, the long, narrow parishes of the chalklands (Fig. 9.6) had ensured that most settlements had some access to woodland, but enclosure changed all that. By the beginning of the eighteenth century the poorer inhabitants of the more open champion and downland country were often hard-pressed to find even fuel wood, turning to the gorse and juniper scrub of the steeper hillsides, or to whatever the hedgerows could provide. On a journey through north Kent some 50 years after Defoe, Young (1769) wrote:

> The beauty of all this country . . . is wretchedly hurt by the abominable custom of stripping up all the trees, insomuch that they look like hop-poles . . .; all the timber is ruined, and a very small quantity of faggot-wood gained.

And from the downs near Burghclere in 1825, Cobbett (1830) describes, with characteristic italics, the farm labourer's family "who had to buy their bread *at the mill*, not being able to bake themselves for *want of fuel*". Even as recently as the opening years of the present century, farmers on Salisbury Plain relied on peat carted from the New Forest for their cooking fires and winter warmth (Hudson, 1910).

In a few places on the Chalk, tracts of ancient woodland survive, largely by historic accident, to the present day, though these are mainly on Clay-with-flints, and all have been much altered by generations of management. Such is the case with the beechwoods of the mid- and south-west Chilterns, which contrast strikingly with the sparsely wooded Dunstable and Berkshire Downs on either hand. Although some of these beechwoods are known to have been planted on seventeenth century farmland (Mansfield, 1952), their history as a whole can be traced back fully 700 years. It is fascinating to

.read (Roden, 1968) of the load of 14 000 bundles of firewood sent from West Wycombe to Marlow and thence by boat down the Thames to Southwark in 1218, but even more so to learn that there was a regular traffic in such commodities at that time. In the late seventeenth century, by which time emphasis was almost exclusively on the maintenance of high forest for timber production, Defoe (1724) recorded "a vast quantity" of beeches sent regularly to London, again via the Thames, from Henley and Maidenhead, as well as from Marlow. Access to these markets is regarded by Roden as "probably the basic reason why extensive private woodlands were enclosed and preserved in the south-western and central Chilterns". The celebrated chair-making industry of High Wycombe further encouraged the enterprise (Massingham, 1939, 1940; Edlin 1949). Elsewhere, as at Cranborne Chase, the continued survival of woodlands already preserved for many centuries through their royal heritage was encouraged by heavy demands for copse-ware, particularly wattle hurdles, made by the sheep-fold (Fig. 9.5).

Changing circumstances in the past few decades have greatly curtailed these as economic enterprises, however. A large proportion of our hard-wood requirements is now imported and many beechwoods are maintained as much for amenity as for profit (see Chapter 10). The demise of the traditional downland sheepflocks and the availability of wire netting and electric fences have diminished the demand for copseware, and apart from isolated localities such as Handley in Dorset, where peep-proof wattle fencing panels are manufactured for the horticultural trade (Prior, 1968), or where management of this description is deliberately practised for educational and amenity purposes (as at Singleton), coppicing is now rarely seen. We are jumping the gun, however, and need now to trace developments in modern forestry from its beginnings in the eighteenth century.

9.4.2 Modern forestry practice on chalk soils

9.4.2.1 Establishment

Attention will be directed mainly to beech, for this has always been by far the most widely used species in chalkland plantations. In his classic mono-graph on British beechwoods, Brown (1953) recognises four distinct situa-tions into which the forester may wish to establish beech:

(i) thriving beechwoods, which may contain other species in addition, but in which the encouragement of natural regeneration of the beech can be expected to meet with some success (see Chapter 8);

(ii) derelict woods in which the intention is to replant without extensive felling;

(iii) areas of former woodland, derelict or otherwise, recently clear-felled;

(iv) chalk downland (i.e. grassland and scrub).

9.4.2.2 Replanting existing woodlands

The actual method chosen for the replanting of existing woods, even with beech, may be strongly influenced by amenity or conservation interests: a derelict wood, for example, may have important ecological or archaeological attributes which preclude wholesale replanting, or there may be restrictions on the size of the area which can be clear-felled at any one time on aesthetic grounds (see Chapter 10). But there are good silvicultural reasons too for underplanting a thinned canopy (the shelterwood system) or planting-up small clearings (group planting). Prolonged exposure to the sun is avoided, and protection provided from wind and frost, ameliorations in topoclimate to which beech responds particularly favourably in its establishment and early growth. Moreover, leaf litter is less likely to dry up and blow away, so that maximum advantage can be taken of the favourable conditions of mineralisation created in the surface layers of the soil following the removal of the mature trees.

Clear-felling is, of course, much more straightforward to carry out and undoubtedly has the economic advantage over more piecemeal methods. However, the larger the area felled, the greater are the extremes of insolation, back radiation and evaporation which are likely to occur. Even with special precautions, such as hardening the young transplants by pre-lifting in the nursery, raising in decomposable pots to minimise stress at transplanting, or interplanting with coniferous nurse species (see below), losses from physical factors alone can be substantial. The growth of herbaceous weeds, brambles and scrub may actually counter these effects to some extent, provided the transplants are not smothered in the process. Mature trees left on the perimeter of any clearing inevitably sustain some damage when felling takes place, but this is seen most markedly around the boundary of a clear-felling. Opening up large gaps in previously closed woodland exposes formerly sheltered trees to the full blast of wind, sun and frost. Swathes of wind-thrown beeches are a common sight in such situations, while the bark of those left standing frequently cracks, and in due course may peel away from top to bottom on the side of the trunk exposed to the sun, jeopardising both the survival and economic value of the afflicted trees.

9.4.2.3 The afforestation of chalk downland

It is with the afforestation of chalk downland, however, as distinct from the replanting of existing woodlands, that we really take up the point raised earlier in this section of replenishing the dwindling woodlands of the chalk country as a whole. The objectives of much of the eighteenth and nineteenth century tree planting on the downs and wolds were as much to provide shade for livestock, shelter for crops and farmsteads, cover for game or simply to enhance the landscape, as to provide timber. The visual impact of these downland plantations (see Fig. 7.6)—variously termed clumps, rings, groves, spinneys,* copses and coverts—is vividly captured in the impressionist landscapes of Paul Nash. Perhaps the most famous example of amenity planting is Chanctonbury Ring, a clump of beeches planted around the rim of an Iron Age hillfort on the crest of the South Downs near Wiston, Sussex, in 1760. In places, more extensive plantations were established in those early years of forestry revival, notably in the Yorkshire Wolds (Wood and Nimmo, 1962). Again, beech was usually the predominant species, though many others were tried (Section 9.4.2.6).

In the present century, the afforestation of large tracts of chalk downland in Hampshire and Sussex formed an early project of the newly created Forestry Commission from the mid-1920s (Brown, 1953; Troup, 1954; Wood and Nimmo, 1962). These were the Buriton Forest south of Petersfield (later named the Queen Elizabeth Forest and now the nucleus of the Queen Elizabeth Country Park—Section 10.3.6.2) and the Friston Forest near Eastbourne. Further extensive planting has been carried out, for example above Brighstone on the Isle of Wight, on the Dorset Downs near Blandford and on parts of Salisbury Plain in Wiltshire. With economically more competitive agricultural enterprises on the one hand, and, on the other, demands to conserve the diminishing remnants of chalk downland and scrub that remain, further large-scale afforestation of the downs seems unlikely, although there is renewed interest in tree planting on marginal land on many of the larger chalkland estates (e.g. University of Reading, 1972; FWAG, 1973).

Regarding practicalities, Wood and Nimmo stress that the problems of planting on chalk downland were fully appreciated as long ago as 1808. So were the means of overcoming them, although these typically involved such operations as ploughing, cultivating and even intercropping with cereals where this was feasible, or setting the trees into carefully dug holes or on to inverted turves where it was not—measures which nowadays would be regarded as more appropriate to horticulture than forestry. But then these were times of abundant labour, when such measures could be, and were, afforded; when, in the words of Miles (1967), there was "a regard for the care of each individual tree". It is well known that the young John Goring regularly tended his beloved Chanctonbury beeches, carrying water to them until they were established. Sadly, such intensive care can rarely be contemplated, at least on a large scale, in modern times.

*According to Rackham (1976), the term spinney strictly applies to hawthorn thickets.

Numerous methods of establishment were tried out during the early years at Queen Elizabeth and Friston Forests, ranging in scale from simply dropping the young trees into slots in the turf (*notching*), with or without prior removal of the mat of surface vegetation with a mattock (*screefing*), to complete cultivation with crawler tractor and various ploughs and subsoilers. Even mechanical planting was successfully carried out at Harting Down in 1956. No strictly systematic comparisons were possible, but the main conclusions were plain enough. The value of at least some degree of cultivation was recognised, both in removing existing vegetation and in providing the young trees with a medium in which their roots could develop freely, and not be bunched up near the surface; adequate consolidation should follow any cultivation. Subsoiling was advantageous, especially where hard chalk or chalky head formed the bedrock, but at the same time it was important not to remove the precious layer of topsoil or bring up raw chalk, either of which could enhance the likelihood of nutritional problems. The return of the sward was regarded as a mixed blessing. In its favour it helped to shelter the transplants from excessive exposure and must have given some protection from the hordes of rabbits then present. On the other hand, the grassy vegetation was likely to compete with the young trees for water and nutrients and to encourage voles (Chapter 6). Moreover, the application of fertilisers, particularly N and K, with the intention of correcting major deficiencies, only served to encourage these more demanding grasses. The use of the herbicide Dalapon, strongly selective against grasses and able to reduce this competition without recourse to soil disturbance, was recognised as a particularly significant step forward, especially on the thinner chalk soils, though it had to be applied with care as it could kill young trees.

Since Wood and Nimmo's report was published there have of course been important developments in equipment for preparing the ground, in methods of propagation and in herbicide technology. Experiments are now in progress on Salisbury Plain to assess the relative merits of such techniques as ripping, a mechanical treatment which loosens hard chalky subsoil without bringing it to the surface, pre-treating tree seedlings and transplants, and applying combinations of N, P and K fertilisers with herbicides such as Paraquat and Atrazine after planting (Fourt, 1973, 1975). Such investigations take time, however, not least in forest research, and it will be some years yet before a detailed assessment of this work is available.

9.4.2.4 Pioneer and nurse crops

The suitability of species other than beech for planting on chalk is discussed briefly at the end of this section, but mention must be made here of the widespread practice of using pioneer and nurse crops both in planting-up clear-felled woodland and in downland afforestation. A pioneer crop is established several years before the main crop; a nurse crop is interplanted simultaneously with the main crop. The latter is the method most commonly adopted. The same species are employed for both purposes, and for convenience the term nurse will be used here to cover both techniques. Ease of establishment and vigorous early growth are the most important attributes of the nurse crop, for it must (i) protect the vulnerable young beeches (or other main-crop trees) from extremes of exposure, (ii) discourage the return of the sward and (iii) provide economic returns over what would otherwise be a lean period. A fourth advantage is that subsequent removal of the nurse trees ensures the correct degree of thinning at a time when the main crop is ready to take advantage of this. Almost exclusively, conifers (though not larch) meet these requirements most satisfactorily.

Species most widely used as nurses are, in fact, characteristically short-lived on chalk. Though a vigorous invader of the downland sward when given the chance (Section 8.2.1), Scots pine (*Pinus sylvestris*), for example, usually begins to die off within 20 years, typically following signs of acute chlorosis. As Wood and Nimmo put it, "The really important question is not whether Scots pine will eventually fail, but whether it can be relied on to do its job before it does". Corsican and Austrian pines (*P. nigra* vars *calabrica* and *austriaca*) are widely grown as nurse species on chalk soils, and have been particularly successful in the south (Fig. 9.17). Corsican is less susceptible than Scots pine to chlorosis, and Austrian pine is quite tolerant of lime: indeed, Anderson (1950) recognised the latter as "one of the very few trees that is quite at home on exposed chalk and limestone soils even where the soil is dry and shallow". Growth of all these may be rather slow, however. Western red cedar (*Thuja plicata*) is unaffected by chlorosis on chalk, but its shallow root system renders it liable to instability on the thinner soils. The same applies to Monterey cypress (*Cupressus macrocarpa*): Professor Somerville's famous plantation of this species at Poverty Bottom may have been aided by dressings of basic slag (Section 9.3.2.4). The generic hybrid *X Cupressocyparis leylandii*, familiar to gardeners as the ultimate in instant hedges, has demonstrated characteristically easy and vigorous establishment on chalk soils, but there is little information yet on its long-term performance. It should always be borne in mind that all conifers growing on calcareous soils are prone to severe attack by the root-rotting fungus *Fomes annosus*. Unusual nurse species which have proved valuable on chalk, not least because of their ability to fix nitrogen, are gorse (*Ulex europaeus*) and Spanish broom (*Spartium junceum*), and the Mediterranean alder *Alnus cordata*, which has grown so well on War Down, in the Queen Elizabeth Forest, as to warrant consideration as a main-crop species.

9.4.2.5 The final crop

This is not the place to review in detail the principles and practice of good silviculture, but since emphasis has so far been laid on planting, we need to assess the potential of the chalklands for timber production, and to pinpoint the main limitations, as far as these are known, to productivity. Five main aspects can be recognised: social, strategic, genetic, edaphic and biotic. In terms of standing crop, the best returns might, theoretically, be expected from a relatively rapid succession of plantations of fast-growing conifers, suitably tolerant of calcareous soils. The first point to make, regarding social

Fig. 9.17. Beech, planted in 1952 with a Corsican pine nurse crop, on a chalk rendzina in the Forestry Commission's Queen Wood, near Watlington, photographed in 1978. Due to slow establishment, Austrian pine (seen in the background) was interplanted in 1961. (C.J.S.)

limitations, is that, at least in the southern chalk country, amenity considerations alone may exclude this possibility (see Chapter 10), so that returns from conifers have to be regarded only as a bonus procured during the course of establishment of beech or other acceptable hardwoods in the manner just noted, for all that the latter method may be less profitable. Secondly—the strategic point—the greater the cropping potential of a soil, the more it is likely to be under arable cultivation rather than under trees. This is one of the principles advocated by the Centre for Agricultural Strategy (Section 10.4), that forestry should be restricted to lower-grade soils unsuitable for intensive arable enterprises, and it ties in closely with edaphic considerations outlined below.

Beech itself, like other tree species, is graded into quality classes on the basis of such attributes as girth and height for age, and straightness and freedom from branching of the main trunk, all of which determine the volume and value of usable timber harvested (Fig. 9.18). Limitations to productivity which may be genetic in origin are well seen in parts of the Chilterns, where native beech of very poor quality can be found, apparently resulting from sustained extraction of the best trees in the so-called Chiltern Selection System. As early as 1769, Arthur Young refers to the beeches "in the tract of country between Wycomb and Tetford" (presumably Tetsworth,

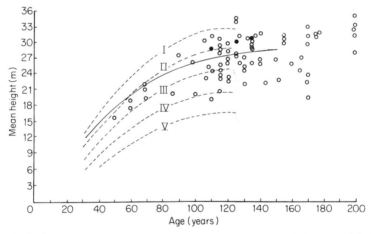

Fig. 9.18. Yield classes for beech (dashed curves I–V) based on Møller's Danish yield tables, against which actual values of height–growth of beech on chalk and limestone soils in Britain are compared. Circles represent data collected by J. M. B. Brown between 1948 and 1950; (o) = one record, (●) = two records; the solid line is a mean of earlier records from temporary sample plots in England, none of which was on a greatly exposed site. From Brown (1953).

just to the north-west of the foot of the escarpment) as being "all small", and appearing "little better than underwood", though this was on soil which was "little else but chalk", and so perhaps as much an effect of substrate (see below) as of bad management. It is, in fact, a large-scale and long-term undertaking to ascertain the extent to which notable traits—good and bad— in forest trees are heritable. It will be interesting to see the results, when they are finally published, of the Forestry Commission's beech provenance trials at Latimer.

Turning, after so many indirect references, to edaphic factors *per se*, it is worth stressing yet again that though beech is almost invariably the dominant species in woods on chalk soils within its climatic range (Chapter 8), it is not particularly favoured by this medium. Growth is sometimes relatively poor, especially on the thinner rendzinas, and greatly enhanced wherever a significant thickness of Clay-with-flints or comparable superficial deposit covers the chalk and contributes to the soil profile (Figs 9.19 and 9.20). Very thin layers of loessal material are sufficient to effect this improvement. The

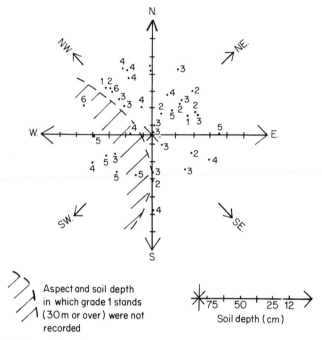

Fig. 9.19. Height of beech at maturity as influenced by aspect and soil depth. Height–range scale: (1) 33 m and over, (2) 30–32·7 m, (3) 27–29·7 m, (4) 24–26·7 m, (5) 21–23·7 m and (6) 18–20·7 m. After J. M. B. Brown, in Wood and Nimmo (1962).

problem does not seem to stem from inadequate mycorrhizal growth (except perhaps on virgin downland), for Harley (1937, 1949) has shown that fungal infection of beech roots is satisfactory on chalk soils; indeed, seedling growth was consistently better on chalk than on the more acid loams tested by Harley. In fact a complex of edaphic and biotic factors seems to account for the relatively poor performance of beech on chalk. According to a critical analysis by Day (1946), water relations and iron supply play a key role, but more recent work invokes N and K deficiency as the major factors (D. F. Fourt, personal communication).

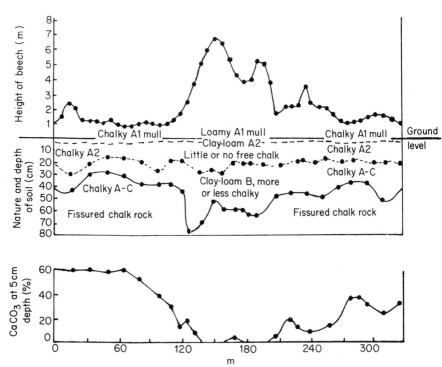

Fig. 9.20. The influence of soil depth and calcium carbonate content on the height of beech at Crawley Down, Hampshire. After D. F. Fourt, in Wood and Nimmo (1962).

It has already been noted that, on the whole, chalk water relations are favourable to tree growth. In the great drought of 1976, there were proportionally fewer losses in Forestry Commission plantations on chalk than from comparable sites on sands and clays. But on steep slopes where the surface soil is thin and the underlying chalk hard, flinty, coarsely fissured or loosened by Pleistocene permafrost, drainage can be excessive and percolating water elusive. Transplants are particularly vulnerable, but even established trees may suffer, if less obviously, for while the root system as a whole may find enough water to maintain a flow to the canopy, individual roots die back, whereupon they are liable to invasion by root-rotting and butt-staining fungi of which *Ustulina vulgaris* is particularly common in chalk soils (Day, 1946). Naturally damaged and diseased roots extract water and mineral nutrients inefficiently, and in exposed situations where litter blows away and soil humus content is low, chlorosis develops which in turn further restricts growth. Hence a vicious circle results. Nor is this all: trees thus weakened are predisposed to attack by the ubiquitous felted beech coccus (*Cryptococcus fagi*), allowing entry of fungi of the genus *Nectria* and consequent development of bark necrosis and canker. Day found beech displaying these combined symptoms of lime-induced chlorosis, root-rot, butt-stain and bark necrosis to be consistently associated with calcareous soils, while trees on adjacent brown earths were largely free of them.

Regarding other biotic factors, the effects on beech seedlings of aphids, slugs and caterpillars, as well as woodland mammals, were noted in Chapter 8. On the woodland fringe and in scrub and short turf, rabbits can do untold damage and, where they occur, are undoubtedly the major cause of death of self-sown and transplanted seedlings and saplings. Moles are sometimes a problem, as they were in the early years at Queen Elizabeth Forest. In longer grass, the short-tailed field vole (Chapter 6) is likely to inflict damage by barking or gnawing off tree stems within the grass mat, and hares may bite off the tips of the main shoots about 30 cm above the ground. Deer cause similar damage and also bark older trees. Although, as already mentioned, the grey squirrel consumes nuts and seedlings, it is between 10 and 15 years after establishment that this animal really comes into its own, stripping bark from trees of all ages, girdling the younger ones and often killing them outright, mutilating the older ones, distorting their growth and exposing them to pathogenic attack as noted above. Only in East Anglia and on the Isle of Wight are chalkland forests completely free from this pest; here the far less destructive red still reigns, and this tends anyway, like the very local *Glis glis* in the Chilterns (Section 8.8.1), to favour the seclusion of coniferous canopies rather than beech. In replanted compartments within existing woodland, wood-mice are sometimes troublesome (Brown, 1959).

As for controlling these pests, rabbits can be kept out by adequate fencing, and their numbers reduced by sustained shooting and gassing of warrens, provided there is not too large a reserve on adjacent land outside the forester's control from which infiltration can continue (Section 9.3.4.8). Fitting plastic collars around young transplants affords additional protection, and not only from rabbits. Where numbers warrant it, the control of deer can form the basis of a profitable link-up between forestry and game conservation. Sentimentality apart, no one would advocate an outright campaign against hares, voles or wood-mice except perhaps in the event of an exceptional plague. After all, these are native species in their natural environment, and much the best means of control from all points of view—economic and ecological—is to encourage a proper balance of natural predators, particularly birds of prey. The grey squirrel is in a class apart, however (Shorten, 1962). The animal has no effective predators in lowland Britain, and the chances of finding an agency as potent and specific as myxomatosis is to rabbits seems remote, even if such measures were deemed morally acceptable. Early, and possibly uneconomic, thinning has been found to greatly retard the progress of squirrels through the canopy; this, though, can have only a very localised effect. Shooting is impossible on the scale which is necessary. Poisoning with Warfarin baits placed in special feeders, as now permitted in lowland Britain under the Squirrel Act, 1973, provided there are no red squirrels in the area, seems to be the only answer, despite the possible threat to other members of the woodland fauna (see Phillipson and Wood, 1976; Phillips and Rowe, 1976). As with the rabbit, the situation is not helped by the appeal of the grey squirrel, despite its ill temper, to the public.

9.4.2.6 Alternatives to beech

Much emphasis has been laid on beech, but in fact a very wide range of timber species has been reported over the years as growing satisfactorily on chalk soils: Pope (1916) grew false acacia (*Robinia pseudoacacia*) for gateposts on his Dorset estate, while Anderson (1950) recommended horse-chestnut (*Aesculus hippocastanum*) as a matchwood substitute for poplars on chalk. At Aston Rowant NNR a fine crack willow (*Salix fragilis*), normally a species of pond-margins and streamsides, thrives on the very crest of the Chiltern scarp, having taken root as a fencing stake in 1923 (T. E. A. Barclay, personal communication). When horticultural species are taken into account the list can be greatly extended, of course (see Stern, 1960; Dyson, 1977), though this is not to say that cultural problems such as those described on the foregoing pages cease to arise. Indeed, some patches of formerly cultivated chalk downland seem incapable of growing anything (Fig. 9.21).

Here, we briefly consider possible alternatives to (or associates with) beech for main-crop, as distinct from short-term use, although there is no hard and fast division between the two categories. Where soil, aspect and a good run of seasons combine favourably, trees originally intended as nurses may do sufficiently well to warrant retention to maturity. Such is frequently the case with both Austrian and Corsican pines (the latter now the most commonly planted conifer in the chalklands), as well as with *Thuja plicata*, and Lawson's and Monterey cypresses. Most of the conifers used more widely in British forestry practice, notably western hemlock (*Tsuga heterophylla*), Douglas fir (*Pseudotsuga menziesii*), lodge-pole pine (*Pinus contorta*) and Sitka spruce (*Picea sitchensis*), require a more acid medium (or heavier rainfall) than is characteristic of chalk soils, though for some quite a thin layer of gravel, sand, clay or loess is sufficient. Others, however, will thrive on the chalkiest of soils as long as they can be coaxed through the precarious stages of initial establishment. Examples are Norway spruce (*Picea abies*) and European larch (*Larix decidua*), illustrated in Figs 9.22a,b. Further details about these species, together with brief notes on many more, can be found in Wood and Nimmo.

Fig. 9.21. Extremely stunted Norway spruce on former arable land near Great Hampden, Buckinghamshire. The trees in the foreground are more than ten years old. (C.J.S.)

Finally, a word or two more about hardwoods, and a little food for thought. Native oaks, birches, limes and elms, ash, field maple, hazel, gean and hornbeam, as well as introduced sycamore, Norway maple, chestnut, certain alders, southern beech (*Nothofagus*) and others, have all been planted in varying quantities throughout the chalk country, alone or in mixtures, with or without beech, in thinned or cleared woodland or on downland. All have shown merit somewhere and at some time, and demonstrate that on the whole, where soil depth is adequate and the topoclimate not too fierce, chalk soils provide a hospitable medium for a large range of hardwood species. It is important to bear this in mind, for although beech traditionally reigns supreme on the Chalk, and it would be unthinkable to propose its outright substitution, it would seem ecologically desirable, not least for the conservation of beech itself, to aim for greater diversity in the final stand than has become customary. Beautiful as they are, we have seen just how vulnerable the aging Chiltern monocultures and their like now are,

Fig. 9.22. Coniferous stands on chalk soils in the Chilterns photographed in 1978. (a, above) Norway spruce in the Forestry Commission's Halton Wood near Wendover planted in 1929, and (b, right) European larch on the National Trust's Bradenham Estate planted in 1961. (C.J.S.)

and it needs little imagination to picture the consequences of a beech-specific analogue of Dutch elm disease, or a continuation of the insidious takeover by the invasive sycamore. Of course, this is only part of the whole complex question of woodland conservation in the broad sense, and the subject is taken up again in Chapter 10.

10
Conservation

10.1 Introduction

It is manifestly obvious from the foregoing account that, ironic as it may be, virtually every feature of the chalklands of ecological and aesthetic interest results from the intervention, intentional or otherwise, by man. As conditions change, so these features are threatened, and without a deliberate attempt to protect and conserve them, they are bound to disappear in one way or another in course of time. The threats come from two major sources, internally (autogenically) from nature herself, and externally (allogenically) from twentieth-century industrial man. In the first instance, we are seeing no more than the natural course of succession which begins immediately cultivation or grazing by farm livestock or rabbits ceases. This has been described in detail in Chapters 6 and 7, and it is clearly fruitless to expect to conserve any community of plants simply by putting a fence around it, especially if it represents an early stage in the seral sequence. This may seem obvious enough now, but in the early days of nature conservation it was surprisingly often a lesson learned the hard way.

The second, external, source of doom—ourselves—includes the processes of industrial and urban expansion and their miscellaneous trappings (Figs 1.38 and 1.39), intensification of agriculture and forestry (Chapter 9), and the effects of pressures on the countryside from ever increasing numbers of people bent on recreation and relaxation. In this chapter, each of these aspects is examined in turn in some detail.

10.2 External Threats to Chalkland Habitats

10.2.1 Industrial and urban expansion

There are plenty of figures illustrating losses of rural land to industry, housing, road-building and so on, but there is no breakdown on a geological basis. Moreover, much of this is agricultural land, and so already lost in the sense under discussion here. However, an indication of the magnitude of this encroachment is given by Duffey (1974a), who notes that between 1933 and 1961 a loss of over 833 000 ha of "common land, heath and moors" in Britain was accompanied by an increase from 1·3 to 1·9 million ha of urban land. More recent statistics from the Centre for Agricultural Strategy (1976) predict a loss of rural land of some 12 000 ha annually to urban development alone, during the closing quarter of the present century.

A special case is chalk quarrying, particularly on the large scale demanded by cement manufacture, which clearly aims ruthlessly and specifically at the chalk countryside (Fig. 9.3). The vegetation, be it grassland, scrub or woodland, is destroyed, and once the quarry is worked out and filled in or used for refuse deposition or for building, the site can be regarded as a total loss. Yet, there is another side to it. Cement companies typically buy up their territory decades before it is actually quarried, and so may maintain for a long time semi-natural communities which may well be of considerable ecological interest. The unusual wood at Darenth, Kent (Section 8.5) is such an example. Another is the floristically rich chalk grassland at Pitstone, Buckinghamshire, in which the rare field fleawort (*Senecio integrifolius*) grows. It is encouraging to learn that negotiations between the Tunnel Cement Company at Pitstone and Buckinghamshire County Council have led to the exchange of this ecologically valuable chalk bank (now managed by the Berks., Bucks. and Oxon. Naturalists' Trust) for an adjacent block of land of lesser conservation interest. Perhaps Darenth Wood may yet be similarly reprieved. The quarries themselves, of course, harbour unusual and interesting habitats of open chalk and calcareous fen (Chapter 4), which would not otherwise occur in the area, while several geological Sites of Special Scientific Interest (SSSIs, see Section 10.3.2) occur in chalk-pits and quarries.

Regarding more general aspects of industrial and urban development, we noted in Chapter 3 that on the whole the chalk country is relatively little affected by aerial pollution. Nor is there any great concern over the contamination of the vast supplies of ground water in the Chalk by residues of cadmium, lead and other toxic metals which affect certain localities elsewhere in Britain. Vast as it may be, however, this water supply is in increasing demand by domestic and industrial users, and wary eyes are being kept on lowering water tables and, in certain coastal regions, the infiltration of marine ground water (see Rodda *et al.*, 1976).

10.2.2 Expansion and intensification of agriculture and forestry

10.2.2.1 Habitat loss and modification

The immediate consequence of the expansion of agriculture and forestry is, as with urbanisation, the loss of habitats. This is most obvious and direct where old grassland is ploughed up and arable crops sown. It is no less obvious but more gradual in its impact in the case of forestry, at least where this entails the planting of spaced saplings in the turf which itself may take more than a decade to lose its identity.

The consequences of intensification of agriculture depend on whether this applies to permanent grassland or to arable land. In the case of grassland, agronomic improvement without cultivation through the use of selective herbicides, fertilisers or sod-seeding techniques, amounts essentially to habitat loss, though the effect is more insidious than with ploughing. Intensification of arable agriculture, in the sense of more specialised and intensive management of croplands already present, may affect adjacent chalk downland indirectly (though not insignificantly), but there are, of course, important consequences for the flora and fauna of the arable land itself, which, as we saw in Chapter 9, is not witł out ecological interest. The same can be said of a change to more intensive woodland management.

10.2.2.2 Loss of chalk downland to agriculture and forestry

As we saw in Chapter 9, the ploughing of chalk downland in historical (as distinct from prehistoric) times has been going on for at least 300 years. It has happened in fits and starts, largely corresponding with fluctuating grain prices. But the situation is now such that there is a real cause for concern. For not only is the total area of semi-natural chalk grassland greatly reduced from its former extent, but it has become more and more fragmented. Useful data are provided on the subject for the chalklands in general by Blackwood and Tubbs (1970), for Wiltshire by Wells (1975) and Wells *et al.* (1976), and for Dorset by Jones (1973).

Blackwood and Tubbs' survey, which was actually carried out in 1966, was necessarily conducted on a scale which demanded some sacrifice of precision. Thus, areas less than about 2 ha in area were disregarded, and estimations of area were made by vertical projection from maps without allowing for angle of slope, so that the total area of chalk grassland must have been underestimated, since virtually all sites were on steep banks.

Table 10.1

The distribution of chalk grassland in England in 1966 [a,b]

County or topographical unit	Area of chalk outcrop (ha)	Area of chalk grassland (ha)	%[c]	Number of fragments	Number of fragments according to area (ha)					
					2–20	21–40	41–80	81–121	122–162	>162
Dorset	95 936	3390	3·5	145	101	28	10	2	1	3
Wiltshire	188 082	29 599	15·7	529	407	51	34	16	5	16
Hampshire	97 103	2116	2·2	119	95	13	7	1	—	3
Berkshire	75 087	637	0·8	59	50	7	2	—	—	—
Isle of Wight	5 686	862	15·2	24	13	6	2	—	2	1
South Downs (Sussex)	90 396	3480	3·8	117	83	13	11	2	1	7
North Downs (Kent and Surrey)	177 293	902	0·5	94	83	10	1	—	—	—
Chiltern Hills (Beds., Berks., Bucks., Herts., Oxon.)	222 782	821	0·4	59	47	7	4	1	—	—
Cambridgeshire	63 423	69	0·1	3	2	1	—	—	—	—
Norfolk	134 719	18	0·01	3	3	—	—	—	—	—
Lincolnshire Wolds	58 028	91	0·2	12	12	—	—	—	—	—
Yorkshire Wolds	106 288	1595	1·5	61	35	15	5	4	2	—
Total	1 314 823	43 580	3·3	1225	931	151	76	26	11	30

[a] From Blackwood and Tubbs (1970).

[b] The authors' original data have been converted to hectares; the last six columns correspond to 5–50, 51–100, 101–200, 201–300, 301–400 and >400 acres.

[c] Percentage of chalk outcrop under "natural" chalk grassland.

However, the authors were well aware of these limitations, which cannot have greatly influenced the overall conclusions (Table 10.1). These are that the majority of surviving areas of chalk grassland are confined to steep slopes and military training areas. A few sites survive because they have been deliberately conserved, but these are in any case on slopes too steep to cultivate. Outstanding is the vast expanse of chalk grassland still to be found in Wiltshire. Rather less than half of this is accounted for by the military ranges, the rest occurs on the county's numerous scarp slopes and valley sides. Of course, not all these grasslands are of high conservation value. Much of the former arable land of the Imber Ranges on Salisbury Plain, for example, is largely covered by tall, floristically poor grasslands dominated by *Arrhenatherum elatius* and *Bromus erectus*. Other ranges, however, such as those at Porton (Wells *et al.*, 1976) include ecologically rich and interesting grasslands (see Chapter 5).

It is difficult to be sure either of the exact original extent of the virgin downland or of the rate at which the present situation has come about. It seems that very large inundations were made, in fact, about the turn of the eighteenth and nineteenth centuries. Blackwood and Tubbs quote work by Naish (1961) suggesting that the total area of chalk grassland in Hampshire had declined to its 1966 value as early as 1840. Jones' (1973) data for Dorset, collected in the summer of 1972, support this idea (Table 10.2) and point again to the problem of fragmentation. Indeed, in Dorset at least, fragmentation of remaining chalk downland has unquestionably proceeded apace even since 1966 (Table 10.3), though bear in mind that the two sets of figures are not directly comparable, for they were acquired in rather different ways. The vital question arises of the minimal area for a viable chalk grassland ecosystem, and this is taken up later on.

10.2.2.3 The effects of intensification of management

The effects of fertilisers and other economic improvements on the ecology of chalk grassland, and of modern arable agriculture on the flora and fauna of the croplands, have been indicated in Chapter 9. But what about the interaction between the croplands and the remaining natural areas? How much of a threat is really posed to surrounding wildlife specifically by the farmer's armoury of insecticides, herbicides, fungicides, desiccants and growth retardants? Of course, some animals are sufficiently mobile to stray into fields during spraying, but too slow or stupefied to escape. Harmless insects from down and verge may be attracted to flowering crops of rape or beans receiving treatment for blossom beetle or blackfly infestations. But, properly conducted, these agencies ought to have a minimal effect on wildlife, for they are acting upon a habitat already transformed by the plough. The emphasis is on responsibility. Strict codes of conduct exist

Table 10.2

A tentative analysis of changes in the distribution of chalk downland in Dorset, 1790–1972 [a]

Year	Number of separate fragments	Total area (ha)	Apparent loss (ha)	Percentage of initial area	Rate of loss (ha year⁻¹)
1793	—	117 450	—	—	—
1811	40	28 314	89 136	76	4952
1815	—	12 665	15 649	55	3912
1934	—	7810	4855	38	41
1956	52	4960	2850	36	130
1966	145	3390	1570	—	—
1967	70	3326	64	33	149
1972	122	2296	1030	31	206

[a] From Jones (1973). See original reference for sources of data and reasons for caution in interpretation.

Table 10.3

Fragmentation of chalk downland in Dorset by 1972 [a,b]

Total number of fragments	Size category (ha)					
	0·4–20	21–40	41–80	81–121	122–162	>162
122	81	19	15	4	2	1

[a] From Jones (1973).

[b] cf. "Dorset" data in Table 10.1, but note that the first category includes fragments >2 ha.

which are no more than plain commonsense. ("If all else fails, read the label.") The same can be said of straw and stubble burning, activities which, practised as they are at the end of the growing season, should have no more effect on the surrounding countryside than the plough itself. Yet here, accidents (if so they can be called) are regrettably frequent, and many a valuable verge or hedge has been destroyed by too casual an attitude to such matters as fire-breaks and wind direction. The ecological interest of the hedgerows may, of course, already have been reduced to a minimum by drastic "trimming", if not outright removal, and it is these practices above all which have led to controversy between farmers and naturalists.

The more widespread the transformation of grassland to arable, the greater is the likelihood that hedgerows will come to be removed, though some grassland systems are no less demanding of wide open spaces. Initially, hedge removal is neither more nor less significant than the ploughing of the turf, but as field size increases, so the hedgerows which are left assume greater ecological importance. Pollard *et al.* (1974) place this emotive subject on a firm footing, and indicate that public concern for hedge removal may to some extent be an over-reaction. Nevertheless, there is absolutely no doubt about the ecological richness and conservation value of an old hedgerow and its associated bank or verge of downland or "tall meadow" plants (Chapters 6 and 7), and this is especially so in the chalklands which have never been more than sparsely hedged. Nowhere is this more dramatically seen than in the green tracks which traverse the more extensive tracts of chalk downland, or the verges of the long-established trunk roads across the arable prairies of Hampshire, Wiltshire and Dorset. They provide a habitat for plants and animals which would otherwise simply have nowhere to live other than among the crop plants themselves. Railway cuttings and embankments serve a similar purpose. They must, too, form vital corridors for the movement and migration of birds, insects and even plants, and so help to counter the effects of fragmentation, although evidence is needed to support this idea. Way (1973) has reviewed the importance of road-verges in biological conservation, and we touch on this later.

Another controversial aspect of the borderline between farming and conservation is the control of larger pests, of which rabbits, foxes and wood-pigeons are obvious examples. Where game-keeping is an important enterprise, the list of "pests" can traditionally be extended almost indefinitely, although to be fair the new generation of keepers seems to display more generally the ecological insight of the enlightened minority of their predecessors. In many places, loss of crops to rabbits is quite as bad as it ever was before myxomatosis (Fig. 9.16), and it is, of course, from hedgerow, scrub and down that the rabbits emanate. With the demise of government support for the rabbit clearance societies, the situation can only worsen. The badger raises a no less serious issue for beef and dairy farmers in those areas where the animal has been found to carry bovine tuberculosis, although so far this problem has missed most of the chalk country; it is almost entirely restricted to the south-western counties, and there mainly to Gloucestershire, Avon and Cornwall (MAFF, 1976a).

10.2.3 The effects of public access

10.2.3.1 Chalk downland

Chalk downland attracts large numbers of visitors for a variety of reasons (see Goldsmith and O'Connor, 1975). Clearly the majority seek the simple joys of attractive scenery and open air. A reasonably impressive or at least pleasing view, and an expanse of short grass on which to walk, cavort, sit or picnic, or from which to fly a kite, are the essentials. If there is a focal point, such as the monuments at Coombe Hill and Box Hill, or the white horse at Uffington, or the cross at Whiteleaf, so much the better. Additional attractions include hang-gliding and grass-skiing (Fig. 10.1). For many visitors,

Fig. 10.1. Grass-skiing is a novel sport for which the chalk country provides ideal slopes. This scene, at Turville Hill, Buckinghamshire, demonstrates the importance of careful management, especially regarding the siting of the ski-lift. (J. Boas)

however, a car park and an ice cream van are sufficient. Writing about the results of a series of surveys conducted at Box Hill, Streeter (1971) observes:

> It is now a recognised fact that the behaviour pattern of the urban visitor to the countryside is such that the majority stay within a relatively small radius of the car park or other focal point. The rest of the area constitutes "the view" into which relatively few intrepid explorers venture.

Burton (1966) has also discussed leisure activities at Box Hill. Of course, this is not to suggest that all the places open to the public are like holiday camps, or that only half-wits visit them. Many people, moreover, seek the solitude of the quieter places (which are often only round the corner or over the brow from the popular ones), and except for collectors (see below) their effects upon the environment are largely confined to wearing down, and perhaps widening, the main footpaths. The purpose of this section is to look more closely at examples of downland which are subject to relatively heavy visitor pressure, to gauge the effects of the visitors, and to focus on those features which are most prone to damage. Only then can schemes be formulated to manage these sites on a basis which ensures both their protection and their continued attraction, though as we shall see, certain sacrifices are often unavoidable.

The effects of public pressure on any particular spot will vary according to the exact nature of the soil and vegetation, but they depend mainly on the number of visitors per unit area, how these are dispersed in time, and what people do while they are there. A great deal of harm can be done in half a day by a single, dedicated collector bent on uprooting orchids or netting blue butterflies. Spooner (1963) records that one private collector was found to have over 770 specimens of the British form of the large blue butterfly in his possession. Happily, whatever can be said in favour of the positive contributions of collectors, and however much the detrimental aspects of their activities have been exaggerated, the basic urge to possess which is at the heart of collecting seems to have diminished (Ratcliffe, 1967; Morris, 1967b), or at least to have been transferred to other pursuits. Attention is nowadays focused more on discouraging the indiscriminate picking of wild flowers, but in the more intensively used areas those species most susceptible to picking are likely to disappear anyway before long (see below) and it seems as unnecessarily drastic to discourage the innocent gathering of the commoner flowers as to ban blackberrying. Fire, of course, is a potential hazard which one careless individual can all too easily unleash, while among the new generation of treasure-seekers, anyone indifferent to plant ecology can do irreparable damage, especially to the rich swards of the old earthworks which it is such a temptation to probe.

Most damage to the flora and fauna comes, however, simply from the action of wheels, hooves, feet and bottoms, either in one short-lived blitz (as in a well-publicised one-day grass-skiing event, or gymkhana), or through less intense but long-sustained usage through all seasons and kinds of weather. An excellent illustration of the short-and-sharp type of impact is provided by Burden and Randerson (1972) (see also Burden, 1970), who studied the effects of an unprecedented volume of pedestrian traffic upon chalk grassland (as well as Clay-with-flints woodland) in what was then the Surrey Naturalists' Trust's Nature Trail at Ranmore on the North Downs near Dorking. The nature of the survey enabled the effects of trampling to be monitored from the very start, and recovery was observed afterwards. Longer-term studies have been conducted by Sankey and Mackworth-Praed (1969) and Streeter (1971) in the North Downs, and by Chappell et al. (1971) in Hampshire.

It is a fact of observation that where trampling is light, the sward may be only minimally affected. Indeed, a major problem of public access is the interference with, or preclusion of, proper grazing management so that away from the main paths and picnic spots, coarse grasses and scrub rapidly develop. In these circumstances, regular pedestrian traffic can be beneficial, maintaining the turf in a short and floriferous state, much as though it was still grazed by sheep. Rabbits may supplement the effects of human trampling by cropping the fringes of the main paths.

But where visitor pressure is greater, the effects become destructive rather than favourable. Duffey (1974b, 1975) has shown that the invertebrate fauna of the field layer is the first to suffer, and its decline may long precede more obvious changes in botanical composition (Table 10.4). The

Table 10.4

The influence of light treading on the numbers of some invertebrate groups in grassland litter [a]

Group	Control (mean of 22 samples)	Trampled (mean of 25 samples) [b]
Annelida	20·2	18·1
Coleoptera	80·1	12·8
Araneae	11·0	1·8
Isopoda	16·5	0·3
Mollusca	13·5	5·2
Diptera (larvae)	25·0	46·2

[a] From Duffey (1974b). For more detailed treatment see Duffey (1975).
[b] Trampling consisted of "one tread by the foot of a 180 lb (75·3 kg) man twice per day for five consecutive days each month over the period of a year".

heavier the trampling, the more seriously compacted both turf and soil become, and in the most excessively worn parts, especially on the slopes where it is impossible not to twist or slide the foot with each step, the turf is destroyed and the soil exposed. Vehicular and equestrian traffic can greatly accelerate this process, but where sufficiently concentrated, foot traffic alone can be spectacularly destructive, as on the cliff-top path at Box Hill. Finally, gulley erosion begins and on chalk in particular this can result in very rapid deterioration of the site (Fig. 4.8). Once the bare chalk is exposed the area of erosion spreads (producing the familiar "giant's footprints"), for wet, bare chalk is extraordinarily slippery, and it is a natural reaction to skirt any such exposure in wet conditions. Similarly, as gulleys develop, horses and vehicles create new tracks.

Exact changes in species composition vary, naturally, from site to site, and there are interactions with season too. Few detailed records are available over any really long period of time, though it has long been known that certain species are characteristic of heavily trampled ground, largely regardless of soil type (see Bates, 1935, 1938, 1950; Davies, 1936). No one seems to have monitored long-term changes of this kind from the start, but some useful data are provided by Chappell *et al.* (1971) from a study they conducted on the effects of pedestrian and vehicular traffic on a floristically rather poor grassland sward on the Hampshire Chalk (actually over calcareous clay loam) at Farley Mount near Winchester. They defined three types of vegetation (Zones I, II and III—see Table 10.5), representative of three levels of visitor pressure, which it is easy to picture. The typical

Table 10.5

Categories of habitat wear on chalk grassland [a]

Zone	Description
I	Lank, tussocky grass (mainly *Dactylis glomerata* and *Festuca rubra*) with occasional invading scrub and accumulating plant litter.
II	Short sward up to 5 cm high, with no significant accumulation of litter, subject to trampling but showing no signs of soil surface disturbance.
III	Most heavily worn areas, regularly used by vehicles as well as being subject to trampling, excessively but not deeply rutted.

[a] From Chappell *et al.* (1971).

grazed-turf species *Plantago lanceolata, Carex flacca* and *Bellis perennis* were most frequent in Zone II, while essentially similar values were recorded in Zones I and II for *Festuca rubra* and *Trifolium repens.* Species which showed a strong decline from Zones I to III were *Poterium sanguisorba* (entirely absent from Zone III), *Leontodon hispidus* and the moss *Rhytidiadelphus squarrosus.* Interestingly, *Dactylis glomerata* gave similarly high frequency values in Zone I (where its normal coarse growth gave it competitive advantage) and in Zone III (where its flat shoots and folded leaves enabled it to survive the heaviest traffic—see Fig. 6.15). Soil compaction was reflected in reduced pore space and aggregate stability, but no significant differences were found in pH, or in the ratios of C:N, ferrous: ferric iron or ammonium: nitrate N, which might have been anticipated. Predictable differences in the invertebrate fauna of the three zones were detected, and the xerophilic snails *Candidula intersecta* and *Pupilla muscorum* were much more in evidence in both Zones II and III than in Zone I.

Yet, there may be more to it than this, especially where pressure from visitors has built-up gradually over a long period. At Box Hill, a popular venue for nearly 300 years, Streeter (1971) has found evidence not only of an increase in species resistant to trampling, but also of eutrophication, with nutritionally demanding species such as *Lolium perenne* and *Plantago major*, not at all normal components of chalk grassland, forming competitive swards, well able to stand up to quite heavy pedestrian traffic (Fig. 10.2). Chemical analysis confirmed the enrichment of the surface soil, particularly in extractable phosphate. These nutrients undoubtedly come from the more obvious sources of biodegradable organic matter, such as canine (and human) excretory wastes, decomposing buns and sandwiches and so on, but, surprisingly, a significant contribution, notably of potassium, is thought to have come from a century of scattered cigarette ash (Table 10.6).

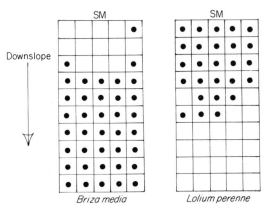

Fig. 10.2. Contrasts in the distribution of *Briza media* and *Lolium perenne* resulting from prolonged visitor pressure in the vicinity of the Saloman Memorial (SM) at Box Hill. Each square represents a 10 m × 10 m quadrat in which the respective species were scored as present (black spot) or absent (absent). Data kindly provided by D. T. Streeter.

10.2.3.2 Public pressures on woodlands

Where they are accessible to the public, woodlands are a great draw, and none more than the chalkland beech hangers. Here, the attraction of the unsurpassed aesthetic beauty of the beeches is augmented for many visitors by the fact that characteristically they do not possess the forbidding aura of deep, extensive forests, or if they do, this feeling can be safely experienced from the edge, where reassuring glimpses of open country are repeatedly encountered. The greatest pressures on woods are seen where these occur alongside, or on the approaches to, an open downland viewpoint, as seen at Box Hill, and on a smaller scale at Whiteleaf Cross, Buckinghamshire (Fig.

Table 10.6

The mineral nutrient content of the ash of a single cigarette [a]

Mineral	mg per cigarette
Nitrogen	1·36
Phosphorus	0·14
Potassium	20·78
Calcium	2·22
Magnesium	2·56
Sodium	0·25

[a] Data from J. Morris (personal communication).

10.3). Here, though the trees themselves may appear to suffer no more than initial-carving by the visitors, the ground around becomes heavily compacted and even churned up in winter, not least by horses and motorcycles, so that percolation by rainwater and satisfactory aeration are impaired. The ground flora, such as it is, is among the most sensitive of all terrestrial herbaceous communities to trampling, and quickly succumbs. Although, as was noted in Chapter 8, regeneration of the trees may be encouraged by a certain degree of trampling, it is but a small step to the stage where this is an unfavourable factor. This is particularly important in view of the aging and diseased state of many amenity beechwoods. The fauna—at least the macro-fauna—is perhaps less affected since, as we have seen, this is not very rich at the best of times, but the destructive grey squirrel is often encouraged, intentionally or otherwise, by visitors, and this can only exacerbate the precarious situation of the ailing beeches.

10.3 Nature Conservation

10.3.1 Attitudes

Before the great upsurge of public and particularly private transport, which resulted in the age of the commuter, access to the countryside was the privilege of the rural population who lived and worked there. By and large, attitudes towards nature were unsentimental: wild plants were used for a whole gamut of household purposes (see Usher, 1974), for some of which huge quantities must have been gathered annually. Certain mammals and birds supplemented the larder, with more providing sport and amusement in customs which were sometimes quite devoid of feeling, if not downright cruel (Sheail, 1976, pp. 3–5). Even Gilbert White thought nothing of shooting down a bird in order to confirm its identity, or to sample its culinary potential:

> On the thirteenth of April [1769] I went to the sheep-down, where the ring-ousels [*Turdus torquatus*] have been observed to make their appearance at spring and fall . . . and was much pleased to see three birds about the usual spot. We shot a cock and a hen; they were plump and in high condition . . . I dressed one . . . and found it juicy and well-flavoured.

Fig. 10.3. Heavy pedestrian traffic on the approach to Whiteleaf Cross, Buckinghamshire, has resulted in bare, compacted soil, a greatly impoverished field layer, and little chance of the regeneration of trees. (C.J.S.)

As recently as the turn of the present century, W. H. Hudson was reporting the annual slaughter of wheatears on the Sussex Downs (Section 6.3.5). Known as ortolans, they were regarded as a great gastronomic delicacy, and were served up by all the fashionable hotels of Brighton and adjacent resorts.

> In July [writes Hudson], the shepherds made their "coops", as their traps were called—a T-shaped trench about fourteen inches long, over which the two long narrow sods cut neatly out of the turf were adjusted, grass downwards. A small opening was left at the end for ingress, and there was room in the passage for the bird to pass through towards the chinks of light coming from the two ends of the cross passage. At the inner end of the passage a horsehair spring was set, by which the bird was caught by the neck as it passed in, but the noose did not as a rule strangle the bird. On some of the high downs near the coast, notably at Beachy Head, at Birling Gap, at Seaford, and in the neighbourhood of Rottingdean, the shepherds made so many coops, placed at small distances apart, that the downs in some places looked as though they had been ploughed. In September, when the season was over, the sods were carefully put back, roots down, in their places, and the smooth green surface was restored to the hills.

Hudson relates the experience of an earlier diarist who recalled catching as many as 156 wheatears in a day, yet who regarded even this as a trifle compared with the claims of trappers of the previous century.

All this is not to say that the beauty of the countryside went largely unappreciated: far from it. But anyone who devoted essentially their full time to botanical, ornithological or entomological pursuits—usually a member of the gentry or clergy—was regarded as at least mildly eccentric. Stokoe (1944) relates the case of Lady Glanville, from whom the Isle of Wight fritillary derives its alternative common name: on her death, Lady Glanville's will was disputed simply because of her passionate interest in butterflies.

10.3.2 The development of the nature conservation movement

This is not the place to review in detail the history of the nature conservation movement, which in any case has been admirably covered by Stamp (1969b) and Sheail (1976). Early milestones were the establishment of the Society (later the Royal Society) for the Protection of Birds (RSPB) in 1889, the Society for the Promotion of Nature Reserves (SPNR) in 1912 (now the Society for the Promotion of Nature Conservation, SPNC), and the first Naturalists' Trust (Norfolk) in 1926. County Naturalists' Trusts which operate in chalkland regions are listed in Table 10.7. In 1949 the Nature Conservancy was founded, an event of enormous significance which marked the recognition of nature conservation as a subject warranting serious scientific research rather than simply a concern of the pressure-groups.

Table 10.7

County Naturalists' Trusts with responsibilities for chalkland reserves

---

Bedfordshire and Huntingdonshire Naturalists' Trust
Berkshire, Buckinghamshire and Oxfordshire Naturalists' Trust
Cambridgeshire and Isle of Ely Naturalists' Trust
Devon Trust for Nature Conservation
Dorset Naturalists' Trust
Essex Naturalists' Trust
Hampshire and Isle of Wight Naturalists' Trust
Hertfordshire and Middlesex Trust for Nature Conservation
Kent Trust for Nature Conservation
Lincolnshire and South Humberside Trust for Nature Conservation
Norfolk Naturalists' Trust
Surrey Trust for Nature Conservation
Sussex Trust for Nature Conservation
Wiltshire Trust for Nature Conservation
Yorkshire Naturalists' Trust

All the Trusts operate under the auspices of the Society for the Promotion of Nature Conservation.

Later came the Council for Nature (1958) and the World Wildlife Fund (1961). An important practical step was the formation in 1959 of the Conservation Corps, at first under the auspices of the Council for Nature, but from 1970 as an organisation in its own right, the British Trust for Conservation Volunteers (BTCV). In 1974 the Nature Conservancy was split into two separate agencies: the Institute of Terrestrial Ecology (ITE), and the Nature Conservancy Council (NCC). ITE, under the auspices of the Natural Environment Research Council (NERC), conducts fundamental research, while NCC, responsible to the Department of the Environment (DoE), is concerned with management aspects of conservation, including the scheduling of Sites of Special Scientific Interest (SSSIs). Though not established specifically for the purpose, the National Trust (1895), and the Forestry Commission (1919) have played an increasing role in nature conservation, as has the National Parks Commission (1949), later (1968) the Countryside Commission, responsible for the National Parks and for designating Areas of Outstanding Natural Beauty (AONBs). The culmination of all these capital-lettered developments has been the recent legislation to protect certain plants and animals which are considered to be in greatest danger of extinction (Table 10.8), though of course this should not detract from the need to ensure the well-being of all our indigenous species and their habitats (see below).

Four categories of nature reserve have emerged: National, Forest, Local and the miscellaneous group which constitutes the rest. National Nature

Table 10.8

Chalkland plants specified in the Conservation of Wild Creatures and Wild Plants Act of 1975 [a]

Daphne mezereum (mezereon)
Cephalanthera rubra (red helleborine)
Epipogium aphyllum (ghost orchid)
Orchis militaris (military orchid)
O. simia (monkey orchid)

---

[a] Under the Act, it is now an offence to uproot any wild plant, and in addition to pick or damage the species listed above (as well as others not found on chalk soils).

Reserves (NNRs) are now graded according to their national, or even international, importance, and those situated on the Chalk are indicated in Table 10.9. These are owned and managed by the Nature Conservancy Council. They are all extensive, some encompassing more than one main kind of habitat and even other soil types, and all are used for research of one kind or another in ecology, pedology, geomorphology and archaeology. Forest Nature Reserves (FNRs) are the responsibility of the Forestry Commission, although none within this category have yet been established on the Chalk. The ambiguously named Local Nature Reserves (LNRs) are strictly the responsibility of district or county authorities. The fourth category includes reserves such as those of the County Naturalists' Trusts, Natural History Societies, colleges and schools. Note that, unless it is within a reserve, the scheduling by NCC of an SSSI demands only that the local authority be informed, lest any planning applications should be made which might threaten the site, and under current legislation the authority can do no more than politely remind the owner or tenant of the interest and value of what he may be about to destroy. Moreover, agricultural and forestry activities, other than the erection of buildings, do not come under the eye of the local authority, and here much depends on the watchfulness and tact of NCC's agents in the field. A useful gazetteer of chalk grassland reserves and SSSIs in southern and south-east England has been compiled by the Kent Trust for Nature Conservation (1973), though of course any such list requires regular updating. This is effected by the Annual Reports of NCC and the Naturalists' Trusts, together with occasional publications such as "BBONT 1959–1969" (Gardiner *et al.*, 1970).

Table 10.9

Chalkland sites of major conservation value listed in the "Nature Conservation Review" (Ratcliffe, 1977)[a]

Region	Site	NCR no. and grade [b]		Comments
Yorkshire Wolds	Bempton and Speeton Cliffs	C63	1*	Chalk cliffs. Botanically unexceptional, but outstanding ornithological interest (Fig. 4.11). Also undercliff areas.
	Waterdale	L130	1	Chalk grassland of the Wolds type.
	Duggleby High Barn Wold	L131	1	
	East Dale	L132	1	Chalk grassland of a more generally typical kind on south facing slope.
Breckland [c]	Weeting Heath, Norfolk	L64	1	NNR. Grassland B, maintained under heavy rabbit grazing and supporting breeding populations of stone-curlew and wheatear. Some arable land with interesting weed flora.
	Lakenheath Warren and Elvedon Heath, Suffolk	L62	1	Grasslands A and B. Lakenheath Warren has been the location for A. S. Watt's classic studies reported in Chapter 6.
Herts. and Beds. Escarpment	Therfield Heath, Herts.	L18	2	Species-rich, typically East Anglian chalk grassland. Renowned entomologically for aberrant forms of the chalk-hill blue butterfly (Lysandra coridon).
	Knocking Hoe, Beds.	L67	1	NNR. Chalk grassland (Festuca ovina type), with rarities Seseli libanotis, Hypochoeris maculata, Senecio integrifolius and Anemone pulsatilla, as well as other notable species.
	Barton Hills, Beds.	L68	1	Chalk grassland (Bromus erectus type), with hawthorn and hazel scrub. The location for important experimental work by T. C. E. Wells and M. G. Morris

Region	Site	Code	No.	Description
Chilterns	Ivinghoe, Steps and Pitstone Hills, Herts./Bucks.	L55	2	Chalk grassland (mainly *Bromus erectus*) and hawthorn scrub (Figs. 1.22 and 7.11).
	Coombe Hill, Bucks.	L56	2	Chalk grassland (*Festuca ovina*) and mixed scrub, including juniper.
	Ellesborough Warren, Bucks.	L38	1	The most extensive stands of apparently native box in Britain (Fig. 8.14). Notable bryophyte flora, as well as the box-specific lichen *Catillaria bouteillii*. Chalk scree and grassland (*Festuca ovina*). Regenerating subspontaneous walnut (*Juglans regia*) a curious feature. Mezereon (*Daphne mezereum*) in adjacent beechwood.
	Windsor Hill, Bucks.	W30	2	Includes beechwood on chalk rendzina. Adjoining juniper scrub in heavily rabbit-infested *Festuca ovina* grassland managed by BBONT (Fig. 7.10).
	Bradenham Woods, Bucks. (including Naphill Common, and Bradenham and Park Woods)	W22	1	Mainly beech and oakwoods on Clay-with-flints, but some chalk grassland, scrub and beechwood.
	Aston Rowant, Bucks./Oxon.	L21/W29	1/2	NNR within L21. Chalk grassland (*Festuca ovina*/*F. rubra*), including the very local grass *Nardurus maritimus*. Some extremely uncommon invertebrates. The location of important grazing experiments. Also scrub and woodlands.
Berkshire Downs	Aston Upthorpe Down	L22	1	Chalk grassland (*Festuca ovina* and *Bromus erectus* types, with some *Brachypodium pinnatum*); herbs include *Anemone pulsatilla*. Good juniper scrub (Fig. 7.6).

Table 10.9 continued

Region	Site	NCR no. and grade[b]		Comments
Hampshire Chalk (excluding western North Downs)	Old Winchester Hill	L25	1	NNR. Chalk grassland (*Festuca ovina* type), scrub (including juniper) and woodland (including yew).
	Burghclere Beacon	L39	1	Chalk grassland with southern-mixed and juniper scrub.
	Rushmore Down	L40	1	
	Stockbridge Down	L54	2	Chalk grassland (*Bromus erectus*) and mixed scrub.
	Martin Down	L24	1	Chalk grassland and several types of chalk heath. Historically well documented.
Wiltshire and Dorset Downs	Bulford Downs, Wilts.	L41	1	Species-rich *Festuca ovina/F. rubra* chalk grassland with regenerating juniper.
	Fyfield Down, Wilts.	L48	2	NNR included. A site dominated by species-poor *Festuca rubra* chalk grassland, but of great archaeological and geological importance.
	Pewsey Downs, Wilts.	L26	1	NNR included. Species-rich chalk grassland (including the rare *Cirsium tuberosum* and its hybrid with *C. acaulon*). Also a small area of interesting woodland.
	Oldbury and Cherhill Downs, Wilts.	L42	1	Chalk grassland (*Festuca ovina*, *Bromus erectus* and *Brachypodium pinnatum* types); another site for *Cirsium tuberosum* and *C. tuberosum* × *acaulon*. Entomologically notable for the rare wartbiter, *Decticus verrucivorus*, a bush cricket (Orthoptera, Tettigoniidae).

Region	Site	Code	Grade	Description
	South-west Wiltshire Downs	L27–36 and others	1/2	Including extremely species-rich chalk grassland of the *Festuca ovina* and *Carex humilis* types, e.g. Wylye Downs, and the notable swards and scrub formations of the Porton Ranges.
	Cranborne Chase, Wilts./Dorset	W32	2	Mixed woodlands on deep chalk soils, including hazel coppice with oak, ash and maple standards.
	Hod and Hambledon Hills, Dorset	L97	1	Species-rich chalk grassland with *Carex humilis*.
	Park Bottom, Higher Houghton, Dorset	L108	2	Chalk grassland. Important archaeological site.
	Eggardon Hill, and Haydon and Askerswell Downs Dorset	L96	1	*Festuca ovina* chalk grassland with good examples of regional and aspect facies, e.g. *Succisa pratensis* and *Leontodon hispidus*.
Dorset and South Devon Coast	Durlston Head–Ringstead Bay	C24	1	Chalk (and other) cliffs and landslips with associated vegetation. Insect fauna includes Lulworth skipper (*Thymelicus acteon*). Classic geological features.
	Axmouth–Lyme Regis Undercliffs	C26/W67	1*	NNR. Chalk (and other) cliffs and landslips with associated vegetation, including ashwood (Fig. 1.24). Notoriously inaccessible. Outstanding examples of natural succession. Classic geological features.
Isle of Wight	Needles–St Catherine's Point	C10		Chalk cliffs, including the Needles (Fig. 1.6), maritime chalk grassland, scrub and undercliff vegetation. Rich lepidopteran fauna, including Isle of Wight wave (*Idaea humiliata*) and fritillary (*Melitaea cinxia*). Landfall for migrants.
	Tennyson Down	L37	1	
	Compton Down	L23		
North Downs	Selborne Hanger, Hants.	W27	1	Beechwood on chalk scarp. Gilbert White's stronghold.

Table 10.9 continued

Region	Site	NCR no. and grade [b]		Comments
	Noar Hill, Hants.	L50	2	Old chalk workings, outstanding for the diversity and abundance of orchids.
	White Downs, Surrey	L12	1	Chalk grassland (mainly *Bromus erectus*); juniper and ex-arable scrub.
	Box Hill–Headley, Surrey	L7	1	Species-rich chalk grassland, scrub and woodland (including box).
	Crookhorn Wood, Kent	W6	1	Beech and other woodlands.
	Halling–Trottiscliffe, Kent	L11	1	Ex-arable chalk grassland; southern-mixed, hawthorn and dog-wood scrub.
	Wouldham–Detling Escarpment, Kent	L10/W7	1	Chalk grassland (mainly *Bromus erectus* and ex-arable swards) and scrub. Beech and other woodlands.
	Purple Hill and Queendown Warren, Kent	L17	2	Herb-rich *Festuca ovina*/*F. rubra* chalk grassland.
	Wye and Crundale Downs, Kent	L3	1	NNR included. Chalk grassland (*Festuca ovina* and *Brachypodium pinnatum* types), scrub and woodland. Numerous rare and local invertebrate species.
	Alkham Valley Woods, Kent	W3	1	Mixed coppice of ash, oak and other species (but no beech except on the margins) on deep calcareous loam over Chalk.
	Folkestone–Etchinghill Escarpment, Kent	L15	2	Chalk grassland (short *Festuca ovina* type and rough grassland including *Brachypodium pinnatum*), scrub and ashwood.

Site	Code	Grade	Description
Folkestone Warren, Kent	C2	1	Chalk cliffs and landslips with associated vegetation. Rich invertebrate fauna.
South Downs (all sites in Sussex)			
Harting Down	L8	1	Chalk grassland (*Festuca ovina*/*F. rubra* with some *Bromus erectus*) and juniper scrub. (Chalk heath on Clay-with-flints capping.)
Kingley Vale	L9/W10	1*	NNR. Chalk grassland and scrub (including juniper), and extensive yew-woods for which the reserve is of international importance.
Heyshott Down	L16	2	Chalk grassland and scrub. Notable for bryophytes of north and north-western distribution in old chalk workings.
Bignor Hill	W8	1	Beechwoods on chalk (and other) soils.
Fulking Escarpment and Newtimber Hill	L19	2	Chalk grassland and scrub; chalk heath and chalk-heath scrub.
Castle Hill	L4	1	Chalk grassland including *Festuca ovina*/*F. rubra* and *Helictotrichon pratense*/*Koeleria cristata* types.
Mount Caburn–Lewes Downs	L5	1	Chalk grassland including *Festuca ovina* and *Bromus erectus* types.
Lullington Heath	L6	1	NNR. Chalk grassland; classic chalk heath and chalk-heath scrub site.
Cuckmere Haven–Beachy Head	C1	1	Chalk cliffs (at 160 m, highest of chalk in Britain) and maritime chalk grassland. Formerly notable for the peregrine falcon colony there.

[a] Some of these sites are National Nature Reserves (NNRs); others are owned or managed by County Naturalists' Trusts; others again are National Trust properties. Note that many of the larger examples include in addition communities on soils derived from strata other than Chalk.

[b] Grade 1* sites are of international importance.

[c] Several other NCR Breckland sites contain calcareous facies.

10.3.3 Factors involved in the establishment of nature reserves

10.3.3.1 Acquiring reserves

Although much care has gone into the establishment of the major nature reserves, many of the smaller ones have been created in a hurry, their plant and animal communities snatched at the eleventh hour from spray-boom, plough or bulldozer, from afforestation, or from irreversible seral change, with little thought of exactly what is to be done with the acquisition. This approach frequently prevails even where matters are less pressing. It has been noted more than once (mainly in Chapters 4–8) that we are still appallingly short of even the most fundamental information on the biology and ecology of our native flora and fauna, so it is hardly surprising that it has been (and still is) common practice to manage nature reserves "by ear". Indeed, there is nothing like acquiring a reserve to turn the spotlights on the gaps in our knowledge. Yet, it is essential to have a clear idea of the role of the reserve from the outset, for only with specific objectives in mind can an appropriate management plan be drawn up. A critical analysis of this subject can be found in the comprehensive "A Nature Conservation Review" (Ratcliffe, 1977), and in the papers presented at the meeting called by the Royal Society to discuss the Review (Smith *et al.*, 1977). Here, we consider the nature and role of the chalkland reserves, the significance of size and diversity, the controversial topic of introductions, and the particular problems associated with managing grassland, scrub and woodland communities for conservation.

10.3.3.2 Types of chalkland nature reserve

It has been common in the past to talk in nature conservation about the naturalness of a community or habitat. On the Chalk, as we have seen, there is little that is truly natural about the grasslands, yet they rank high in conservation value because they represent an immensely rich and interesting biome which has been drastically reduced in extent, and which is still threatened on all sides. Scrub is less artificial in the sense that it reflects the natural course of secondary succession, but since normally it is essentially transient, scrub which is deliberately held at a particular stage of development is not strictly natural. Again, however, the importance of scrub in conservation hardly needs emphasis. Most types of chalk grassland and scrub are important, moreover, for the diversity and rarity of the species they contain. Regarding the woodlands, although these represent the natural climax vegetation, we saw in Chapter 8 that very few tracts can have escaped modification by man, and that the majority have been planted anyway. Here, a rather involved argument has developed, from which the concepts of original-, future-, potential- and present-naturalness have emerged (Table 10.10). As far as the more strictly artificial communities are concerned, the possible need to conserve rare arable weeds was noted in Chapter 9, but this is a project most appropriately undertaken in the botanic garden.

Table 10.10
Categories of naturalness [a]

Original-naturalness	"Actually existed in the past before man became a significant ecological factor."
Future-naturalness	"Would develop if man's influence were completely and permanently removed."
Present-naturalness	"Would prevail now if man had not become a significant ecological factor."
Potential-naturalness	"Would develop if man's influence were removed and the resulting succession were completed in a single instant."

[a] After Tüxen and Küchler (see Peterken, 1977).

10.3.3.3 Size, shape and isolation

The marked decline and fragmentation of our semi-natural chalkland habitats has already been noted. Clearly there must be a limit to the extent to which a community can be thus reduced yet remain stable and viable as a whole. In his study of the dwindling Dorset heaths, Moore (1962) concluded that the smallest viable size of a habitat is that which can support a viable population of its weakest key species. The concept of key species is less easily applied to the more complex communities of Chalk and limestone, but the principle is demonstrated by observations on blue butterfly populations in the Cotswolds (Muggleton, 1973, 1975; Muggleton and Benham, 1975). The large blue (*Maculinea arion*—on the brink of extinction, it will be recalled, and no longer found on the Chalk) is extremely susceptible to habitat fragmentation for reasons which are indicated in the next section, whereas the chalk-hill and small blues (*Lysandra coridon* and *Cupido minimus*) are able to maintain viable populations on very small areas indeed. The same seems to apply to the Duke of Burgundy "fritillary" (*Hamearis lucina*), of which a colony has persisted for many years at BBONT's Park Wood reserve on a patch of herb-rich chalk grassland little more than 20 m² in extent.

Other crucial and strongly interconnected factors are the length of time over which a population can maintain itself, and the extent to which it is isolated from other populations. Continued reinforcement maintains both numbers and genetic diversity (Berry, 1971). The shape of a reserve may also be important here: one which is compact and approaches circularity in shape has a small edge-to-area ratio (perimeter ratio), and is less likely to be encountered by, say, a migrating butterfly than is a more straggly tract. Compactness may be an advantage, however, if invasion, as of chalk grassland by scrub, is not required (see Bradley, 1968). The point has already been made several times that there are marked contrasts between species in their capacity to disperse. The possible importance of hedgerows and verges as corridors of dispersal has also been noted. The subject has strong affinities with island biogeography (Elton, 1958; MacArthur and Wilson, 1967) and the siting of nature reserves has been discussed in this context in general terms by Hooper (1971) and Usher (1973), and in relation to juniper conservation by Ward and Lakhani (1977).

10.3.3.4 Richness, diversity and stability

A common source of confusion is the distinction between species richness and diversity. Species richness is simply the number of different kinds of species in a given area. Species diversity, on the other hand, expresses the way in which numbers of individuals are dispersed between the species present. Many formulae for diversity have been devised, but one which is now widely accepted is the Shannon–Wiener function derived from information theory and first applied to ecology by Margalef (1958). Expressed in terms of species diversity , D, the formula can be written

$$D = - \; \Sigma \; [p_i \, (\log_2 \, p_i)],$$

where p_i is the proportion of the total sample represented by the ith species. Logs to base 2 reflect the original application of the formula to binary data; natural logarithms are sometimes substituted. The same principle can be applied to habitat diversity, an important factor which again is well illustrated by the large blue butterfly: in addition to the larval food-plant and suitable ant host (Section 6.4.4.1), the large blue appears to require gorse bushes and rush spikes for diurnal shelter and nocturnal roosting, respectively (Usher, 1973). It is an axiom of theoretical ecology that diversity and stability are strongly positively correlated, and that both increase towards the climax (though see Section 7.7.1 and 8.9).

Morris (1971a) applied an alternative index of diversity to his extremely detailed analysis of frog- and leaf-hoppers (Auchenorrhyncha) of the grazed and ungrazed chalk grassland plots in Bedfordshire already described in

Section 6.5.2. In fact, he identified no less than 45 species. Using the expression

$$D = \frac{1}{N} \log_2 \frac{N!}{N_1! \, N_2! \, N_3! \ldots N_s!}$$

where N = the total number of individuals and $N_1, N_2 \ldots N_s$ = the number of individuals in each species, s being the total number of species (see Pielou, 1969), he found that the diversity of hoppers was greater in the ungrazed than the grazed plots, but that peak values occurred, respectively, in September and August, when actual numbers both of species and individuals were well past their maxima.

Little more can be said about this aspect of conservation until more data obtained on the same basis are available for comparison, but its application depends very much on the size and role of the reserve. Considerations of diversity (and stability) may not matter very much where regular and carefully planned management is practised, and may indeed be of very limited relevance in small reserves which exist mainly to cater for a very few species which are threatened with extinction (or in some types of woodland—Section 10.3.6). Perhaps this is where the arguable distinction can be drawn between conservation and preservation. There is little that is natural, for example, about the monkey orchid populations of Kent and Oxfordshire, which are painstakingly hand-pollinated every spring. Long may these plants flourish, but their well-being results as much from skilled horticulture as from conservation at its most sophisticated. Much the same can be said regarding the intensive care of the critical egg, larval or pupal stages of rare or local insects by caging and supplementary feeding.

10.3.3.5 Introduction of species and restoration of habitats

It is a small step from the examples noted in the previous paragraph to intentional seeding or transplanting, or to the releasing of insects or even birds (such as the Porton Down bustards). Yet this is a controversial aspect of nature conservation. The deliberate introduction of alien species or alien races of native species is, of course, a separate matter, and emphasis here is on the transfer of native plants or animals, either raised in captivity or collected from a location where they are relatively abundant, to a site where they are declining, or have disappeared, or even where they have not been known to occur but might reasonably be expected to have done (see Streeter, 1967; Frazer, 1967). Transfers of this kind have met with mixed success, however, and have been mainly confined to populations threatened by destruction or seral change. In Buckinghamshire, for example, about 400 fragrant orchids (*Gymnadenia conopsaea*), together with a number of junipers, were rescued from an old chalk pasture scheduled for ploughing and successfully transplanted in the adjacent BBONT reserve at Kimble Rifle Range. In a similar exercise, a dwindling population of the rare meadow clary (*Salvia pratensis*) was removed in its entirety to the Dancers End Reserve, though here the transplants are barely holding their own and have never flowered in seven years. It is essential, of course, that the Biological Records Centre and county museums are informed of moves of this kind. In an unusual and very successful operation in Oxfordshire, cuttings were taken from the many junipers in the path of the M40 motorway extension, and these were subsequently returned as rooted plants to an area of Aston Rowant NNR very close to their original site.

Notwithstanding the initiative and sheer hard work with which projects such as these have been tackled, a lobby which is gaining strength suggests an altogether bolder strategy in which the wholesale restoration of floristically rich communities, and even the creation of new ones, are proposed (see Bradshaw, 1977). Emphasis would be on maintaining species which are well within their geographical range. Ideal situations for this approach are to be found in the reseeding of worked-out quarries, road workings and verges, and sites restored from a state of serious degradation (Sections 4.6, 6.6 and 10.2.3.1; see also Goldsmith and O'Connor, 1975; Streeter, 1977). Plants and seeds would need to come from known provenance, the latter ideally from seed-banks of the kind envisaged by Thompson (1974).

10.3.4 The management of chalk grassland for conservation

10.3.4.1 Grazing

The main objective in managing chalk grassland is to maintain the plagioclimax and control the invasion of scrub. Ideally, this is effected by continuing to graze the turf, preferably with sheep. A flock of non-breeding Border Leicester–Cheviot ewes has been run on the Aston Rowant NNR for more than a decade and much has been learned about grazing habits of the sheep and the effects of stocking at different rates and times of year (Hawes, 1971; Duffey *et al.*, 1974; Lowday and Wells, 1977). Stock should be moved around the reserve if the area is sufficiently large in relation to the size of herd or flock; a put-and-take system has to operate if the reserve is smaller. In the latter case it is often possible to borrow stock as appropriate from a neighbouring farm. Stocking rate must be adjustable to allow for seasonal fluctuations in the availability of herbage, but this will vary according to the height at which it is required to maintain the sward. Internal fencing and water supplies are essential in management of this kind, and it is not always appreciated how extraordinarily costly fencing is, especially if, as is usually the case, it is necessary to include rabbit-proof netting, which has to be buried and turned under. The latter is, of course, to keep rabbits in, rather than out.

It is sometimes possible to broaden the interest and run sheep of rare or unusual breeds on reserves. Jacob sheep graze the National Trust's Coombe Hill property. Soays and St Kildas have been run on Aston Rowant. Here, an interesting project was undertaken jointly by the Grassland Research Institute and the Nature Conservancy Council, to investigate the feasibility of using Soays specifically for the conservation of chalk downland (Large, 1976; Large and King, 1978). The animals are hardy, they are fleet of foot and have horns to defend themselves, and they lose their fleece in an annual moult which precludes the need for shearing. An additional advantage is that they appear to browse leaves and buds from woody vegetation such as brambles and thorn bushes, so that they could be used to control even quite advanced scrub, as well as maintaining a close sward.

A further step was envisaged which involved crossing the Soays with a more economically viable breed of downland sheep in order to obtain a saleable lamb crop to offset costs. In preliminary trials at Hurley, Soay ewes were put to either a Soay or Dorset Horn ram. Fears that the 23 kg Soay ewes might have difficulty in producing lambs sired by a 75 kg Dorset Horn ram were allayed by reassurances that dam, not sire, controls lamb size and, as it turned out, the cross-bred lambs were only about half as heavy again as the pure-breds; although the foetal burden was undoubtedly heavy, there were no lambing difficulties. Fed on a high plane of nutrition for four months from weaning, the lambs produced a small but "useful and acceptable" carcase.

Trials on the exposed slopes of Aston Rowant, which unfortunately had to be terminated after only two years, confirmed that the scheme is feasible but drew attention to several problems. The first was the high foetal burden in ewes carrying cross-bred lambs. This might be offset by cutting down stocking rates, by providing vitamin and mineral blocks, by supplementary feeding, or even by removing the ewes from the down altogether during this critical period. It also proved necessary to dose the flock with a drench against helminthic worms, to spray them with a formulation of Crotoxyphos as a protection against head-fly (*Hydrotaea irritans*) which had caused the ewes to rub themselves raw with their horns, and to undertake regular dagging (removal of accumulated dung from the ewe's tails).

Where space permits and concern is for maximum diversity of species, a good arrangement is to have a combination of both closely and more laxly grazed swards within the same area, or at least in adjacent blocks. Some possible schemes are suggested in Chapter 8 of Duffey *et al.* (1974). In moderation, rabbits can help to maintain the more closely-grazed mosaics, but they should be securely enclosed within the reserve, and populations should be prevented from building up to damaging numbers. Shooting rights are reserved in some agreements, and this may cover rabbit control. Weeting Heath NNR in Breckland is an interesting example where a heavy rabbit population has been deliberately encouraged in order to maintain the unique botanical composition of Grassland B (Section 5.3.2). The restoration of its characteristically open structure has encouraged wheatears and stone-curlews back into the area. In smaller reserves it is usually necessary to select either close or more lax grazing regimes according to management priorities, although it ought to be possible to arrange clusters of reserves to serve separate but complementary purposes.

Cattle can be put to graze chalk grassland and are especially useful if an accumulation of rank growth needs clearing off, but as noted in Chapter 6, they are best not used on a prolonged basis and care must be exercised if ragwort or yew are present. The St John's worts (*Hypericum*) present similar problems on grassland of former arable land. New Forest ponies and donkeys have been used at Kingley Vale NNR (R. L. C. Williamson, personal communication), and goats at Dancers End.

10.3.4.2 Mowing

Where it is impossible or impracticable to use livestock, grassland management must be effected by mowing. How this is done will clearly depend on the nature and extent of the terrain. On large areas of relatively flat grassland, the tractor-mounted mower or even flail forage harvester may be practicable. On smaller reserves, or where the slope is very steep, or strewn with flints or ant-hills, lighter machinery such as the motor-scythe or rotary mower, or even clearance by hand with scythe or ripping hook, may be necessary. Where coarse grasses or scrub are present, the cut herbage and toppings should be removed in order to reverse the trend of nutrient enrichment and eutrophication, though this is not always possible; it may result, moreover, in the removal of ripening seeds of desirable species. It is likewise arguable whether cutting accurately simulates grazing: a well-sharpened mower chops rather than pulls off the herbage, for example, and where lighter, air-cushion machines are used, an unconsolidated mat may steadily accumulate. Moreover, though some motor-scythes are notoriously temperamental they are rarely quite so selective about what they remove as are livestock. Nevertheless, it is common practice to employ machinery and to disregard its relatively minor drawbacks.

10.3.4.3 Burning

Although fire is a natural hazard on heath and moorland, where it has long been a traditional method of mangement, it is less commonly associated with grassland communities, at least in temperate climates. Duffey *et al.* (1974)

point out, however, that an annual spring burn (swale) of accumulated litter in Cotswold grasslands dominated by *Brachypodium pinnatum* and *Bromus erectus* is an important factor in maintaining a species-rich turf. This is generally confirmed by Lloyd's (1968) work on limestone grasslands in Derbyshire, although Morris (1975) notes that fire can have more damaging effects on the invertebrate fauna. It was noted in Chapter 9 that burning has proved effective in checking the invasion of chalk grasslands in the South Downs by *B. pinnatum*, and similar results have been obtained on the Wye and Crundale Downs NNR on the North Downs of Kent (see Plate 30 of Duffey *et al.*).

10.3.4.4 Timing of operations
The importance of choosing the right time of year to burn an undergrazed pasture is obvious, but timing is no less important where the sward is grazed or mown. Reference to Section 6.2.2 will indicate the time of year at which particular species would be most prone to damage. Work by T. C. E. Wells (e.g. 1967a, 1971) bears out these generalities. Thus, when domestic live-stock were allowed access in September to chalk grassland where the autumn ladies tresses orchid (*Spiranthes spiralis*) was growing, predictable damage ensued (Table 10.11). Similarly, mowing *Bromus erectus* during its period of peak growth caused it a major setback and enabled certain species which would otherwise have succumbed to its bulk to grow and flower much more vigorously. This was particularly true for *Briza media* and *Hippocrepis comosa*, and the significance of the latter for the blue butterflies needs no re-emphasis. Later in the year, however, *Bromus* proved much more res-ilient to cutting, especially on deeper soils (see Wells, 1971, Figs 3 and 4; Duffey, 1974b, Fig. 11.3). It might be noted in passing that the removal of flowering shoots, at least from perennials, is not necessarily a bad thing. Assimilates may be diverted to enhance vegetative growth, or to build-up food reserves in storage organs, and either way the performance of a population may be boosted in the following year. Whether such treatment can be effected without jeopardising the quality of the sward as a whole, or the invertebrates which depend on it, is another matter however. This is a point which illustrates well the problems of objective reserve management.

10.3.4.5 Scrub control in grassland management
All the foregoing treatments help to control scrub, but more drastic measures may be necessary where encroachment threatens to get ahead of mower or livestock. This is a classic assignment for volunteer forces, either locally organised or arranged through BTCV (Section 10.3.2). Smaller bushes and saplings may be grubbed up or winched out, though this results in considerable soil disturbance which may materially alter the floristic com-

Table 10.11

Damage to inflorescences of Spiranthes spiralis caused by grazing animals [a]

	Grazing animals and year										
	Rabbits (1962)	None (1963)	Sheep (1964)	Bullocks with some rabbits							Heavy bullocks (1972)
				(1965)	(1966)	(1967)	(1968)	(1969)	(1970)	(1971)	
Total number of inflorescences	122	449	313	250	773	640	557	—	—	—	—
Number damaged	12	3	94	56	66	98	90	—	—	—	—
Percentage damaged	10	1	30	22	9	15	16	23	14	20	39

[a] Based on data in Wells (1973).

position of the sward (Chapter 5) and is in any case impracticable with the larger bushes. These are best sawn off at ground level, and the stumps treated with a suitable chemical such as 2,4,5-T or, better, ammonium sulphamate to prevent regeneration. Dogwood is particularly troublesome in this respect, for it suckers vigorously and without chemical control can only be held in check by further cutting (or grazing by donkeys, as at Kingley Vale). Once this control is relaxed, ever more impenetrable thickets result. Curiously, if dogwood bushes are left alone, they often die spontaneously at the age of perhaps 15–20 years, though by this time the botanical interest of the turf is likely to have been lost.

Scrub clearance is a winter job so that growing plants and breeding birds are not disturbed, and the trimmings can be safely burned on the site with minimal damage to the turf and its fauna. Nervertheless, obvious precautions have to be taken. Bonfires must be as restricted in area and number as possible, care must be taken over flocks of overwintering birds such as redwings and fieldfares, and livestock should be prevented from licking treated stumps, which cattle in particular are wont to do.

In the management of chalk grassland, it is arguable whether it is necessary or even desirable to remove every last bush. A scattering of shrubs can provide shelter, resting places and a source of nectar for grassland invertebrates, and the removal of clumps of older bushes is pointless if the result is merely to encourage the growth of eutrophic species of wood-edge, glade and bare ground.

10.3.4.6 Chalk grassland for recreation and amenity

A problem repeatedly encountered in nature conservation is whether or not public access should be allowed. From what has been said so far in this chapter it is obvious that on the whole limited access is better than none (provided it does not interfere with grazing regimes or experiments, or endanger rare species or fragile habitats), but that excessive damage results from heavy human traffic. Moreover, the result is frequently a mosaic of under- and over-trampled vegetation. There is, of course, an entirely different aspect to this problem: to what extent conservation is feasible on areas to which the public have full access and which are not managed primarily as nature reserves. In fact, such areas frequently retain a remarkable wealth of wild plants and animals, but in addition they provide two vital services: (i) they attract people away from areas which may well be more delicate and prone to irreversible damage; and (ii) properly appointed, they provide the best possible means of educating both school children and the public at large in countryside ethics and conservation. The Queen Elizabeth Country Park is a superb example within the chalk country of how informal recreation and education can be thus combined (see Section 10.3.6.2). The special problems of managing grassland in country parks, and the extent to which conservation can be encouraged, are lucidly reviewed by Lowday and Wells (1977).

10.3.5 The management of scrub

The deliberate management of scrub in its own right is one of the most challenging aspects of practical conservation, not least because, apart from hedge-trimming and possibly coppicing, the deliberate maintenance of such a community has no equivalent in traditional rural economies, and so is very much in its infancy as an aspect of applied ecology. Unlike grassland, scrub cannot be regularly trimmed, nor, except in rare instances, can it be left to itself as might be appropriate to a climax community. Dense scrub, more-over, lacks the general aesthetic appeal of either grassland or woodland, and is a favourite target of the amenity societies. Even naturalists haggle over whether scrub should be cleared in favour of the grassland species it threatens to eclipse (Section 7.2.2), or left to attract a rich bird fauna. Scrub conservation above all demands clear objectives in reserve management. A useful chapter on the management of scrub can be found in Duffey *et al.* (1974) so it is necessary here only to touch briefly on the subject.

A scrubland reserve may cater primarily for one species in particular, or it may contain a mixture of species. In either case, of course, it may form part of a more comprehensive system including open grassland or woodland communities. Juniper is an example of essentially single-species scrub (see Chapter 7). So is box, though this is one of the few species which appears to be able to maintain itself: hence its treatment as a climax community in Chapter 8. Mixed-scrub reserves are more habitat-orientated in the sense that the actual species represented are of secondary importance either to physiognomy or to the herbaceous element, though if relatively uncommon species happen to be included, such as spindle or guelder rose, so much the better. In either of these situations, unwanted bushes and trees can be selectively removed as desired. However, where space permits, a much more satisfactory arrangement is to aim for a system in which mosaics of scrub and open grassland alternate with one another, for not only is maximum ecological interest assured, but management is simplified in that a rotation can be devised of the kind shown in Fig. 10.4. The time-scale will vary according to site and species, and to how advanced a stage of scrub is required: juniper would be given a relatively long rotation, for example, as would a mixed system in which a pioneer woodland stage was required. In practice, other factors, some inevitably unforeseen, arise to influence the best laid out of plans. An example of such a plan is shown for BBONT's Park Wood Reserve in Fig. 10.4b.

(a)

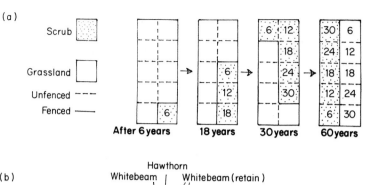

Scrub ▢   Grassland ▢   Unfenced ---   Fenced —

After 6 years   18 years   30 years   60 years

(b)

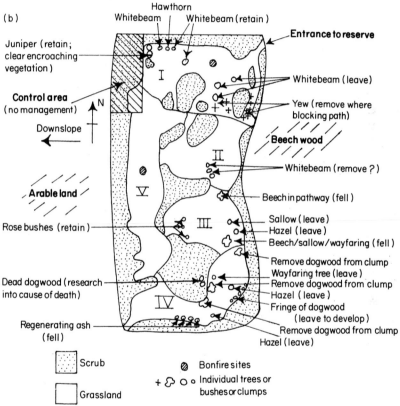

Hawthorn

Whitebeam — Whitebeam (retain)

Entrance to reserve

Juniper (retain; clear encroaching vegetation)

Control area (no management)

N

Downslope

Arable land

Rose bushes (retain)

Dead dogwood (research into cause of death)

Regenerating ash (fell)

Whitebeam (leave)

Yew (remove where blocking path)

Beech wood

Whitebeam (remove?)

Beech in pathway (fell)

Sallow (leave)
Hazel (leave)
Beech/sallow/wayfaring (fell)

Remove dogwood from clump
Wayfaring tree (leave)
Remove dogwood from clump
Hazel (leave)
Fringe of dogwood
(leave to develop)
Remove dogwood from clump
Hazel (leave)

▢ Scrub   ⊘ Bonfire sites

▢ Grassland   + 🜚 O o Individual trees or bushes or clumps

Fig. 10.4. (a) Hypothetical 30-year scrub rotation using 10 blocks of land on a 6-year cycle over 60 years. Numbers indicate the length of time (in years) that each block will be allowed to progress to scrub, or will be maintained under grass after scrub clearance. See Duffey *et al.* (1974), Chapter 11. (b) Preliminary (1972–1976) management plan for BBONT's 1·25 ha chalk/grassland/scrub reserve at Park Wood, Buckinghamshire. Shaded areas represent whitebeam and mature mixed scrub to be managed on a long-term basis, while compartments I–V were to be cleared of scrub after 1, 2, 3, 4 and 5 years, so that an overall mosaic of grassland, young and advanced scrub would be achieved. Encroachment of coppiced dogwood threatens to jeopardise this scheme, however.

10.3.6 Woodland conservation

10.3.6.1 Objectives

The conservation of woodland, more than any other biome, is susceptible to a variety of interpretations. Four main approaches can be identified: economic, aesthetic, historic and ecological. The first concerns the proper management of woodland as a renewable resource, and should form the basis of any good forest enterprise. Spencer (1972) makes this point very nicely in his reminder that "in most forest services in the world, including our own, it is no accident that the senior professional officers are given the title of Conservator". Economic aspects of woodland management have already been outlined in Chapter 9, but are briefly returned to here in relation to the multiple use of woodlands. The aesthetic approach reflects our appreciation of trees and woodlands in the landscape, and our concern over such matters as unimaginative block-planting, particularly of conifers, indiscriminate felling, and the continuing ravages of beech-bark and Dutch elm diseases. This is the recreation-and-amenity department, but of this, too, more shortly.

The subject of woodland history has recently been put on firm foundations in an absorbing review by Rackham (1976). As has already been noted in earlier chapters, however, virtually all the chalklands have long been cleared of their original forest cover, and even where existing woods do have a long history, they have been robbed of most of their ancient features, either by generations of management for large timber or, more drastically, by replanting with species more responsive to intensive methods. Coppicing survives precariously in a few localities: it would be pleasing to see the practice expand again, not only as a demonstration of the disappearing skills of woodcraft, but in the interests of the rich flora which develops under coppice management.

Ecological reasons for conserving woodlands are various, but include the maintenance of continuity, naturalness and diversity, the protection of rare species, and the study of succession and climax. The first four are closely interrelated in that a "natural" wood (though not easy to define—Table 10.10) ideally contains, as well as mature trees, a proportion of over-mature, dying and dead ones, with gaps and glades, areas of scrub, and invading pioneers. Such structural complexity ensures maximum diversity of woody, herbaceous, epiphytic and heterotrophic plants, mammals, birds and invertebrates, rare and local species included. Yet, as we saw for chalk grassland, the smaller and more fragmented our remaining woodlands become, the greater is the difficulty of maintaining this diversity. As Peterken (1977) puts it, ". . . non-woodland habitats, which were formerly islands in a woodland matrix, now form the matrix for the wooded islands". Consequently the most insignificant shelterbelt, grove or overgrown hedge may warrant conservation on ecological grounds, if only as a bridge between woodland fragments.

Finally, almost any piece of woodland may be of value in providing sore-needed data on the dynamics of climax ecosystems and on mechanisms and rates of succession—*vide* Wytham Wood and Broadbalk Wilderness, respectively. A rare example of natural woodland succession observed from scratch is seen on the very fringe of the chalk country in the ashwoods of the Axmouth–Lyme Regis undercliff (Fig. 1.24), but much can be learned from careful observation even in plantations. Roberts Wood, a beechwood established on former arable land near Kings Ash, Buckinghamshire, in 1937, and generously donated to BBONT in 1975, has already surprised local naturalists with its dense carpets of yellow birds' nest (*Monotropa hypopitys*). A long-term study of permanent 10 m^2 quadrats (located on National Grid intersections) has recently been set in train in a joint venture between Oxford University and the Forestry Commission, in the Commission's Wendover Woods.

10.3.6.2 The multiple use of woodlands

There are plainly many interests, some apparently conflicting, to be satisfied in woodland conservation, but it is now generally accepted that, with forethought and care, these can largely be reconciled. Indeed, woodlands have always served a wider range of functions than the purely utilitarian, and forestry and wildlife conservation in particular have long been broadly synonymous. Williamson (1970) has shown how quite simple measures can enrich the bird-life of plantations (Fig. 10.5); merely installing nestboxes can effect a remarkable transformation once the birds have got used to them (Williamson, 1968; Baron, 1971). These and other aspects of catering for wildlife in managed woodlands are discussed by Steele (1972b,c) and others (see Steele, 1972a).

Regarding recreation and amenity, which of course includes the observation of wild animals and plants, there is now a greater demand than ever before for our forests to provide relaxation and enjoyment for the public. Although originally established to attend to the nation's strategic reserves of timber, the Forestry Commission soon found itself involved in wider interests, and the first Forest Park was established in 1936 (Sheail, 1976). It is in the last 15 years, however, that the most rapid developments have taken place, the Countryside Act of 1968 having provided an additional push. Facilities range from modest picnic sites and forest walks through provision for riding and camping, to ambitious projects of which the Queen Elizabeth Country Park, launched jointly by the Forestry Commission and Hampshire County

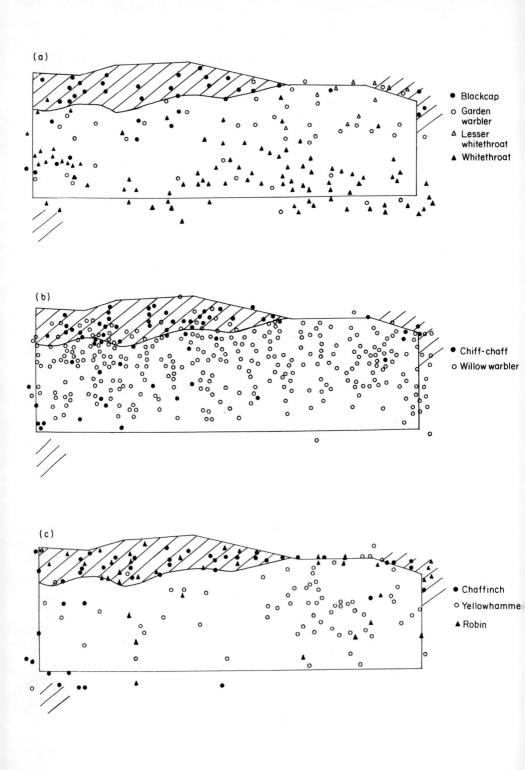

(a)

- ● Blackcap
- ○ Garden warbler
- △ Lesser whitethroat
- ▲ Whitethroat

(b)

- ● Chiff-chaff
- ○ Willow warbler

(c)

- ● Chaffinch
- ○ Yellowhammer
- ▲ Robin

Council in 1976, is a prime example to be found in the chalk country. As already noted in relation to open downland, such undertakings serve a vital role in environmental education (see below), as well as guiding the majority of the visitors away from more sensitive areas on the honeypot principle. Most developments of this kind have, obviously, to fall to public bodies such as the Forestry Commission, the National Trust and county and district councils, as well as the newly formed Woodland Trust. Private owners wishing to open their woodlands to the public are eligible for grant aid under the Countryside Act, but these are often better able to cater for more specialised requirements such as shooting (again, see Steele, 1972a).

As far as aesthetics are concerned, almost all practitioners of forest management are under some sort of obligation to maintain pleasing landscapes, particularly if these fall into areas designated as National Parks (of which none have so far been declared on the Chalk) or Areas of Outstanding Natural Beauty (AONBs). In the Chilterns AONB, for example, restrictions are placed on the proportion of softwood species planted in relation to hardwoods, as well as on the size and shape of clear-fellings and the frequency and timing of such operations (Chilterns Standing Conference, 1971).

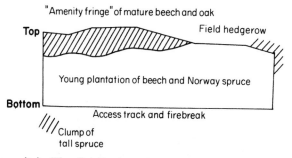

Fig. 10.5 (opposite). The distribution of breeding bird pairs in a plantation of Norway spruce and beech on Upper Chalk at Homefield Wood, Buckinghamshire, in 1969. Each point represents a record of territorial bird-song, and song-registrations for all visits are combined. The importance of both the "amenity fringe" of mature beeches and the grassy ride for species diversity is very clear. Features of vegetation are shown in the small diagram above. The area of the plot was about 10 ha. After Williamson (1970).

10.4 The Future

In the past few years, the approach to conservation has become more objective and less naïve. It has been accepted as a discipline warranting serious scientific study, though it has to be admitted that a wealth of research even in basic natural history, let alone applied ecology, waits to be done. It is also recognised that there is more to conservation than managing nature reserves and that the subject cannot be isolated from politics, economics and sociology (Warren and Goldsmith, 1974). This broad approach highlights the need for an overall strategy for rural resource management which, while encouraging maximum efficiency and productivity in agriculture, forestry and the extractive industries, permits, through adequate financial support, the maintenance of aesthetically pleasing and ecologically rich landscapes, in which protection from and access by the general public are appropriately balanced (Countryside Commission, 1974; Centre for Agricultural Strategy, 1976; Nature Conservancy Council, 1977).

Fig. 10.6. Children from Walters Ash Middle School engaged in simple ecological studies near Park Wood, Buckinghamshire, on the National Trust's Bradenham Estate. Properly organised and conducted, activities of this kind are of inestimable value in developing skills of observation and expression and in fostering informed and sympathetic attitudes towards the rural environment and its conservation. (C.J.S.)

In the final analysis, however, everything hangs on attitudes. There can be little doubt that the situation has improved since the turn of the century (Section 10.3.1) and that a strong conservationist ethic exists, particularly among the young. Following the lead of the Field Studies Council, our educational system has at last come round to recognising this. Environmental studies now form a significant component of middle and secondary school curricula (Fig. 10.6), while an increasing range of certificate, diploma and degree courses in conservation is being mounted by polytechnics and universities, through which better integration of the so-called land-linked disciplines can be anticipated (Newbould, 1974; Ireland, 1975). In the field of adult and community education, regular television programmes on natural history, ecology and conservation, of which the quality and educational value are unsurpassed, reawaken latent feelings for the countryside, prompting further enquiry and active involvement. These in turn are provided for by such facilities as evening classes, nature trails, forest and country parks, as well as voluntary work of the kind already mentioned. Farm holiday schemes help to keep the working countryside in perspective: they encourage a realistic attitude towards conservation, and can do much to restore the traditional good relations between the general public and the farming community. Farmers, after all, will always be the guardians of the wider countryside, and in their turn are by no means as indifferent to conservation as is sometimes suggested (MAFF, 1976b). Those of us in a position to do so must ensure that all these encouraging trends are sustained. For only an ecologically informed population can begin to face the infinitely greater issues with which the human race has now to come to terms.

References

Abeyakoon, K. F. and Pigott, C. D. (1975). The inability of *Brachypodium sylvaticum* and other species to utilise apatite or organically bound phosphate in calcareous soils. *New Phytologist* **74**, 147–154.

Adamson, R. S. (1921). The woodlands of Ditcham Park, Hampshire. *Journal of Ecology* **9**, 114–219.

Allison, F. E. (1955). The enigma of soil nitrogen balance sheets. *Advances in Agronomy* **7**, 213–250.

Allorge, P. (1921–1922). Les associations végétales du Vexin français. *Revue Générale Botanique* **33, 34** (16 separate papers). See Stott (1970) for details.

Anderson, M. L. (1950). "The Selection of Tree Species". Oliver and Boyd, Edinburgh.

Anderson, Violet L. (1927). Studies of the vegetation of the English Chalk. V. The water economy of the chalk flora. *Journal of Ecology* **15**, 72–129.

Anslow, R. C. and Green, J. O. (1967). The seasonal growth of pasture grasses. *Journal of Agricultural Science, Cambridge* **68**, 109–122.

Arber, Muriel A. (1940). The coastal landslips of south-east Devon. *Proceedings of the Geologists' Association* **51**, 257–271.

Arber, Muriel A. (1973). Landslips near Lyme Regis. *Proceedings of the Geologists' Association* **84**, 121–133.

Archibald, E. E. A. (1949). The specific character of plant communities. I. Herbaceous communities. *Journal of Ecology* **37**, 260–273.

Arnold, P. W. (1958). Potassium uptake by cation exchange resins from soils and minerals. *Nature, London* **182**, 1594–1595.

Arnold, P. W. (1960). Potassium-supplying power of some British soils. *Nature, London* **187**, 436–437.

Ash, J. E. and Barkham, J. P. (1976). Changes and variability in the field layer of a coppiced woodland in Norfolk. *Journal of Ecology* **64**, 697–712.

Ashcroft, W. (1901). Report of an experiment for ascertaining the influence of various manures upon the production of mutton. *Journal of the Bath and West and Southern Counties Society, 4th series* **11**, 141–143.

Ashcroft, W. (1910). Report of an experiment for ascertaining the influence of various manures upon the production of mutton. *Journal of the Bath and West and Southern Counties Society, 5th series* **4**, 78–89.

Atkinson, R. J. C. (1957). Worms and weathering. *Antiquity* **31**, 219–233.

Aubrey, J. (1685). "Memoires of Naturall Remarques in the County of Wilts." Later (1847) incorporated into "The Natural History of Wiltshire" (John Button, ed.), published by the Wiltshire Topographical Society. Reprinted 1969, with an introduction by K. G. Ponting, by David and Charles Reprints, Newton Abbot.

Austin, M. P. (1968a). Pattern in a *Zerna erecta* dominated community. *Journal of Ecology* **56**, 197–218.

Austin, M. P. (1968b). An ordination study of a chalk grassland community. *Journal of Ecology* **56**, 739–757.

Avebury, Lord (1902). "England's Scenery". Macmillan, London.

Avery, B. W. (1964). "The Soils and Land Use of the District around Aylesbury and Hemel Hempstead", Memoirs of the Soil Survey of Great Britain. HMSO London.

Avery, B. W. (1973). Soil classification in the Soil Survey of England and Wales. *Journal of Soil Science* **24**, 324–338.

Baden-Powell, D. F. W. (1948). The chalky boulder clays of Norfolk and Suffolk. *Geological Magazine* **85**, 279–296.

Bailey, E. B. (1924). The desert shores of the Chalk Seas. *Geological Magazine* **61**, 102–116.

Bakker, D. (1960). A comparative life-history study of *Cirsium arvense* (L.)Scop. and *Tussilago farfara* L., the most troublesome weeds in the newly reclaimed polders of the former Zuiderzee. *In* Harper (1960), 205–222.

Ballif, J. L. and Dutil, P. (1975). Le réchauffement des sols de craie par films plastiques. Mesures et bilans thermiques. *Annales Agronmiques* **26**, 159–167.

Barber, D. (1970). "Farming and Wildlife: a Study in Compromise." Royal Society for the Protection of Birds, Sandy, Bedfordshire.

Barber, K. E. (1976). History of vegetation. *In* Chapman (1976a), 5–83.

Barnard, T. and Banner, F. T. (1953). Arenaceous foraminifera from the Upper Cretaceous of England. *Quarterly Journal of the Geological Society* **109**, 173–216.

Baron, W. M. M. (1971). "Nature Conservation". Methuen, London.

Bates, G. H. (1935). The vegetation of footpaths, sidewalks, cart-tracks and gateways. *Journal of Ecology* **23**, 470–489.

Bates, G. H. (1938). Life-forms of pasture plants in relation to treading. *Journal of Ecology* **26**, 452–454.

Bates, G. H. (1950). Track making by man and domestic animals. *Journal of Animal Ecology* **19**, 21–28.

Batey, T. and Davies, D. B. (1971). Soil structure and the production of agricultural crops. *Journal of the Royal Agricultural Society of England* **132**, 106–122.

Baver, L. D., Gardner, W. H. and Gardner, W. R. (1972). "Soil Physics", 4th Edition. Wiley–Interscience, New York.

Berry, R. J. (1971). Conservation aspects of the genetical constitution of populations. *In* Duffey and Watt (1971), 177–206.

Beven, G. (1964). The feeding sites of birds in grassland with thick scrub. Some comparisons with dense oakwood. *The London Naturalist* No. 43, 86–109.

Bhadresa, R. (1977). Food preferences of rabbits *Oryctolagus cuniculus* L. at Holkham sand dunes, Norfolk. *Journal of Applied Ecology* **14**, 287–291.

Bibby, J. S. and Mackney, D. (1969). "Land Use Capability Classification". Soil Survey of Great Britain Technical Monograph No. 1. Soil Survey, Harpenden.

Bilham, E. G. (1938). "The Climate of the British Isles". Macmillan, London.

Birch, H. F. (1960). Nitrification in soils after different periods of dryness. *Plant and Soil* 12, 81–96.

Black, M. (1965). Coccoliths. *Endeavour* 24, 131–137.

Black, M. (1971). The systematics of coccoliths in relation to the palaeontological record. *In* Funnell and Riedel (1971), 611–624.

Black, Rhona M. (1970). "The Elements of Palaeontology". Cambridge University Press, Cambridge.

Blackman, G. E. and Rutter, A. J. (1946). Physiological and ecological studies of the plant environment. I. The light factor and the distribution of the bluebell (*Scilla non-scripta*) in woodland communities. *Annals of Botany, New Series* 10, 361–390.

Blackman, G. E. and Rutter, A. J. (1947). Physiological and ecological studies of the plant environment. II. The interaction between light intensity and mineral nutrient supply in the growth and development of the bluebell (*Scilla non-scripta*). *Annals of Botany, New Series* 11, 125–158.

Blackman, G. E. and Rutter, A. J. (1948). Physiological and ecological studies of the plant environment. III. The interaction between light intensity and mineral nutrient supply in leaf development and in the net assimilation rate of the bluebell (*Scilla non-scripta*). *Annals of Botany, New Series* 12, 1–26.

Blackman, G. E. and Rutter, A. J. (1949). Physiological and ecological studies of the plant environment. IV. The interaction between light intensity and mineral nutrient supply on the uptake of nutrients by the bluebell (*Scilla non-scripta*). *Annals of Botany, New Series* 13, 453–489.

Blackman, G. E. and Rutter, A. J. (1950). Physiological and ecological studies in the analysis of the plant environment. V. An assessment of the factors controlling the distribution of the bluebell (*Scilla non-scripta*) in different communities. *Annals of Botany, New Series* 14, 487–520.

Blackwood, J. W. and Tubbs, C. R. (1970). A quantitative survey of chalk grassland in England. *Biological Conservation* 3, 1–5.

Blunden, J. (1975). "The Mineral Resources of Britain. A Study in Exploitation." Hutchinson, London.

Boddington, M. A. B. (1972). Economics of outdoor pigs. *Agriculture* 79, 6–12.

Bourne, R. (1931). "Regional Survey, and its Relation to Stocktaking of the Agricultural and Forest Resources of the British Empire", Oxford Forestry Memoirs No. 13.

Bourne, R. (1942). A note on beech regeneration in southern England. *Quarterly Journal of Forestry* 36, 42–49.

Bowen, H. C. (1961). "Ancient Fields". British Association for the Advancement of Science, London.

Bowen, H. J. M. (1965). Sulphur and the distribution of British plants. *Watsonia* 6, 114–119.

Bowen, H. J. M. (1970). Air pollution and its effects on plants. *In* Perring (1970), 119–127.

Boycott, A. E. (1934). Habitats of land Mollusca in Britain. *Journal of Ecology* 22, 1–38.

Brade-Birks, S. G. and Furneaux, B. S. (1930). Soil survey of the College Farm. *Journal of the South-East Agricultural College, Wye* No. 27, 252–254.

Bradley, P. N. (1968). "Small and Large-scale Studies of Scrub Invasion along the Chiltern Escarpment". M.Sc. Thesis, University of London.

Bradshaw, A. D. (1977). Conservation problems in the future. *In* Smith *et al.* (1977), 77–96.

Braun-Blanquet, J. (1932). "Plant Sociology. The Study of Plant Communities" (English translation). McGraw-Hill, New York.

Braun-Blanquet, J. and Moor, M. (1938). "Prodromus der Pflanzengesellschaften Fasc. 5. Verband des Bromion erecti". Comité International du Prodrome Phytosociologique, Leiden.

Bremner, J. M. (1959). Determination of fixed ammonium in soils. *Journal of Agricultural Science, Cambridge* **52**, 147–160.

Bremner, J. M. and Nelson, D. W. (1968). Chemical decomposition of nitrite in soils. *Proceedings of the 9th International Congress of Soil Science* **2**, 495–503.

Brenchley, Winifred E. (1911). The weeds of arable land in relation to the soils on which they grow. *Annals of Botany* **25**, 155–165.

Brenchley, Winifred E. (1912). The weeds of arable land in relation to the soils on which they grow. II. *Annals of Botany* **26**, 95–109.

Brenchley, Winifred E. (1913). The weeds of arable land. III. *Annals of Botany* **27**, 141–166.

Brenchley, Winifred E. (1924). "Manuring of Grassland for Hay". Longman, London.

Brenchley, Winifred E. and Adam, Helen (1915). Recolonisation of cultivated land allowed to revert to natural conditions. *Journal of Ecology* **3**, 193–210.

British Ecological Society (1944). Nature conservation and nature reserves. *Journal of Ecology* **32**, 45–82.

Bromley, R. G. (1967). Some observations on burrows of thalassinidian Crustacea in chalk hardgrounds. *Quarterly Journal of the Geological Society* **123**, 157–182.

Brown, J. M. B. (1953). "Studies on British Beechwoods", Forestry Commission Bulletin No. 20. HMSO, London.

Brown, J. M. B. (1959). Forest ecology. *In* "Report on Forest Research", year ending March 1959, 61–62. HMSO. London.

Bull, A. J. (1940). Cold conditions and land forms in the South Downs. *Proceedings of the Geologists' Association* **51**, 63–71.

Bunting, A. H. and Elston, J. (1966). Water relations of crops and grass on chalk soil. *Scientific Horticulture* **18**, 116–120.

Burden, R. F. (1970). Some effects of people on the Nature Trail at Ranmore, Surrey. *In* "The Surrey Naturalist Annual Report 1969", 14–22.

Burden, R. F. and Randerson, P. F. (1972). Quantitative studies of the effects of human trampling on vegetation as an aid to the management of semi-natural areas. *Journal of Applied Ecology* **9**, 439–457.

Burges, A. (1960). Time and size as factors in ecology. *Journal of Ecology* **48**, 273–285.

Burges, A. (1967). The soil system. *In* Burges and Raw (1967), 1–13.

Burges, A. and Raw, F. (Eds) (1967). "Soil Biology". Academic Press, London and New York.

Burges, A., Hurst, H. M., Walkden, S. B., Dean, F. M. and Hirst, M. (1963). Nature of humic acids. *Nature, London* **199**, 696–697.

Burton, T. (1966). A day in the country. A survey of leisure activity at Box Hill in Surrey. *Chartered Surveyor* **98**, 378–380.

Caborn, J. M. (1965). "Shelterbelts and Windbreaks". Faber and Faber, London.

Cain, A. J. and Currey, J. D. (1963a). Area effects in *Cepaea. Philosophical Transactions of the Royal Society of London Series B* **246**, 1–81.

Cain, A. J. and Currey, J. D. (1963b). Area effects in *Cepaea* on the Larkhill artillery ranges, Salisbury Plain. *Journal of the Linnaean Society (Zoology)* **45**, 1–15.

Cameron, R. A. D. (1970a). The survival, weight-loss and behaviour of three species of land-snail in conditions of low humidity. *Journal of Zoology* **160**, 143–157.

Cameron, R. A. D. (1970b). The effect of temperature on the activity of three species of Helicid snail (Mollusca: Gastropoda). *Journal of Zoology* **162**, 303–315.

Carter, E. S. (1970). The agriculture of Lincolnshire. *Journal of the Royal Agricultural Society of England* **131**, 56–68.

Castle, M. E. and Holmes, W. (1960). The intensive production of herbage for crop drying. 7. The effect of further continued massive applications of nitrogen with and without phosphate and potash on the yield of grassland herbage. *Journal of Agricultural Science, Cambridge* **55**, 251–260.

Cates, R. G. and Orians, G. H. (1975). Successional status and the palatability of plants to generalized herbivores. *Ecology* **56**, 410–418.

Centre for Agricultural Strategy (1976). "Land for Agriculture", Report No. 1. Centre for Agricultural Strategy, Reading.

Chandler, T. J. and Gregory, S. (Eds) (1976). "The Climate of the British Isles". Longman, London.

Chapman, H. D. and Liebig, G. F. (1952). Field and laboratory studies of nitrite accumulation in soils. *Proceedings of the Soil Science Society of America* **16**, 276–282.

Chapman, S. B. (Ed.) (1976a). "Methods in Plant Ecology". Blackwell, Oxford.

Chapman, S. B. (1976b). Production ecology and nutrient budgets. *In* Chapman (1976a), 157–228.

Chappell, H. G., Ainsworth, J. F., Cameron, R. A. D. and Redfern, M. (1971). The effect of trampling on a chalk grassland ecosystem. *Journal of Applied Ecology* **8**, 869–882.

Chatwin, C. P. (1960). "The Hampshire Basin and Adjoining Areas", 3rd Edition, British Regional Geology. HMSO, London.

Chatwin, C. P. (1961). "East Anglia and Adjoining Areas", 4th Edition, British Regional Geology. HMSO, London.

Chilterns Standing Conference (1971). "A Plan for the Chilterns". Chilterns Standing Conference, Aylesbury.

Chinery, M. (1973). "A Field Guide to the Insects of Britain and Northern Europe". Collins, London.

Christy, M. (1924). The hornbeam (*Carpinus betulus* L.) in Britain. *Journal of Ecology* **12**, 39–94.

Clapham, A. R., Tutin, T. G. and Warburg, E. F. (1962). "Flora of the British Isles", 2nd Edition. Cambridge University Press, Cambridge.

Clark, F. E. (1967). Bacteria in soil. *In* Burges and Raw (1967), 15–49.

Clark, J. G. D. (1952). "Prehistoric Europe. The Economic Basis". Methuen, London.

Clark, M. J., Lewin, J. and Small, R. J. (1967). The sarsen stones of the Marlborough Downs and their geomorphological implications. *Southampton Research Series in Geography* No. 4, 3–40.

Clement, C. R. (1958). "The Comparative Significance of Water-stable Aggregation and the Nitrogen Status of the Soil as Factors in Crop Production Following Diverse Leys". Ph.D. Thesis, University of Reading.

Clement, C. R. and Williams, T. E. (1964). Leys and soil organic matter. I. The accumulation of organic carbon in soils under different leys. *Journal of Agricul-*

tural Science, Cambridge **63**, 377–383.

Clements, F. E. (1916). "Plant Succession: an Analysis of the Development of Vegetation", Publication 242. Carnegie Institute, Washington.

Clymo, R. S. (1962). An experimental approach to part of the calcicole problem. *Journal of Ecology* **50**, 707–731.

Cobbett, W. (1830). "Rural Rides". William Cobbett, London. Various more recent editions, e.g. Penguin English Library, 1967.

Cohen, O. P. and Sharabani, N. (1964). Moisture extraction by grape-vines from chalk. *Israel Journal of Agricultural Research* **14**, 179–185.

Colquhoun, M. K. and Morley, A. (1941). The density of downland birds. *Journal of Animal Ecology* **10**, 35–46.

Cooke, G. W. (1967). "The Control of Soil Fertility". Crosby Lockwood, London.

Coombe, D. E. (1957). The spectral composition of shade light in woodlands. *Journal of Ecology* **45**, 823–830.

Cope, D. W. (1976). "Soils in Wiltshire. I.", Soil Survey Record No. 32. Soil Survey, Harpenden.

Coppock, J. T. (1961). Agricultural changes in the Chilterns 1875–1900. *Agricultural History Review* **9**, 1–16.

Coppock, J. T. (1971). "An Agricultural Geography of Great Britain. Bell, London.

Corbet, G. B. and Southern, H. N. (Eds) (1977). "The Handbook of British Mammals", 2nd Edition. Blackwell, Oxford.

Cornish, M. W. (1954). The origin and structure of the grassland types of the central North Downs. *Journal of Ecology* **42**, 359–374.

Countryside Commission (1974). "New Agricultural Landscapes". Countryside Commission, Cheltenham.

Cousens, J. (1974). "An Introduction to Woodland Ecology". Oliver and Boyd, Edinburgh.

Coward, T. A. (revised by A. W. Boyd) (1950). "The Birds of the British Isles and Their Eggs", Series III. Warne, London.

Cowling, D. W. (1966). The response of grass swards to nitrogenous fertilizer. *Proceedings of the 10th International Grassland Congress* 204–209.

Cowling, D. W. and Jones, L. H. P. (1970). A deficiency in soil sulfur supplies for perennial ryegrass in England. *Soil Science* **110**, 346–354.

Cox, C. B., Healey, I. N. and Moore, P. D. (1976). "Biogeography. An Ecological and Evolutionary Approach", 2nd Edition. Blackwell, Oxford.

Crawford, O. G. S. and Keiller, A. (1928). "Wessex from the Air". Clarendon Press, Oxford.

Croney, D. and Coleman, J. D. (1954). Soil structure in relation to soil suction (pF). *Journal of Soil Science* **5**, 75–84.

Cross, D. A. E. (1967). The great Till flood of 1841. *Weather* **22**, 430–433.

Crowe, P. R. (1971). "Concepts in Climatology". Longman, London.

Cuanalo de la C., H. E. (1966). "A Study of the Short-range Soil Variation Associated with Intricate Image-patterns on Air Photographs". Diploma Thesis, University of Oxford.

Curry, J. P. (1969). The decomposition of organic matter in soil. I. The role of the fauna in decaying grassland herbage. *Soil Biology and Biochemistry* **1**, 253–258.

Curtis, L. F., Courtney, F. M. and Trudgill, S. (1976). "Soils in the British Isles". Longman, London.

Darby, H. C. (Ed.) (1969). "An Historical Geography of England before A.D. 1800". Cambridge University Press, Cambridge.

Darlington, A. (1969). "The Ecology of Refuse Tips". Heinemann, London.

Daubenmire, R. (1968). "Plant Communities. A Textbook of Plant Synecology." Harper and Row, New York.

Davies, A. M. and Barnes, A. H. J. (1953). A preliminary survey of the sarsen and puddingstone blocks of the Chilterns. *Proceedings of the Geologists' Association* **64**, 1–9.

Davies, W. (1936). Vegetation of grass verges and other excessively trodden habitats. *Journal of Ecology* **26**, 38–49.

Davis, B. N. K. (1977). The *Hieracium* flora of chalk and limestone quarries in England. *Watsonia* **11**, 345–351.

Davis, W. E. and Rose, R. C. (1912). The effects of external conditions upon the after-ripening of the seeds of *Crataegus mollis*. *Botanical Gazette* **54**, 49–62.

Davy, A. J. and Taylor, K. (1974a). Water characteristics of contrasting soils in the Chiltern Hills and their significance for *Deschampsia caespitosa* (L.) Beauv. *Journal of Ecology* **62**, 367–378.

Davy, A. J. and Taylor K. (1974b). Seasonal patterns of nitrogen availability in contrasting soils in the Chiltern Hills. *Journal of Ecology* **62**, 793–807.

Davy, A. J. and Taylor, K. (1975). Seasonal changes in the inorganic nutrient concentrations in *Deschampsia caespitosa* (L.) Beauv. in relation to its tolerance of contrasting soils in the Chiltern Hills. *Journal of Ecology* **63**, 27–39.

Day, W. R. (1946). The pathology of beech on chalk soils. *Quarterly Journal of Forestry* **40**, 72–82.

De Crespigny, E. C. (1877). "A New London Flora". Clowes, London.

Defoe, D. (1724). "A Tour Through the Whole Island of Great Britain". Strahan, London. Various more recent editions, e.g. Everyman's Library, 2nd Edition, 1962.

De Silva, B. L. T. (1934). The distribution of "calcicole" and "calcifuge" species in relation to the content of the soil in calcium carbonate and exchangeable calcium, and to soil reaction. *Journal of Ecology* **22**, 532–553.

De Wit, C. T. (1960). On competition. *Verslagen van Landbouwkundige Onderzoekingen* No. 66.8.

Dimbleby, G. W. and Evans, J. G. (1974). Pollen and land-snail analysis of calcareous soils. *Journal of Archaeological Science* **1**, 117–133.

Donisthorpe, H. St J. K. (1927). "British Ants, their Life History and Classification", 2nd Edition. Routledge, London.

Dony, J. G. (1967). "Flora of Hertfordshire". Urban District Council, Hitchin.

Dony, J. G. (1976). "Bedfordshire Plant Atlas". Borough Museum and Art Gallery, Luton.

Double, I. S. (1927). The microscopic character of certain horizons of the British Chalk. *Journal of the Royal Microscopical Society* **47**, 226–231.

Dowse, G. A. (1967). Cereal rotations on the Chalk. The problems in Wiltshire. *Agriculture* **74**, 368–370, 485–488.

Druce, G. C. (1926). "The Flora of Buckinghamshire". Buncle, Arbroath.

Druce, G. C. (1927). "The Flora of Oxfordshire", 2nd Edition. Clarendon Press, Oxford.

Duchaufour, P. (1968). "L' Évolution des Sols". Masson, Paris.

Duchaufour, P. (1970). "Précis de Pédologie", 3rd Edition. Masson, Paris.

Duffey, E. (1956). Aerial dispersal in a known spider population. *Journal of Animal Ecology* **25**, 85–111.

Duffey, E. (1962a). A population study of spiders in limestone grassland. Description of study area, sampling methods and population characteristics. *Journal of Animal Ecology* **31**, 571–599.

Duffey, E. (1962b). A population study of spiders in limestone grassland. The field-layer fauna. *Oikos* **13**, 15–34.

Duffey, E. (Ed.) (1967). "The Biotic Effects of Public Pressures on the Environment", Monks Wood Experimental Station Symposium No. 3. Natural Environment Research Council, London.

Duffey, E. (1974a). "Nature Reserves and Wildlife". Heinemann, London.

Duffey, E. (1974b). Lowland grassland and scrub: management for wildlife. *In* Warren and Goldsmith (1974), 167–183.

Duffey, E. (1975). The effects of human trampling on the fauna of grassland litter. *Biological Conservation* **7**, 255–274.

Duffey, E. and Morris, M. G. (1966). The invertebrate fauna of the Chalk and its scientific interest. *In* "Handbook and Annual Report of the Society for the Promotion of Nature Reserves 1966", 83–94.

Duffey, E. and Watt, A. S. (Eds) (1971). "The Scientific Management of Animal and Plant Communities for Conservation". Blackwell, Oxford.

Duffey, E., Morris, M. G., Sheail, J., Ward, L. K., Wells, D. A. and Wells, T. C. E. (1974). "Grassland Ecology and Wildlife Management". Chapman and Hall, London.

Duncan, R. (1972). Farming in Kent. *Journal of the Royal Society of England* **133**, 55–65.

Dunning, R. A., Mathias, P. L., and Chwarszczynska, Daphne M. (1977). Spiral nematodes associated with the problem of establishing sugar beet on calcareous soils. *Institut International de Recherches Betteravieres, 40th Winter Congress* 323–332.

Dyson, R. (1977). "Gardening on Chalk and Lime". Dent, London.

Eckardt, F. E. (Ed.) (1968). Functioning of terrestrial ecosystems at the primary production level. UNESCO, Paris.

Edlin, H. L. (1949). "Woodland Crafts in Britain" (reprinted 1973). David and Charles, Newton Abbot.

Edlin, H. L. (1955). "Trees, Woods and Man". Collins, London.

Edmonds, E. A., McKeown, M. C. and Williams, M. (1969). "South-west England", 3rd Edition, British Regional Geology. HMSO, London.

Edney, E. B. (1954). Woodlice and the land habitat. *Biological Reviews* **29**, 185–219.

Ellis, W. (1745). "Chiltern and Vale Farming Explained", 2nd Edition. Osborne, London.

Elston, J. (1963). "An Investigation into the Plant/Soil Inter-relationships of Crops Growing on the Chalk". Ph.D. Thesis, University of Reading.

Elton, C. S. (1958). "The Ecology of Invasions". Methuen, London.

Elton, C. S. (1966). "The Pattern of Animal Communities". Methuen, London.

Elton, C. S. and Miller, R. S. (1954). The ecological survey of animal communities, with a practical system of classifying habitats by structural characters. *Journal of Ecology* **42**, 160–196.

Evans, J. G. (1972). "Land Snails in Archaeology". Seminar Press, London.

Evans, J. G. (1975). "The Environment of Early Man in the British Isles". Elek, London.

Farrow, E. P. (1917). On the ecology of the vegetation of Breckland. III. General effects of rabbits on the vegetation. *Journal of Ecology* **5**, 1–18.

Fenner, M. (1975). "Factors Limiting the Distribution of Strict Calcicoles". Ph.D.

Thesis, University of Cambridge.

Fitter, A. (1968). The present distribution of juniper (*Juniperus communis*) in the Chilterns. *In* "Proceedings and Report of the Ashmolean Natural History Society, Oxford 1968", 16–23.

Flaig, W. (1968). Einwirkung von organischen Bodenbestandteilen auf das Pflanzenwachstum. *Landwirtschaftliche Forschung* **21**, 103–127.

Floyd, C. (1965). Conservation of chalk grassland. Winter grazing or summer grazing or mowing of chalk downland on nature reserves. *In* "Handbook and Annual Report of the Society for the Promotion of Nature Reserves 1965", 59–61.

Ford, E. B. (1957). "Butterflies", 2nd Edition. Collins, London.

Ford, E. B. (1972). "Moths", 3rd Edition. Collins, London.

Ford, J. (1937). Fluctuations in natural populations of Collembola and Acarina. *Journal of Animal Ecology* **6**, 98–111.

Fourt, D. F. (1973). Studies on chalk soils. *In* "Report on Forest Research", year ending March 1973, 61–62. HMSO, London.

Fourt, D. F. (1975). Studies on calcareous soils in the lowlands. *In* "Report on Forest Research", year ending March 1975, 24–25. HMSO, London.

Frame, J. (1975). A comparison of herbage production under cutting and grazing (including comments on deleterious factors such as treading). *In* Hodgson and Jackson (1975), 39–49.

Fraps, G. S. and Sterges A. J. (1935). "Availability of nitrous nitrogen to plants", Texas Agricultural Experiment Station Bulletin 515.

Frazer, J. F. D. (1967). Insect introductions. *In* Duffey (1967), 9–15.

Fryer, J. D. and Chancellor, R. J. (1970). Herbicides and our changing weeds. *In* Perring (1970), 105–118.

Fryer, J. D. and Makepiece, R. J. (1977). "Weed Control Handbook", Vols I "Principles", 6th Edition, and II "Recommendations", 8th Edition. Blackwell, Oxford.

Funnell, B. M. and Riedel, W. R. (Eds) (1971). "The Micropalaeontology of Oceans." Cambridge University Press, Cambridge.

FWAG (Farming and Wildlife Advisory Group) (1973). "The Chalkland Exercise: Chalkland Farming and Wildlife Conservation". MAFF/ADAS, London.

Gallois, R. W. (1965). "The Wealden District", 4th Edition, British Regional Geology. HMSO, London.

Gardiner, Phyllis, Fitter, R. S. R. and Campbell, B. (1970). "BBONT 1959–1969". Berks., Bucks. and Oxon. Naturalists' Trust, Oxford.

Gardner, H. W. (1959). Manuring of barley on Chalk. *Agriculture* **66**, 396–397.

Gardner, H. W. (1967). "A Survey of the Agriculture of Hertfordshire", County Agricultural Surveys No. 5. Royal Agricultural Society of England, London.

Garrad, G. H. (1954). "A Survey of the Agriculture of Kent", County Agricultural Surveys No. 1. Royal Agricultural Society of England, London.

Gasser, J. K. R. (1969). Some processes affecting nitrogen in the soil. *In* MAFF (1969), 15–29.

Geiger, R. (1965). "The Climate Near the Ground", 4th Edition (English Translation). Harvard University Press, Cambridge, Massachusetts.

Gigon, A. and Rorison, I. H. (1972). The response of some ecologically distinct plant species to nitrate- and ammonium-nitrogen. *Journal of Ecology* **60**, 93–102.

Gilbert, E. W. (1969). The human geography of Roman Britain. *In* Darby (1969), 30–87.

Gilbert, N. and Wells, T. C. E. (1966). Analysis of quadrat data. *Journal of Ecology*

54, 675–685.

Gillham, M. E. (1956). Ecology of the Pembrokeshire islands. IV. Effects of treading and burrowing by birds and animals. *Journal of Ecology* **44**, 51–82.

Gilpin, W. (1804). "Observations on the Coasts of Hampshire and Sussex". Cadell and Davies, London.

Gittins, R. (1965a). Multivariate approaches to a limestone grassland community. I. A stand ordination. *Journal of Ecology* **53**, 385–401.

Gittins, R. (1965b). Multivariate approaches to a limestone grassland community. II. A direct species ordination. *Journal of Ecology* **53**, 403–409.

Gittins, R. (1965c). Multivariate approaches to a limestone grassland community. III. A comparative study of ordination and association analysis. *Journal of Ecology* **53**, 411–425.

Gittins, R. (1969). The application of ordination techniques. *In* Rorison (1969), 37–66.

Godwin, H. (1975). "The History of the British Flora", 2nd Edition. Cambridge University Press, London.

Goldsmith, F. B. (1973a). The vegetation of exposed sea-cliffs at South Stack, Anglesey. I. The multivariate approach. *Journal of Ecology* **61**, 787–818.

Goldsmith, F. B. (1973b). The vegetation of exposed sea-cliffs at South Stack, Anglesey. II. Experimental studies. *Journal of Ecology* **61**, 819–829.

Goldsmith, F. B. (1974). Ecological effects of visitors in the countryside. *In* Warren and Goldsmith (1974), 217–231.

Goldsmith, F. B. and Harrison, C. M. (1976). Description and analysis of vegetation. *In* Chapman (1976a), 85–155.

Goldsmith, F. B. and O'Connor, F. B. (1975). "Ivinghoe Beacon Experimental Restoration Project". University College, London. Since published by the Countryside Commission.

Good, R. (1928). Notes on a comparision of the angiosperm floras of Kent and Pas de Calais. *Journal of Botany* **66**, 253–265.

Good, R. (1936). On the distribution of the lizard orchid (*Himantoglossum hircinum* Koch.). *New Phytologist* **35**, 142–170.

Good, R. (1948). "A Geographical Handbook of the Dorset Flora". Dorset Natural History and Field Club, Dorchester.

Good, R. (1974). "The Geography of the Flowering Plants", 4th Edition. Longman, London.

Goodden, R. (1978). "British Butterflies". David and Charles, Newton Abbot.

Gough, H. C. and Dunnett, F. W. (1950). Rabbit damage to winter corn. *Agriculture* **57**, 374–378.

Graham, C. W. and Stone, L. E. W. (1975). Field experiments on the cereal cyst nematode (*Heterodera avenae*) in south-east England, 1967–72. *Annals of Applied Biology* **80**, 61–73.

Granville, A. B. (1841). "Spas of England", Vol. 2 "The Midlands and South". Colburn, London.

Green, B. H. (1972). The relevance of seral eutrophication and plant competition to the management of successional communities. *Biological Conservation* **4**, 378–384.

Green, R. D. and Fordham, S. J. (1973). "Soils in Kent. I.", Soil Survey Record No. 14. Soil Survey, Harpenden.

Greenwood, D. J. (1961). The effect of oxygen concentration on the decomposition of organic materials in soil. *Plant and Soil* **24**, 360–376.

Greig-Smith, P. (1964). "Quantitative Plant Ecology", 2nd Edition. Butterworth, London.

Griffiths, M., Thomas, J. F. H. and Line, R. (1951). "Reclaiming Land for Agriculture". Crosby Lockwood, London.

Grime, J. P. (1974). Vegetation classification by reference to strategies. *Nature, London* **250**, 26–31.

Grime, J. P. and Blythe, G. M. (1969). An investigation of the relationships between snails and vegetation at the Winnats Pass. *Journal of Ecology* **57**, 45–66.

Grime, J. P. and Hodgson, J. G. (1969). An investigation of the ecological significance of lime-chlorosis by means of large-scale comparative experiments. *In* Rorison (1969), 67–99.

Grime, J. P. and Hutchinson, T. C. (1967). The incidence of lime-chorosis in the natural vegetation of England. *Journal of Ecology* **55**, 557–566.

Grime, J. P., MacPherson-Stewart, S. F. and Dearman, R. S. (1968). An investigation of leaf palatability using the snail *Cepaea nemoralis* L. *Journal of Ecology* **56**, 405–420.

Grose, D. (1957). "The Flora of Wiltshire". Wiltshire Archaeological and Natural History Society, Devizes.

Grubb, P. J. (1976). A theoretical background to the conservation of ecologically distinct groups of annuals and biennials in the chalk grassland ecosystem. *Biological Conservation* **10**, 53–76.

Grubb, P. J. and Suter, M. B. (1971). The mechanism of acidification of soil by *Calluna* and *Ulex* and the significance for conservation. *In* Duffey and Watt (1971), 115–133.

Grubb, P. J., Green, H. E. and Merrifield, R. C. J. (1969). The ecology of chalk heath: its relevance to the calcicole–calcifuge and soil acidification problems. *Journal of Ecology* **57**, 175–212.

Gunn, R. E. (1968). Break crops: grain maize. *Agriculture* **75**, 66–70.

Guyot, A. L. (1957). Les microassociations végétales au sein du Brometum erecti. *Vegetatio* **7**, 321–354.

Hagen, C. E. and Hopkins, H. T. (1955). Ionic species in orthophosphate absorption by barley roots. *Plant Physiology, Lancaster* **30**, 193–199.

Hall, A. D. and Russell, E. J. (1911). "A Report on the Agriculture and Soils of Kent, Surrey and Sussex". HMSO, London.

Hall, J. B. (1967). "Some Aspects of the Ecology of *Brachypodium pinnatum*". Ph.D. Thesis, University of London.

Hall, J. B. (1971). Pattern in a chalk grassland community. *Journal of Ecology* **59**, 749–762.

Hallsworth, E. G. and Crawford, D. V. (Eds) (1965). "Experimental Pedology". Butterworth, London.

Hancock, J. M. (1975). The petrology of the Chalk. *Proceedings of the Geologists' Association* **86**, 499–535.

Hancock, J. M. and Kennedy, W. J. (1967). Photographs of hard and soft chalks taken with a scanning electron microscope. *Proceedings of the Geological Society* **1643**, 249–252.

Harberd, D. J. (1961). Observations on population structure and longevity of *Festuca rubra*. *New Phytologist* **60**, 184–206.

Harley, J. L. (1937). Ecological observations on the mycorrhiza of beech. *Journal of Ecology* **25**, 421–423.

Harley, J. L. (1949). Soil conditions and the growth of beech seedlings. *Journal of*

Ecology **37**, 28–37.

Harley, J. L. (1969). "The Biology of Mycorrhiza", 2nd Edition. Hill, London.

Harmsen, G. W. and van Schreven, D. A. (1955). Mineralization of organic nitrogen in soil. *Advances in Agronomy* **7**, 299–398.

Harper, J. L. (1957). Ecological aspects of weed control. *Outlook on Agriculture* **1**, 197–205.

Harper, J. L. (Ed.) (1960). "The Biology of Weeds". Blackwell, Oxford.

Harper, J. L. and Wood, W. A. (1957). Biological flora of the British Isles. *Senecio jacobaea* L. *Journal of Ecology* **45**, 617–637.

Hauck, R. D. and Bremner, J. M. (1969). Significance of the nitrification reaction in nitrogen balances. *In* University of California (1969), 31–39.

Havill, D. C., Lee, J. A. and Stewart, G. R. (1974). Nitrate utilization by species from acidic and calcareous soils. *New Phytologist* **73**, 1221–1231.

Hawes, P. T. J. (1971). "Changes in the Botanical Composition of Chalk Grassland Resulting from Dietary Selection by Sheep". M.Phil. Thesis, Council for National Academic Awards, London.

Hawke, E. L. (1933). Extreme diurnal ranges of air temperature in the British Isles. *Quarterly Journal of the Royal Meteorological Society* **59**, 261–265, 401–403.

Hawke, E. L. (1944). Thermal characteristics of a Hertfordshire frost hollow. *Quarterly Journal of the Royal Meteorological Society* **70**, 23–48.

Hawkins, H. L. (1931). The significance of the Chalk. *Transactions of the South-eastern Union Science Section 1931*, 29–43.

Hawkins, H. L. (1952). A pinnacle of chalk penetrating the Eocene on the floor of a buried river channel at Ashford Hill, near Newbury, Berkshire. *Quarterly Journal of the Geological Society* **108**, 233–260.

Hepburn, I. (1952). "Flowers of the Coast". Collins, London.

Heslop-Harrison, J. (1953). "New Concepts in Flowering Plant Taxonomy". Heinemann, London.

Heywood, G. S. P. (1933). Katabatic winds in a valley. *Quarterly Journal of the Royal Meteorological Society* **59**, 47–58.

Hodge, C. A. H. and Seale, R. S. (1966). "The Soils of the District around Cambridge", Memoirs of the Soil Survey of Great Britain: England and Wales. Soil Survey, Harpenden.

Hodgson, J. and Jackson, D. K. (Eds) (1975). "Pasture Utilisation by the Grazing Animal", Occasional Symposium No. 8. British Grassland Society, Hurley.

Hodgson, J. M. (1967). "Soils of the West Sussex Coastal Plain", Bulletin of the Soil Survey of Great Britain: England and Wales. Soil Survey, Harpenden.

Hodgson, J. M. (1974). "Describing and Sampling Soil Profiles", Soil Survey Field Handbook, Technical Monograph No. 5. Soil Survey, Harpenden.

Hooper, L. J. and Davies, D. B. (1968). Melanism and associated symptoms in wheat grown on copper-responsive chalkland soils. *Journal of the Science of Food and Agriculture* **19**, 733–739.

Hooper, M. D. (1970a). Dating hedges. *Area* **4**, 63–65.

Hooper, M. D. (1970b). The botanical importance of our hedgerows. *In* Perring (1970), 58–62.

Hooper, M. D. (1971). The size and surroundings of nature reserves. *In* Duffey and Watt (1971), 555–561.

Hope-Simpson, J. F. (1938). A chalk flora on the Lower Greensand: its use in interpreting the calcicole habit. *Journal of Ecology* **26**, 218–235.

Hope-Simpson, J. F. (1940a). On the errors in the ordinary use of subjective frequency estimations in grassland. *Journal of Ecology* **28**, 193–209.

Hope-Simpson, J. F. (1940b). Studies on the vegetation of the English Chalk. VI. Late stages in succession leading to chalk grassland. *Journal of Ecology* **28**, 386–402.

Hope-Simpson, J. F. (1940c). The utilisation and improvement of chalk down pasture. *Journal of the Royal Agricultural Society of England* **100**, 44–49.

Hope-Simpson, J. F. (1941a). Studies of the vegetation of the English Chalk. VII. Bryophytes and lichens in chalk grassland, with a comparison of their occurrence in other calcareous grasslands. *Journal of Ecology* **29**, 107–116.

Hope-Simpson, J. F. (1941b). Studies on the vegetation of the English Chalk. VIII. A second survey of the chalk grasslands of the South Downs. *Journal of Ecology* **29**, 217–267.

Hope-Simpson, J. F. (1955). Review of Salisbury (1952). *Journal of Ecology* **43**, 311–315.

Hosier, A. J. and Hosier, F. H. (1951). "Hosier's Farming System". Crosby Lockwood, London.

Hoskins, W. G. (1955). "The Making of the English Landscape". Hodder and Stoughton, London.

Howarth, S. E. and Williams, J. T. (1968). Biological flora of the British Isles. *Chrysanthemum leucanthemum* L. *Journal of Ecology* **56**, 585–595.

Howe, J. A. (1910). "The Geology of Building Stones". Arnold, London.

Hubbard, C. E. (1968). "Grasses", 2nd Edition. Pelican, London.

Hudson, A. A. (1930). Planting on chalk soil. *Quarterly Journal of Forestry* **24**, 175–186.

Hudson, W. H. (1900). "Nature in Downland". Longman, London.

Hudson, W. H. (1910). "A Shepherd's Life. Impressions of the South Wiltshire Downs". Methuen, London.

Hughes, A. P. (1959). Effects of the environment on leaf development in *Impatiens parviflora* DC. *Journal of the Linnaean Society (Botany)* **56**, 161–165.

Hughes, A. P. (1965). Plant growth and the aerial environment. IX. A synopsis of the autecology of *Impatiens parviflora*. *New Phytologist* **64**, 399–413.

Hutchings, M. J. and Barkham, J. P. (1976). An investigation of shoot interactions in *Mercurialis perennis* L., a rhizomatous perennial herb. *Journal of Ecology* **64**, 723–743.

Hutchinson, J. N. (1969). A reconsideration of the coastal landslides at Folkestone Warren, Kent. *Géotechnique* **19**, 6–38.

Imms, A. D. (1971). "Insect Natural History", 3rd Edition. Collins, London.

Ireland, D. C. (1975). "Curriculum Change: the Development of Environmental Education in Primary/Middle Schools in Buckinghamshire". M.Ed. Thesis, University of Reading.

Jackson, W. A. (1967). Physiological effects of soil acidity. *In* Pearson and Adams (1967), 43–124.

Jarvis, M. G. (1973). "Soils of the Wantage and Abingdon District", Memoirs of the Soil Survey of Great Britain: England and Wales. Soil Survey, Harpenden.

Jarvis, R. A. (1968). "Soils of the Reading District', Memoirs of the Soil Survey of Great Britain: England and Wales. Soil Survey, Harpenden.

Jeans, C. V. (1968). The origin of the montmorillonite of the European Chalk with special reference to the Lower Chalk of England. *Clay Mineralogy* **7**, 311–329.

Jefferies, J. R. (1879). "Wild Life in a Southern County". Smith, Elder and Co.; London. Various more recent editions, e.g. Cape, 1940.

Jefferies, R. L. and Willis, A. J. (1964). Studies on the calcicole–calcifuge habit. II. The influence of calcium on the growth and establishment of four species in soil and sand culture. *Journal of Ecology* **52**, 691–707.

Jenkins, J. Geraint (Ed.) (1972). "The Wool Textile Industry in Great Britain". Routledge and Kegan Paul, London.

Jermy, A. C. and Stott, P. A. (Eds) (1973). "Chalk Grassland. Studies on its Conservation and Management in South-east England". Kent Trust for Nature Conservation, Maidstone.

Jermy, A. C. and Tutin, T. G. (1968). "British Sedges". Botanical Society of the British Isles, London.

Jesse, R. H. B. (1960). "A Survey of the Agriculture of Sussex", County Agricultural Surveys No. 2. Royal Agricultural Society of England, London.

Johnson, P. A., Boyd, D. A. and Sparrow, P. E. (1977). Manurial experiments with spring barley on chalk soils in southern England. *Experimental Husbandry* No. 32, 8–18.

Jones, Carys A. (1973). "The Conservation of Chalk Downland in Dorset". Dorset County Council, Dorchester.

Jones, E. L. (1967). "Agriculture and Economic Growth in England, 1650–1815". Methuen, London.

Jones, J. O. and Dermott, W. (1951). The potash manuring of lucerne. *Agriculture* **57**, 507–509.

Jones, P. J. (1967). Cereals on the Chalk. *Agriculture* **74**, 2–5.

Jones, P. J. (Ed.) (1976). Bridgets Experimental Husbandry Farm. Annual Review No. 16, 1–4.

Jukes-Browne, A. J. and Hill W. (1904). "The Cretaceous Rocks of Britain", Vol. III "The Upper Chalk of England", Memoirs of the Geological Survey of the United Kingdom. HMSO, London.

Kay, F. F. (1934). "A Soil Survey of the Eastern Portion of the Vale of the White Horse", University of Reading Bulletin 48. University of Reading.

Kay, F. F. (1940). "Survey around Carstens and Partridges Farms, Hampshire'. University of Reading, unpublished.

Kay, Q. O. N. (1971a). Biological flora of the British Isles. *Anthemis cotula* L. *Journal of Ecology* **59**, 623–636.

Kay, Q. O. N. (1971b). Biological flora of the British Isles. *Anthemis arvensis* L. *Journal of Ecology* **59**, 637–648.

Kellaway, G. A. (1971). Glaciation and the stones of Stonehenge. *Nature, London* **233**, 30–35.

Kennet of the Dene (1940). Chalk landscape. *Nature, London* **145**, 466.

Kerney, M. P. (1972). The British distribution of *Pomatias elegans* (Müller). *Journal of Conchology* **27**, 359–361.

Kerney, M. P., Levy, J. F. and Oakley, K. P. (1963). Late-glacial deposits on the Chalk of south-east England. *Philosophical Transactions of the Royal Society of London, Series B* **246**, 203–254.

Kerney, M. P., Brown, E. H. and Chandler, T. J. (with Carreck, J. N., Lambert, C. A., Levy, J. F. and Millman, A. P.) (1964). The late-glacial and post-glacial history of the chalk escarpment near Brook, Kent. *Philosophical Transactions of the Royal Society of London, Series B* **248**, 135–204.

Kerridge, E. (1967). "The Agricultural Revolution". Allen and Unwin, London.

Kerridge, E. (1972). Wool growing and wool textiles in mediaeval and early modern times. *In* Jenkins (1972), 19–33.

Kerridge, E. (1973). "The Farmers of Old England". Allen and Unwin, London.

Kershaw, K. (1973). "Quantitative and Dynamic Plant Ecology", 2nd Edition. Arnold, London.

Kevan, D. K. McE. (1962). "Soil Animals". Witherby, London.

King, T.J. (1972). "The Plant Ecology of Ant-hills in Grassland". D.Phil. Thesis, University of Oxford.

King, T. J. (1974). Fruit dispersal in pioneer and climax trees. *School Science Review* **55**, 519–521.

King, T. J. (1975). Inhibition of seed germination under leaf canopies in *Arenaria serpyllifolia, Veronica arvensis* and *Cerastium holosteoides. New Phytologist* **75**, 87–90.

King, T. J. (1977a). The plant ecology of ant-hills in calcareous grasslands. I. Patterns of species in relation to ant-hills in southern England. *Journal of Ecology* **65**, 235–256.

King, T. J. (1977b). The plant ecology of ant-hills in calcareous grasslands. II. Succession on the mounds. *Journal of Ecology* **65**, 257–278.

King, T. J. (1977c). The plant ecology of ant-hills in calcareous grasslands. III. Factors affecting the population sizes of selected species. *Journal of Ecology* **65**, 279–315.

Kirkby, E. A. and Mengel, K. (1967). Ion uptake in different tissues of the tomato plant in relation to NO_3, urea or ammonium nutrition. *Plant Physiology, Lancaster* **42**, 6–14.

Kloet, G. S. and Hincks, W. D. I. (1964–1972). "A Check List of British Insects", 1st and 2nd Editions. Royal Entomological Society, London.

Knipe, P. R. and Maycock, R. (in preparation). "Flora of Buckinghamshire."

Köppen, W. (1931). "Grundriss der Klimakunde". de Gruyten, Berlin.

Kubiena, W. L. (1953). "The Soils of Europe". Murby, London.

Kuroiwa, S. (1968). A new calculation method for total photosynthesis of a plant community under illumination consisting of direct and diffused light. *In* Eckardt (1968), 391–398.

Kydd, D. D. (1964). The effect of different systems of cattle grazing on the botanical composition of permanent downland pasture. *Journal of Ecology* **52**, 139–149.

Lack, D. (1937). The psychological factor in bird distribution. *British Birds* **31**, 130–136.

Lack, D. and Venables, L. S. V. (1939). The habitat distribution of British woodland birds. *Journal of Animal Ecology* **8**, 39–71.

Lamb, H. H. (1950). Tornadoes of May 21 1950. *Meteorological Magazine* **79**, 245–256.

Lamb, H. H. (1966). "The Changing Climate. Selected Papers". Methuen, London.

Lamb, H. H., Collison, P. and Ratcliffe, R. A. S. (1973). Northern Hemisphere monthly and annual mean-sea-level pressure distribution for 1951–1966, and changes of pressure and temperature compared with those of 1900–1939. *Geophysical Memoirs* **16**, No. 118.

Lange, M. and Hora, F. B. (1965). "Collins' Guide to Mushrooms and Toadstools", 2nd Edition. Collins, London.

Large, R. V. (1976). "The Use of Soay Sheep in the Integrated Use of Land for Agriculture and Amenity Purposes". Grassland Research Institute, Hurley, unpublished.

Large, R. V. and King, N. (1978). "The Integrated Use of Land for Agricultural and Amenity Purposes", Technical Report No. 25. Grassland Research Institute,

Hurley.

Larsen, S. and Gunary, D. (1962). Ammonia loss from ammoniacal fertilisers applied to calcareous soils. *Journal of the Science of Food and Agriculture* **13**, 566–572.

Lee, I. K. and Monsi, M. (1963). Ecological studies on *Pinus densiflora* forest. I. Effects of plant substances on the floristic composition of the undergrowth. *Botanical Magazine, Tokyo* **76**, 400–413.

Lee, J. A. and Greenwood, Barbara (1976). The colonisation by plants of calcareous wastes from the salt and alkali industry in Cheshire, England. *Biological Conservation* **10**, 53–76.

Lee, J. A. and Woolhouse, H. W. (1969a). A comparative study of bicarbonate inhibition of root growth of certain grasses. *New Phytologist* **68**, 1–11.

Lee, J. A. and Woolhouse, H. W. (1969b). Root growth and dark fixation of carbon dioxide in calcicoles and calcifuges. *New Phytologist* **68**, 247–255.

Lewis, B. G. (1970). Effects of water potential on the infection of potato tubers by *Streptomyces scabies* in soil. *Annals of Applied Biology* **66**, 83–88.

Liger, J. (1952). Études sur la végétation des falaises calcaires de la Basse Seine. Bulletin des Amis des Sciences Naturelles et du Muséum de Rouen **88**, 17–54.

Liger, J. (1956). Aperçu sur la végétation des falaises littorales du Pays de Caux. *Reveue des Sociétés Savantes de Haute Normandie. Sciences* No. 1, 37–69.

Linton, D. L. (1963). The forms of glacial erosion. *Transactions of the Institute of British Geographers* **33**, 1–28.

Litav, M. (1965). Mycorrhizal association in dwarf shrub species growing in soft Cretaceous rocks. *Journal of Ecology* **53**, 147–151.

Litav, M. and Orshan, G. (1963). Ecological studies on some sub-lithophytic communities in Israel. *Israel Journal of Botany* **12**, 41–54.

Littleton, E. J. (1968). "Studies of the Scrub Vegetation of the South-western Chiltern Hills". B.Sc. Thesis, University of Reading.

Lloyd, P. S. (1968). The ecological significance of fire in limestone grassland communities of the Derbyshire Dales. *Journal of Ecology* **56**, 811–826.

Lloyd, P. S. and Pigott, C. D. (1967). The influence of soil conditions on the course of succession on the Chalk of southern England. *Journal of Ecology* **55**, 137–146.

Locket, G. H. (1946a). Observations on the colonisation of bare chalk. *Journal of Ecology* **33**, 205–209.

Locket, G. H. (1946b). A preliminary investigation of the availability to plants of water in chalk. *Journal of Ecology* **33**, 222–229.

Locket, G. H., Millidge, A. F. and Merrett, P. (1951–1974). "British Spiders", Vols I–III. The Ray Society, London.

Lockley, R. M. (1964). "The Private Life of the Rabbit". Deutsch, London.

Lousley, J. E. (1950). "Wild Flowers of Chalk and Limestone". Collins, London. Second edition 1969.

Lousley, J. E. (Ed.) (1953). "The Changing Flora of Britain". Botanical Society of the British Isles, Arbroath.

Lousley, J. E. (Ed.) (1957). "Progress in the Study of the British Flora". Buncle, Arbroath.

Lowday, J. E. and Wells, T. C. E. (1977). "The Management of Grassland and Heathland in Country Parks". Countryside Commission, Cheltenham.

Ludwig, J. W. and Harper, J. L. (1958). The influence of the environment on seed and seedling mortality. VIII. The influence of soil colour. *Journal of Ecology* **46**, 381–389.

MacArthur, R. H. and Wilson, E. O. (1967). "Island Biogeography". Princeton

University Press, Princeton.

MacConnell, J. T. and Bond, G. (1957). Nitrogen fixation in wild legumes. *Annals of Botany, New Series* **21**, 185–192.

MAFF (Ministry of Agriculture, Fisheries and Food) (1968). "A Century of Agricultural Statistics". HMSO, London.

MAFF (1969). "Nitrogen and Soil Organic Matter", Technical Bulletin 15. HMSO, London.

MAFF (1970). "Modern Farming and the Soil", Report of the Agricultural Advisory Council. HMSO, London.

MAFF (1971). "Insect Pests of *Brassica* Seed Crops", Advisory Leaflet 576. MAFF (Publications), Pinner.

MAFF (1972). "Sugar Beet Pests", Bulletin 162. HMSO, London.

MAFF (1973). "Cereal Aphids", Advisory Leaflet 586. MAFF (Publications), Pinner.

MAFF (1975). "The Wild Rabbit", Advisory Leaflet 534. MAFF (Publications), Pinner.

MAFF (1976a). "Bovine Tuberculosis in Badgers". MAFF, London.

MAFF (1976b). "Wildlife Conservation in Semi-natural Habitats on Farms". HMSO, London.

Manley, G. (1952). "Climate and the British Scene". Collins, London.

Mansfield, A. J. (1952). "The Historical Geography of the Woodlands of the Southern Chilterns". M.Sc. Thesis, University of London.

Margalef, R. (1958). Information theory in ecology. *General Systems* **3**, 36–71.

Margary, I. D. (1967). "Roman Roads in Britain", 2nd Edition. Baker, London.

Marples, M. (1949). "White Horses and Other Hill Figures". Country Life, London.

Martin, W. Keble (1965). "The Concise British Flora in Colour". Ebury Press and Michael Joseph, London.

Massingham, H. J. (1936). "English Downland". Batsford, London.

Massingham, H. J. (1939). "Country Relics". Cambridge University Press, Cambridge. Reprinted 1974 by E. P. Publishing.

Massingham, H. J. (1940). "Chiltern Country". Batsford, London.

Matthews, J. R. (1955). "Origin and Distribution of the British Flora". Hutchinson, London.

Matthews, L. H. (1952). "British Mammals". Collins, London.

McClintock, D. (1977). J. E. Lousley and plants alien in the British Isles. *Watsonia* **11**, 287–290.

McLaren, A. D. and Petersen, G. H. (Eds) (1967). "Soil Biochemistry". Arnold, London.

McRae, S. G. (1973). Agriculture and horticulture. *In* McRae and Burnham (1973), 99–125.

McRae, S. G. and Burnham, C. P. (Eds) (1973). "The Rural Landscape of Kent". Wye College, Ashford, Kent.

Meidner, H. and Mansfield, T. A. (1968). "Physiology of Stomata". McGraw-Hill, London.

Mellanby, K. (1971). "The Mole". Collins, London.

Merton, L. F. H. (1970). The history and status of woodlands of the Derbyshire limestone. *Journal of Ecology* **58**, 723–744.

Meteorological Office (1952). "Climatological Atlas". Meteorological Office, London.

Middleton, A. D. (1937). Whipsnade ecological survey 1936–1937. *Proceedings of the Zoological Society of London Series A* **107**, 471–481.

Miles, R. (1967). "Forestry in the English Landscape". Faber and Faber, London.

Mirghani, M. A. (1965). "Studies on the Effects of Agronomic Treatments on Water Loss from Chalk Turf". M.Sc. Thesis, University of Reading.

Monsi, M. and Saeki, T. (1953). Uber der Lichtfaktor in den Pflanzengesellschaften und seine Bedeutung fur die Stoffproduktion. *Japan Journal of Botany* **14**, 22–52.

Monteith, J. L. (1973). "Principles of Environmental Physics". Arnold, London.

Montford, H. M. (1970). The terrestrial environment during Upper Cretaceous and Tertiary times. *Proceedings of the Geologists' Association* **81**, 181–204.

Moore, D. R. E. and Waid, J. S. (1971). The influence of washings of living roots on nitrification. *Soil Biology and Biochemistry* **3**, 69–83.

Moore, N. W. (1962). The heaths of Dorset and their conservation. *Journal of Ecology* **50**, 369–391.

Morgan, R. (1975). Breeding bird communities on chalk downland in Wiltshire. *Bird Study* **22**, 71–83.

Morris, M. G. (1967a). Differences between the invertebrate faunas of grazed and ungrazed chalk grassland. I. Responses of some phytophagous insects to cessation of grazing. *Journal of Applied Ecology* **4**, 459–474.

Morris, M. G. (1967b). Insect collecting with special reference to nature reserves. *In* Duffey (1967), 20–24.

Morris, M. G. (1968). Differences between the invertebrate faunas of grazed and ungrazed chalk grassland. II. The fauna of sample turves. *Journal of Applied Ecology* **5**, 601–611.

Morris, M. G. (1969). Differences between the invertebrate faunas of grazed and ungrazed chalk grassland. III. The heteropterous fauna. *Journal of Applied Ecology* **6**, 475–487.

Morris, M. G. (1971a). Differences between the invertebrate faunas of grazed and ungrazed chalk grassland. IV. Abundance and diversity of Homoptera–Auchenorhyncha. *Journal of Applied Ecology* **8**, 37–52.

Morris, M. G. (1971b). The management of grassland for the conservation of invertebrate animals. *In* Duffey and Watt (1971). 527–552.

Morris, M. G. (1975). Preliminary observations on the effects of burning on the Hemiptera (Heteroptera and Auchenorhyncha) of limestone grassland. *Biological Conservation* **7**, 311–319.

Muggleton, J. (1973). Some aspects of the history and ecology of blue butterflies in the Cotswolds. *Proceedings and Transactions of the British Entomological Society* **6**, 77–84.

Muggleton, J. (1975). Observations on *Lysandra coridon* Poda (Lep. Lycaenidae) colonies at two sites in Gloucestershire using mark, release, recapture methods. *Proceedings and Transactions of the British Entomological Society* **8**, 73–82.

Muggleton, J. and Benham, B. R. (1975). Isolation and the decline of the large blue butterfly (*Maculinea arion*) in Great Britain. *Biological Conservation* **7**, 119–128.

Mukerji, S. K. (1936). Contributions to the autecology of *Mercurialis perennis* L. *Journal of Ecology* **24**, 38–81, 317–339.

Mundy, E. J. and Roebuck, J. F. (1966). Production of seed potatoes on the Yorkshire Wolds. *Experimental Husbandry* No. 14, 55–65.

Myerscough, P. J. and Whitehead, F. H. (1966). Comparative biology of *Tussilago farfara* L., *Chamaenerion angustifolium* (L.) Scop., *Epilobium montanum* L., and *Epilobium adenocaulon* Hausskn. I. General biology and germination. *New Phytologist* **65**, 192–210.

Naish, M. C. (1961). "The Historical Geography of the Hampshire Chalklands". M.A. Thesis, University of London.

Nature Conservancy Council (1977). "Nature Conservation and Agriculture". Nature Conservancy Council, London.

Newbould, P. J. (1974). Conservation in education. *In* Warren and Goldsmith (1974), 437–451.

Niel, E. (1886). Les variétés de *Daucus carota. Bulletin de l'Association des Sciences Naturelles de Rouen* 1886, 63.

Nishida, K. (1963). Studies on stomatal movement of Crassulacean plants in relation to the acid metabolism. *Physiologia Plantarum* 16, 281–298.

Noel, Denise (1970). "Coccolithes Crétacés". Éditions du Centre Nationale de la Recherche Scientifique, Paris.

Norman, M. J. T. (1956). Intervals of superphosphate application to downland permanent pasture. *Journal of Agricultural Science, Cambridge* 47, 157–171.

Norman, M. J. T. (1957). The influence of various grazing treatments upon the botanical composition of a downland permanent pasture. *Journal of the British Grassland Society* 12, 246–256.

Norman, M. J. T. and Green, J. O. (1957a). The renovation of downland permanent pasture, I. Surface cultivation and seeding. *Journal of the British Grassland Society* 12, 30–38.

Norman, M. J. T. and Green, J. O. (1957b). The renovation of downland permanent pasture. II. Herbicidal treatments, surface cultivation and seeding. *Journal of the British Grassland Society* 12, 74–80.

Norman, M. J. T. and Green, J. O. (1958). The local influence of cattle dung and urine upon the yield and botanical composition of permanent pasture. *Journal of the British Grassland Society* 13, 39–45.

North, F. J. (1930). "Limestones. Their Origins, Distribution and Uses". Murby, London.

Ogden, J. (1974). The reproductive strategy of higher plants. II. The reproductive strategy of *Tussilago farfara* L. *Journal of Ecology* 62, 291–324.

Ollier, C. D. and Thomasson, A. J. (1957). Asymmetrical valleys of the Chiltern Hills. *Geographical Journal* 123, 71–80.

Olson, R. A. and Lucas, R. E. (1966). Fertility requirements. *In* Pierre *et al.* (1966), 285–330.

Owen, T. R. (1976). "The Geological Evolution of the British Isles". Pergamon, Oxford.

Pack, D. A. (1921). After-ripening and germination of *Juniperus* seeds. *Botanical Gazette* 71, 32–60.

Packham, J. R. and Willis, A. J. (1977). The effects of shading on *Oxalis acetosella. Journal of Ecology* 65, 619–642.

Painter, R. B. (1971). A hydrological classification of the soils of England and Wales. *Proceedings of the Institute of Civil Engineers* 48, 93–95.

Pallant, D. (1969). The food of the grey field slug (*Agriolimax reticulatus* (Müller)) in woodland. *Journal of Animal Ecology* 38, 391–397.

Pallant, D. (1972). The food of the grey field slug (*Agriolimax reticulatus* (Müller)) on grassland. *Journal of Animal Ecology* 41, 761–769.

Parkes, G. D. (Ed.) (1956). "Mellor's Modern Inorganic Chemistry", Vol. III. Longman, London.

Pawsey, R. (1973). Amenity tree pathology: a neglected science. *New Scientist* 60, 532–534.

Pearson, R. W. and Adams, F. (Eds) (1967). Soil acidity and liming. *Agronomy* 12.

Pedgley, D. E. (1971). Some weather patterns in Snowdonia. *Weather* 26, 412–444.

Pedro, G. (1972). Les sols développés sur roches calcaires. *Science du Sol, Supplement au Bulletin de l'Association Française pour l'Étude du Sol* 1972, No. 1, 5–18.

Penistan, M. J. (1974). The silviculture of beech woodland. *Forestry*, supplement, 71–78.

Pennington, Winifred (1974). "The History of British Vegetation", 2nd Edition. The English Universities Press, London.

Perrin, R. M. S. (1956). Nature of "chalk heath" soils. *Nature, London* 178, 31–32.

Perrin, R. M. S. (1965). The use of drainage water analyses in soil studies. *In* Hallsworth and Crawford (1965), 73–96.

Perring, F. (1958). A theoretical approach to the study of chalk grassland. *Journal of Ecology* 46, 665–679.

Perring, F. (1959). Topographical gradients of chalk grassland. *Journal of Ecology* 47, 447–482.

Perring, F. (1960). Climatic gradients of chalk grassland. *Journal of Ecology* 48, 415–442.

Perring, F. (Ed.) (1968). "Critical Supplement to the Atlas of the British Flora". Botanical Society of the British Isles, London.

Perring, F. (Ed.) (1970). "The Flora of a Changing Britain." Botanical Society of the British Isles and Pendragon Press, Hampton.

Perring, F. and Walters, S. M. (1976). "Atlas of the British Flora", 2nd Edition. Botanical Society of the British Isles and Nelson, London.

Perring, F., Sell, P. D., Walters, S.M. and Whitehouse, H. L. K. (1964). "A Flora of Cambridgeshire". Cambridge University Press, Cambridge.

Perry, P. J. (1974). "British Farming in the Great Depression. 1870–1914". David and Charles, Newton Abbot.

Peterken, G. F. (1974). A method of assessing woodland flora for conservation using indicator species. *Biological Conservation* 6, 239–245.

Peterken, G. F. (1977). Habitat conservation priorities in British and European woodlands. *Biological Conservation* 11, 223–236.

Pfitzenmeyer, C. D. C. (1962). Biological flora of the British Isles. *Arrhenatherum elatius* (L.) J. & C. Presl. (*A. avenaceum* Beauv.). *Journal of Ecology* 50, 235–245.

Phillips, D. and Rowe, Judy (1976). Warfarin: a question of balance. *New Scientist* 70, 400–401.

Phillipson, J. and Wood, D. (1976). Poisoning grey squirrels. *New Scientist* 70, 398–399.

Pickering, S. U. (1917). The effect of one plant on another. *Annals of Botany* 31, 181–187.

Pielou, E. C. (1969). "An Introduction to Mathematical Ecology". Wiley-Interscience, New York.

Pierre, W. H., Aldrich, S. A. and Martin, W. P. (Eds) (1966). "Advances in Corn Production". State University Press, Iowa.

Pigott, C. D. (1955). Biological flora of the British Isles. *Thymus* L. *Journal of Ecology* 43, 365–387.

Pigott, C. D. (1968). Biological flora of the British Isles. *Cirsium acaulon* (L.) Scop. *Journal of Ecology* 56, 597–612.

Pigott, C. D. (1969). The status of *Tilia cordata* and *T. platyphyllos* on the Derbyshire limestone. *Journal of Ecology* 57, 491–504.

Pigott, C. D. (1970). The response of plants to climate and climatic change. *In*

Perring (1970), 32–44.

Pigott, C. D. and Taylor, K. (1964). The distribution of some woodland herbs in relation to the supply of nitrogen and phosphorus in the soil. *Journal of Ecology* **55**, supplement, 175–185.

Pigott, C. D. and Walters, S. M. (1953). Is the box-tree a native of England? *In* Lousley (1953), 184–187.

Pigott, C. D. and Walters, S. M. (1954). On the interpretation of the discontinuous distributions shown by certain British species of open habitats. *Journal of Ecology* **42**, 95–116.

Pinchemel, P. (1954). "Les Plaines de Craie du Nord-ouest du Bassin Parisien et du Sud-est du Bassin de Londres et Leurs Bordures". Libraire Armand Colin, Paris.

Piozzi, Hesther L. (1786). "Anecdotes of the Late Samuel Johnson, LL.D., During the Last Twenty Years of His Life". London. Various more recent editions, e.g. S. C. Roberts (1925), Cambridge University Press.

Pollard, E., Hooper, M. D. and Moore, N. W. (1974). "Hedges". Collins, London.

Pontin, A. J. (1955). "Colony Foundation and Competition between Ants". D. Phil. Thesis, University of Oxford.

Pontin, A. J. (1962). A method for quick comparison of the total solar radiation incident on different microhabitats. *Ecology* **43**, 740–741.

Pontin, A. J. (1963). Further considerations of competition and the ecology of the ants *Lasius flavus* (F.) and *L. niger* L. *Journal of Animal Ecology* **32**, 565–574.

Poore, M. E. D. (1956). The use of phytosociological methods in ecological investigations. IV. General discussion of phytosociological problems. *Journal of Ecology* **44**, 28–50.

Poore, M. E. D. (1962). The method of successive approximation in descriptive ecology. *Advances in Ecological Research* **1**, 35–68.

Pope, A. (1916). Plantations on Dorset Downs. *Quarterly Journal of Forestry* **10**, 294–297.

Pratt, D. J., Greenaway, P. J. and Gwynne, M. D. (1966). A classification of East African rangeland, with an appendix on terminology. *Journal of Applied Ecology* **3**, 369–382.

Prime, C. T. (1960). "Lords and Ladies". Collins, London.

Prior, R. (1968). "The Roe-deer of Cranborne Chase". Oxford University Press, Oxford.

Proctor, M. and Yeo, P. (1973). "The Pollination of Flowers". Collins, London.

Proudfoot, V. B. (1965). The study of soil development from the construction and excavation of experimental earthworks. *In* Hallsworth and Crawford (1965), 282–294.

Pugsley, A. J. (1939). "Dewponds in Fable and Fact". Country Life, London.

Pyatt, E. C. (1973?). "Chalkways of South and South-east England". David and Charles, Newton Abbot.

Rackham, O. (1976). "Trees and Woodland in the British Landscape". Dent, London.

Radet, E. (1958). Fertilisation des sols calcaires. *Bulletin de l'Association Française pour l'Etude du Sol* **6**, 304–338.

Radet, E. and Mantelet, C. (1938). Étude pédologique de la Champagne Crayeuse. *Bulletin de l'Associaton Française pour l'Etude du Sol* **4**, 279–340.

Ratcliffe, D. A. (1967). Conservation and the collector. *In* Duffey (1967), 16–19.

Ratcliffe, D. A. (Ed.) (1977). "A Nature Conservation Review", Vols I and II. Cambridge University Press, London.

Raunkiaer, C. (1934). "The Life Forms of Plants and Statistical Plant Geography" (English translation). Clarendon Press, Oxford.

Ravikovitch, S. and Pines, F. (1967). Mountain rendzina soils in Israel. *Anales de Edafologia y Agrobiologia, Madrid* **26**, 573–584.

Rawson, P. F., Curry, D., Dilley, F. C., Hancock, J. M., Kennedy, W. J., Neale, J. W., Wood, C. and Worssam, B. C. (1978). "A Correlation of Cretaceous Rocks in the British Isles", Special Report No. 9. Geological Society of London.

Richardson, H. L. (1938). The nitrogen cycle in grassland soils with special reference to the Rothamsted Park Grass experiment. *Journal of Agricultural Science, Cambridge* **28**, 73–121.

Ridley, H. N. (1930). "The Dispersal of Plants Throughout the World." Reeve, Ashford, Kent.

Roberts, M. J. G. (1968). Protection against lightning damage. *Agriculture* **75**, 350–351.

Robinson, K. L. (1948). The soils of Dorset. *In* Good (1948), 19–28.

Rodda, J. C., Downing, R. A. and Law, F. M. (1976). "Systematic Hydrology". Newnes–Butterworth, London.

Roden, D. (1968). Woodland and its management in the medieval Chilterns. *Forestry* **41**, 59–71.

Rogers, P. F. (1969). Organic manuring for potato scab control and its relation to soil manganese. *Annals of Applied Biology* **63**, 371–378.

Rorison, I. H. (1960a). Some experimental aspects of the calcicole–calcifuge problem. I. The effects of competition and mineral nutrition upon seedling growth in the field. *Journal of Ecology* **48**, 585–599.

Rorison, I. H. (1960b). The calcicole–calcifuge problem. II. The effects of mineral nutrition on seedling growth in solution culture. *Journal of Ecology* **48**, 679–688.

Rorison, I. H. (1968). The response to phosphorus of some ecologically distinct plant species. I. Growth rates and phosphorus absorption. *New Phytologist* **67**, 913–923.

Rorison, I. H. (Ed.) (1969). "Ecological Aspects of the Mineral Nutrition of Plants". Blackwell, Oxford.

Rose, F. (1957). The importance of the study of disjunct distributions to progress in understanding the British flora. *In* Lousley (1957), 61–78.

Rose, F. (1972). Floristic connections between south-east England and north France. *In* Valentine (1972), 363–379.

Rose, F. (1973). The refugium hypothesis and the flora of the chalk grasslands of south-east England. *In* Jermy and Stott (1973), 6–9.

Rosenberg, N. J. (1974). "Microclimate. The Biological Environment". Wiley, New York.

Ross, R. (1936). The ecology of hawthorn scrub in south-west Cambridgeshire. Unpublished. See Tansley (1939).

Rovira, A. D. and McDougall, Barbara M. (1967). Microbiological and biochemical aspects of the rhizosphere. *In* McLaren and Petersen (1967), 417–463.

Russell, E. J. and Keen, B. A. (1921). The effect of chalk on the cultivation of heavy land. *Journal of the Ministry of Agriculture* **28**, 419–422.

Russell, E. W. (1973). "Soil Conditions and Plant Growth", 10th Edition. Longman, London.

Salisbury, E. J. (1916). The oak–hornbeam woodlands of Hertfordshire. *Journal of Ecology* **4**, 83–117.

Salisbury, E. J. (1918a). The oak–hornbeam woods of Hertfordshire. Parts III and IV. *Journal of Ecology* **6**, 14–52.

Salisbury, E. J. (1918b). The ecology of scrub in Hertfordshire: a study in colonisation. *Transactions of the Hertfordshire Natural History Society* **17**, 53–64.

Salisbury, E. J. (1921). The significance of the calcicolous habit. *Journal of Ecology* **8**, 202–215.

Salisbury, E. J. (1924). The influence of earthworms on soil reaction and stratification of undisturbed soils. *Journal of the Linnaean Society (Botany)* **46**, 415–425.

Salisbury, E. J. (1942). "The Reproductive Capacity of Plants". Bell, London.

Salisbury, E. J. (1952). "Downs and Dunes". Bell, London.

Salisbury, E. J. (1964). "Weeds and Aliens", 2nd Edition. Collins, London.

Salisbury, E. J. (1976). A note on shade tolerance and vegetative propagation of woodland species. *Proceedings of the Royal Society of London Series B* **192**, 257–258.

Sankey, J. (1966). "Chalkland Ecology". Heinemann, London.

Sankey, J. and Mackworth-Praed, H. (1969). Headley Warren Reserve in 1968. The Surrey Naturalist Annual Report for 1968, 14–21.

Sarukhán, J. and Harper, J. L. (1973). Studies on plant demography: *Ranunculus repens* L., *R. bulbosus* L., and *R. acris* L. I. Population flux and survivorship. *Journal of Ecology* **61**, 675–716.

Schinas, S. and Rowell, D. L. (1977). Lime-induced chlorosis. *Journal of Soil Science* **28**, 351–368.

Scott, G. A. M. (1963). Biological flora of the British Isles. *Glaucium flavum* Crantz. *Journal of Ecology* **51**, 743–754.

Seale, R. S. (1975). "Soils of the Ely District", Memoirs of the Soil Survey of Great Britain: England and Wales. Soil Survey, Harpenden.

Seddon, B. (1971). "Introduction to Biogeography". Duckworth, London.

Sharrock, J. T. R. (1976). "The Atlas of Breeding Birds in Britain and Ireland". British Trust for Ornithology, Tring.

Sheail, J. (1971). "Rabbits and their History". David and Charles, Newton Abbot.

Sheail, J. (1976). "Nature in Trust. The History of Nature Conservation in Britain". Blackie, Glasgow.

Shepherd, W. (1972). "Flint. Its Origin, Properties and Uses". Faber and Faber, London.

Sherlock, R. L., Casey, R., Holmes, S. C. A. and Wilson, V. (1960). "London and Thames Valley", 3rd Edition, British Regional Geology. HMSO, London.

Shimwell, D. W. (1968). "The Phytosociology of Calcareous Grasslands in the British Isles". Ph.D. Thesis, University of Durham.

Shimwell, D. W. (1971a). "The Description and Classification of Vegetation". Sidgwick and Jackson, London.

Shimwell, D. W. (1971b). Festuco-Brometea Br-Bl. & R. Tx 1943 in the British Isles: the phytogeography and phytosociology of limestone grasslands. I. General introduction; Xerobromion in England. *Vegetatio* **23**, 1–28.

Shimwell, D. W. (1971c). Festuco-Brometea Br-Bl. & R. Tx 1943 in the British Isles: the phytogeography and phytosociology of limestone grasslands. Eu-Mesobromion in the British Isles. *Vegetatio* **23**, 29–60.

Shimwell, D. W. (1973). An introduction to the geography and ecology of chalk grassland. *In* Jermy and Stott (1973), 1–5.

Shorten, Monica (1962). "Squirrels, their Biology and Control", MAFF Bulletin 184. HMSO, London.

Side, K. C. (1955). A study of the insects living on the wayfaring tree. *Bulletin of the Amateur Entomologists' Society* **14**, 3–50.

Simms, E. (1971). "Woodland Birds". Collins, London.

Simonson, R. W. (1968). Concept of soil. *Advances in Agronomy* **20**, 1–47.

Simpson, J. E. (1967). Aerial and radar observations of some sea-breeze fronts. *Weather* **22**, 306–316.

Slatyer, R. O. (1967). "Plant–Water Relationships". Academic Press, London and New York.

Sledge, W. A. (Ed.) (1971). "The Naturalists' Yorkshire". Dalesman, Clapham, Lancaster.

Small, R. J. (1958). The origin of Rake Bottom, Butser Hill. *Proceedings of the Hampshire Field Club* **21**, 22–30.

Small, R. J. (1970). "The Study of Landforms". Cambridge University Press, Cambridge.

Smith, A. G. (1970). The influence of Mesolithic and Neolithic man on British vegetation: a discussion. *In* Walker and West (1970), 81–96.

Smith, C. J. (1980). "The Natural History of Newland Park" (in preparation).

Smith, C. J. and Bunting, A. H. (1967–70). Unpublished internal reports. University of Reading, Reading.

Smith, C. J., Elston, J. and Bunting, A. H. (1971). The effects of cutting and fertiliser treatments on the yield and botanical composition of chalk turf. *Journal of the British Grassland Society* **26**, 213–219.

Smith, J. E., Clapham. A. R. and Ratcliffe, D. A. (Eds) (1977). "Scientific Aspects of Nature Conservation in Great Britain". The Royal Society, London.

Smith, K. A. and Russell, R. S. (1969). Occurrence of ethylene, and its significance, in anaerobic soil. *Nature, London* **222**, 769–771.

Smith, L. P. (1968). "Seasonable Weather". Allen and Unwin, London.

Smith, L. P. (1976). "The Agricultural Climate of England and Wales", MAFF Technical Bulletin 35. HMSO, London.

Snaydon, R. W. (1962a). Microdistribution of *Trifolium repens* L. and its relation to soil factors. *Journal of Ecology* **50**, 133–143.

Snaydon, R. W. (1962b). The growth and competitive ability of contrasting natural populations of *Trifolium repens* L. on calcareous and acid soils. *Journal of Ecology* **50**, 439–447.

Snaydon, R. W. and Bradshaw, A. D. (1961). Differential response to calcium within the species *Festuca ovina* L. *New Phytologist* **60**, 219–234.

Somerville, W. (1911). Some notes on the Society's experiments at Sevington, and on the improvement of poor pastures on chalk near Shoreham (Applesham). *Journal of the Bath and West and Southern Counties Society, 5th series* **5**, 90–98.

Somerville, W. (1914). Accumulated fertility in grassland in consequence of phosphatic manuring. *Journal of the Board of Agriculture* **21**, 481–492.

Somerville, W. (1916). Accumulated fertility in grassland in consequence of phosphatic manuring. Second report. *Journal of the Board of Agriculture* **22**, 1201–1209.

Somerville, W. (1918). Poverty Bottom: An experiment in increased food production. *Journal of the Board of Agriculture* **24**, 1186–1202.

South, R. (1906). "The Butterflies of the British Isles". Warne, London.

Southern, H. N. (1940). The ecology and population dynamics of the wild rabbit (*Oryctolagus cuniculus*). *Annals of Applied Biology* **27**, 509–526.

Sowter, F. A. (1949). Biological flora of the British Isles. *Arum maculatum* L. *Journal of Ecology* **37**, 207–219.

Sparks, B. W. (1960). "Geomorphology". Longman, London.

Sparks, B. W. and Lewis, W. V. (1957). Escarpment dry valleys near Pegsdon, Hertfordshire. *Proceedings of the Geologists' Association* **68**, 26–38.

Spencer, J. A. (1972). The Forestry Commission. *In* Steele (1972a), 29–34.

Spooner, G. M. (1963). On causes of the decline of *Maculinea arion* L. (Lep. Lycaenidae) in Britain. *The Entomologist* **96**, 199–210.

Spradbery, J. P. (1973). "Wasps". Sidgwick and Jackson, London.

Stamp, L. D. (1967). "Britain's Structure and Scenery", 6th Edition. Collins, London.

Stamp, L. D. (1969a). "Man and the Land", 3rd Edition. Collins, London.

Stamp, L. D. (1969b). "Nature Conservation in Britain". Collins, London.

Stapledon, R. G. (1912). Pasture problems: drought resistance. *Journal of Agricultural Science* **5**, 129–151.

Stapledon, R. G. and Davies, W. (1940). "Grassland Survey of England and Wales". Grassland Improvement Station, Stratford-on-Avon.

Staples, M. J. C. (1970). A history of box in the British Isles. *Boxwood Bulletin* **10**, 18–23, 34–37, 55–60.

Stearn, L. F. (Ed.) (1975). "Supplement to the Flora of Wiltshire". Wiltshire Archaeological and Natural History Society, Devizes.

Steele, R. C. (Ed.) (1972a). "Lowland Forestry and Wildlife Conservation", Monks Wood Experiment Station Symposium No. 6. Institute of Terrestrial Ecology and Natural Environment Research Council, Monks Wood.

Steele, R. C. (1972b). Wildlife conservation: needs and problems. *In* Steele (1972a), 138–148.

Step, E. (1932). "Bees, Wasps, Ants and Allied Insects of the British Isles". Warne, London.

Stern, F. C. (1960). "A Chalk Garden". Nelson, London.

Stern, W. R. and Donald, C. M. (1962). Light relationships in grass–clover swards. *Australian Journal of Agricultural Research* **13**, 599–614.

Stevenson, Catherine M. (1968). An analysis of the chemical composition of rainwater and air over the British Isles and Eire for the years 1959–1964. *Quarterly Journal of the Royal Meteorological Society* **94**, 56–70.

Stokoe, W. J. (1944). "The Caterpillars of the British Butterflies, including the Eggs, Chrysalids and Food Plants". Warne, London.

Stott, P. A. (1970). The study of chalk grassland in northern France. An historical review. *Biological Journal of the Linnaean Society* **2**, 173–207.

Stout, J. D. (1963). Some observations on the Protozoa of some beechwood soils on the Chiltern Hills. *Journal of Animal Ecology* **32**, 281–287.

Stratton, L. W. (1963). An ecological study. *Journal of Conchology* **25**, 174–179.

Streeter, D. T. (1965). Some considerations relating to woodland reserve management. Society for the Promotion of Nature Reserves, Handbook and Annual Report 1965, 42–47.

Streeter, D. T. (1967). Conservation policy and introductions to nature reserves. *In* Duffey (1967), 2–8.

Streeter, D. T. (1971). The effects of public pressure on the vegetation of chalk downland at Box Hill, Surrey. *In* Duffey and Watt (1971), 459–468.

Streeter, D. T. (1977). Gulley restoration on Box Hill. *Countryside Recreation Review* **2**, 38–40.

Summerhayes, V. S. (1951). "Wild Orchids of Britain". Collins, London.

Sutherland, A. M. (1967). "The Yorkshire Wolds", Farming Cameo Series 3, No. 51. *Agriculture* **74**, 198–199.

Sutton, C. D. and Gunary, D. (1969). Phosphate equilibria in soils. *In* Rorison (1969), 127–134.

Sutton, S. L. (1972). "Woodlice". Ginn, London.

Tansley, A. G. (Ed.) (1911). "Types of British Vegetation". Cambridge University Press, Cambridge.

Tansley, A. G. (1917). On competition between *Galium saxatile* L. (*G. hercynicum* Weig.) and *Galium sylvestre* Poll. (*G. asperum* Schreb.) on different types of soil. *Journal of Ecology* **5**, 173–179.

Tansley, A. G. (1922). Studies on the vegetation of the English Chalk. II. Early stages of redevelopment of woody vegetation on chalk grassland. *Journal of Ecology* **10**, 168–177.

Tansley, A. G. (1939). "The British Islands and Their Vegetation", Vols I and II. Cambridge University Press, Cambridge.

Tansley, A. G. (Ed. M. C. F. Proctor) (1968). "Britain's Green Mantle", 2nd Edition. Allen and Unwin, London.

Tansley, A. G. and Adamson, R. S. (1925). Studies of the vegetation of the English Chalk. III. The chalk grasslands of the Hampshire–Sussex border. *Journal of Ecology* **13**, 177–223.

Tansley, A. G. and Adamson, R. S. (1926). Studies of the vegetation of the English Chalk. IV. A preliminary survey of the chalk grasslands of the Sussex Downs. *Journal of Ecology* **14**, 1–32.

Tansley, A. G. and Rankin, W. M. (1911). The plant formation of calcareous soils. B. The sub-formation of the Chalk. *In* Tansley (1911), 161–186.

Tarling, D. H. and Tarling, M. P. (1971). "Continental Drift". Bell, London.

Tarr, W. A. (1925). Is the Chalk a chemical deposit? *Geological Magazine* **62**, 252–264.

Taylor, J. A. and Yates, R. A. (1967). "British Weather in Maps", 2nd Edition. Macmillan, London.

Temple, M. S. (1929). "A Survey of the Soils of Buckinghamshire", University of Reading Bulletin 38. University of Reading.

Thomas, A. S. (1959). Sheep paths: observations on the variability of chalk pastures. *Journal of the British Grassland Society* **14**, 157–164.

Thomas, A. S. (1960). Changes in vegetation since the advent of myxomatosis. *Journal of Ecology* **48**, 287–306.

Thomas, A. S. (1962a). Botany and geology of the downs of southern England. *Nature, London* **193**, 214–217.

Thomas, A. S. (1962b). Anthills and termite mounds in pastures. *Journal of the British Grassland Society* **17**, 103–108.

Thomas, A. S. (1963). Further changes in vegetation since the advent of myxomatosis. *Journal of Ecology* **51**, 151–186.

Thomas, A. S., Rawes, M. and Banner, W. J. L. (1957). The vegetation of the Pewsey Vale escarpment, Wiltshire. *Journal of the British Grassland Society* **12**, 39–48.

Thomas, J. F. H. (1951). Land reclamation on the Chalk uplands. *In* Griffiths *et al.* (1951), 57–82.

Thomas, R. N. (1930). Flora of paper-mill lime waste dumps near Glasgow. *Journal of Ecology* **18**, 333–351.

Thompson, L. M., Black, C. A. and Zoellner, J. A. (1954). Occurrence and mineralisation of organic phosphorus in soils, with particular reference to associations with nitrification, carbon and pH. *Soil Science* **77**, 185–196.

Thompson, P. A. (1974). The use of seedbanks for conservation of populations of species and ecotypes. *Biological Conservation* **6**, 15–19.

Timperley, H. W. and Brill, Edith (1965). "Ancient Trackways of Wessex". Phoenix House, London.

Tittensor, Ruth M. (1976). "The Ecology and History of Yew (*Taxus baccata* L.) Woodlands of the South Downs". Royal Society, London, unpublished.

Trewartha, G. (1968). "An Introduction to Climate", 4th Edition. McGraw–Hill, New York.

Troughton, A. (1957). "The Underground Organs of Herbage Grasses", Commonwealth Bureau of Pastures and Field Crops. Bulletin 44. CBPFC, Hurley.

Troup, L. C. (1954). The afforestation of chalk downland. *Forestry* **27**, 135–144.

Turner, R. C. (1958). A theoretical treatment of the pH of calcareous soils. *Soil Science* **86**, 32–34.

Turrill, W. B. (1948). "British Plant Life". Collins, London.

University of California (1969). "Biology and Ecology of Nitrogen", Proceedings of a Conference. National Academy of Sciences.

University of Reading, 1972. "Farming and Wildlife: Churn Estate". MAFF/ADAS, Reading.

Usher, G. (1974). "A Dictionary of Plants Used by Man". Constable, London.

Usher, M. B. (1973). "Biological Management and Conservation". Chapman and Hall, London.

Valentine, D. H. (Ed.) (1972). "Taxonomy, Phytogeography and Evolution". Academic Press, London and New York.

Van Andel, J. (1976). Ecological aspects of mineral nutrition of some plant species characteristic of clearings. Paper read to the Winter Meeting of the British Ecological Society, University College, Cardiff, January 1976.

Van der Drift, J. (1950). Analysis of the animal community in a beech forest floor. *Tijdschrift voor Entomologie* **94**, 1–168. See also an alternative publication of this same monograph cited by Kevan (1962).

Van der Meulen, F. and Wiegers, J. (1972). "A Phytosociological Research of some Chalk Grasslands in Southern England". Instituut voor Systematische Plantkunde, Utrecht.

Venables, L. S. V. (1939). Bird distribution on the South Downs and a comparison with that of Surrey Greensand heaths. *Journal of Animal Ecology* **8**, 227–237.

Wagar, J. A. (1964). The insolation grid. *Ecology* **45**, 636–639.

Walker, D. and West, R. G. (Eds) (1970). "Studies in the Vegetational History of the British Isles". Cambridge University Press, Cambridge.

Walsh, P. T. (1966). Cretaceous outliers in south-west Ireland and their implications for Cretaceous palaeogeography. *Quarterly Journal of the Geological Society* **122**, 63-84.

Walter, H. and Lieth, H. (1967). "Klimadiagramm-Weltatlas". Fischer, Jena.

Walter, H., Harnickell, Elisabeth and Mueller-Dombois, D. (1975). "Climate-diagram Maps of the Individual Continents and the Ecological Climatic Regions of the Earth". Springer-Verlag, New York.

Ward, Lena K. (1973). The conservation of juniper. I. Present status of juniper in southern England. *Journal of Applied Ecology* **10**, 165–188.

Ward, Lena K. (1977). The conservation of juniper: the associated fauna with special reference to southern England. *Journal of Applied Ecology* **14**, 81–120.

Ward, Lena K. and Lakhani, K. H. (1977). The conservation of juniper: the fauna of food-plant island sites in southern England. *Journal of Applied Ecology* **14**,

121–135.

Wardle, P. (1959). The regeneration of *Fraxinus excelsior* in woods with a field layer of *Mercurialis perennis*. *Journal of Ecology* **37**, 483–497.

Wardle, P. (1961). Biological flora of the British Isles. *Fraxinus excelsior* L. *Journal of Ecology* **49**, 739–751.

Warne, L. G. G. (1934). Intensive treatment of a Wiltshire Down pasture: effect on the botanical composition. *Journal of the Ministry of Agriculture* **41**, 470–475.

Warren, A and Goldsmith, F. B. (Eds) (1974). "Conservation in Practice". Wiley, London.

Waterhouse, F. L. (1955). Microclimatological profiles in grass cover in relation to biological problems. *Quarterly Journal of the Royal Meteorological Society* **81**, 63–71.

Watson, E. V. (1960). A quantitative study of the bryophytes of chalk grassland. *Journal of Ecology* **48**, 397–414.

Watson, E. V. (1968). "British Mosses and Liverworts", 2nd Edition. Cambridge University Press, Cambridge.

Watson, W. C. R. (1958). "Handbook of the Rubi of Great Britain and Ireland". Cambridge University Press, Cambridge.

Watt, A. S. (1923). On the ecology of British beechwoods with special reference to their regeneration. I. The causes of failure of natural regeneration of the beech (*Fagus sylvatica* L.). *Journal of Ecology* **11**, 1–48.

Watt, A. S. (1924). On the ecology of British beechwoods with special reference to their regeneration. II. The development and structure of beech communities on the Sussex Downs. *Journal of Ecology* **12**, 145–204.

Watt, A. S. (1925). On the ecology of British beechwoods with special reference to their regeneration. II (continued). The development and structure of beech communities on the Sussex Downs. *Journal of Ecology* **13**, 27–73.

Watt, A. S. (1926). Yew communities of the South Downs. *Journal of Ecology* **14**, 282–316.

Watt, A. S. (1934). The vegetation of the Chiltern Hills with special reference to the beechwoods and their seral relationships. *Journal of Ecology* **22**, 230–270, 445–507.

Watt, A. S. (1940). Studies in the ecology of Breckland. IV. The grass heath. *Journal of Ecology* **28**, 42–70.

Watt, A. S. (1947). Pattern and process in the plant community. *Journal of Ecology* **35**, 1–22.

Watt, A. S. (1957). The effect of excluding rabbits from Grassland B (Mesobrometum) in Breckland. *Journal of Ecology* **45**, 861–878.

Watt, A. S. (1962). The effect of excluding rabbits from Grassland A (Xerobrometum) in Breckland, 1936–60. *Journal of Ecology* **50**, 181–198.

Watt, A. S. (1974). Senescence and rejuvenation in ungrazed chalk grassland (Grassland B) in Breckland: the significance of litter and moles. *Journal of Applied Ecology* **11**, 1157–1171.

Watt, A. S. and Fraser, G. K. (1933). Tree roots and the field layer. *Journal of Ecology* **21**, 404–414.

Way, J. M. (1973). "Road Verges on Rural Roads", Monks Wood Experimental Station Occasional Report No. 1. Nature Conservancy, Monks Wood.

Webley, D. M. and Duff, R. B. (1965). The incidence in soils and other habitats of micro-organisms producing 2-ketogluconic acid. *Plant and Soil* **22**, 307–313.

Webster, R., Hodge, C. A. H., Draycott, A. P. and Durrant, M. J. (1977). The effect of soil type and related factors on sugar beet yield. *Journal of Agricultural Science, Cambridge* **88**, 455–469.

Weir, A. H. and Catt, J. A. (1965). The mineralogy of some Upper Chalk samples from the Arundel area, Sussex. *Clay Mineralogy* **6**, 97–110.

Wells, P. A. (1971). A map of effective precipitation. *Weather* **26**, 397–405.

Wells, T. C. E. (1967a). Changes in a population of *Spiranthes spiralis* (L.) Chevall. at Knocking Hoe National Nature Reserve, Bedfordshire, 1962–65. *Journal of Ecology* **55**, 83–99.

Wells, T. C. E. (1967b). Changes in the botanical composition of a sown pasture on the Chalk in Kent, 1956–64. *Journal of the British Grassland Society* **22**, 277–281.

Wells, T. C. E. (1969). Botanical aspects of conservation management of chalk grasslands. *Biological Conservation* **2**, 36–44.

Wells, T. C. E. (1971). A comparison of the effects of sheep grazing and mechanical cutting on the structure and botanical composition of chalk grassland. *In* Duffey and Watt (1971), 497–515.

Wells, T. C. E. (1973). Botanical aspects of chalk grassland management. *In* Jermy and Stott (1973), 10–15.

Wells, T. C. E. (1975). The floristic composition of chalk grassland in Wiltshire. *In* Stearn (1975), 99–125.

Wells, T. C. E. and Barling, D. M. (1971). Biological flora of the British Isles. *Pulsatilla vulgaris* Mill. (*Anemone pulsatilla* L.). *Journal of Ecology* **59**, 275–292.

Wells, T. C. E., Sheail, J., Ball, D. F. and Ward, L. K. (1976). Ecological studies on the Porton Ranges: relationships between vegetation, soils and land-use history. *Journal of Ecology* **64**, 589–626.

West, R. G. (1968). "Pleistocene Geology and Biology". Longman, London.

Wetselaar, R. (1968). Soil organic nitrogen mineralisation as affected by low soil water potentials. *Plant and Soil* **29**, 9–17.

White, G. (1788). "The Natural History of Selborne". Bensley, London. Various more recent editions.

Whitehead, D. C. (1964). Soil and plant nutrition aspects of the sulfur cycle. *Soils and Fertilisers* **27**, 1–8.

Whitehead, D. C. (1970). Carbon, nitrogen, phosphorus and sulphur in herbage plant roots. *Journal of the British Grassland Society* **25**, 236–241.

Williams, C. H. (1968). Seasonal fluctuation in mineral sulphur under subterranean clover pasture in southern New South Wales. *Australian Journal of Soil Research* **6**, 131–139.

Williams, J. T. (1969). Mineral nitrogen in British grassland soils. I. Seasonal patterns in simple models. *Oecologia Plantarum* **4**, 307–320.

Williams, O. B., Wells, T. C. E. and Wells, D. A. (1974). Grazing management of Woodwalton Fen: seasonal changes in the diet of cattle and rabbits. *Journal of Applied Ecology* **11**, 499–516.

Williamson, K. (1967). Some aspects of the scientific interest and management of scrub on nature reserves. *In* Duffey (1967), 94–100.

Williamson, K. (1968). A breeding bird survey of Queen Wood in the Chilterns, Oxon. *Quarterly Journal of Forestry* **62**, 118–131.

Williamson, K. (1970). Birds and modern forestry. *Bird Study* **17**, 167–176.

Williamson, K. (1975). The breeding bird community of chalk grassland scrub in the Chiltern Hills. *Bird Study* **22**, 59–70.

Williamson, K. and Williamson, R. (1973). The bird community of yew woodland at Kingley Vale, Sussex. *British Birds* **66**, 12–23.

Williamson, P. (1976). Above-ground primary production of chalk grassland allowing for leaf death. *Journal of Ecology* **64**, 1059–1075.

Wilson, Joan F. (1968). "The Control of Density in Some Woodland Plants". Ph.D Thesis, University of Lancaster.

Wilson, M. (1911). Plant distribution in the woods of north-east Kent. *Annals of Botany* **25**, 857–902.

Wilson, V. (1948). "East Yorkshire and Lincolnshire", British Regional Geology. HMSO, London.

Woldendorp, J. W. (1968). Losses of soil nitrogen. *Stikstof* No. 12, 32–46.

Wood, R. F. and Nimmo, M. (1962). "Chalk Downland Afforestation", Forestry Commission Bulletin 34. HMSO, London.

Wooldridge, S. W. and Kirkaldy, J. F. (1937). The geology of the Mimms valley. *Proceedings of the Geologists' Association* **48**, 307–315.

Wooldridge, S. W. and Linton, D. L. (1933). The loam terrains of south-east England and their relation to its early history. *Antiquity* **7**, 297–310.

Wooldridge, S. W. and Linton, D. L. (1955). "Structure, Surface and Drainage in South-east England", 2nd Edition. Philip, London.

Woolhouse, H. W. (1966a). Comparative physiological studies on *Deschampsia flexuosa, Holcus mollis, Arrhenatherum elatius and Koeleria gracilis* in relation to growth on calcareous soils. *New Phytologist* **65**, 22–31.

Woolhouse, H. W. (1966b). The effect of bicarbonate on the uptake of iron in four related grasses. *New Phytologist* **65**, 372–375.

Wyatt, J. A. (1971). "East Sussex", Farming Cameo Series 4, No. 43. *Agriculture* **78**, 84–86.

Yaalon, D. H. (1957). Problems of soil testing on calcareous soils. *Plant and Soil* **8**, 275–288.

Yapp, W. B. (1955). A classification of the habitats of British birds. *Bird Study* **2**, 111–121.

Young, A. (1769). "A Six Weeks Tour Through the Southern Counties of England and Wales", 2nd Edition. Strahan, London.

Young, A. (1813). "General View of the Agriculture of Hertfordshire". Nichol, London.

Zeuner, F. E. (1959). "The Pleistocene Period". Hutchinson, London.

Zinke, P. J. (1962). The pattern of influence of individual forest trees on soil properties. *Ecology* **43**, 130–133.

Index

Italicised numbers refer to illustrations

H